Benchmark Papers in Geology

Series Editor: Rhodes W. Fairbridge
Columbia University

RIVER MORPHOLOGY / Stanley A. Schumm
SLOPE MORPHOLOGY / Stanley A. Schumm and M. Paul Mosley
SPITS AND BARS / Maurice L. Schwartz
BARRIER ISLANDS / Maurice L. Schwartz
ENVIRONMENTAL GEOMORPHOLOGY AND LANDSCAPE CONSERVATION,
    VOLUME I: Prior to 1900 / VOLUME II: Urban Areas / VOLUME III: Non-Urban
    Regions / Donald R. Coates
TEKTITES / Virgil E. Barnes and Mildred A. Barnes
GEOCHRONOLOGY: Radiometric Dating of Rocks and Minerals / C. T. Harper
MARINE EVAPORITES: Origins, Diagenesis, and Geochemistry / Douglas W. Kirkland
    and Robert Evans
GLACIAL ISOSTASY / John T. Andrews
GLACIAL DEPOSITS / Richard P. Goldthwait
PHILOSOPHY OF GEOHISTORY: 1785-1970 / Claude C. Albritton, Jr.
GEOCHEMISTRY OF GERMANIUM / Jon N. Weber
GEOCHEMISTRY AND THE ORIGIN OF LIFE / Keith A. Kvenvolden
GEOCHEMISTRY OF WATER / Yasushi Kitano
GEOCHEMISTRY OF IRON / Henry Lepp
GEOCHEMISTRY OF BORON / C. T. Walker
SEDIMENTARY ROCKS: Concepts and History / Albert V. Carozzi
METAMORPHISM AND PLATE TECTONIC REGIMES / W. G. Ernst
SUBDUCTION ZONE METAMORPHISM / W. G. Ernst
PLAYAS AND DRIED LAKES: Occurrence and Development / James T. Neal
PLANATION SURFACES: Peneplains, Pediplains, and Etchplains / George Adams
SUBMARINE CANYONS AND DEEP-SEA FANS: Modern and Ancient / J. H. McD.
    Whitaker
ENVIRONMENTAL GEOLOGY / Frederick Betz, Jr.
LOESS: Lithology and Genesis / Ian J. Smalley
PERIGLACIAL DEPOSITS / Cuchlaine A. M. King
LANDFORMS AND GEOMORPHOLOGY: Concepts and History / Cuchlaine A. M. King
METALLOGENY AND GLOBAL TECTONICS / Wilfred Walker
HOLOCENE TIDAL SEDIMENTATION / George deVries Klein
PALEOBIOGEOGRAPHY / Charles A. Ross
MECHANICS OF THRUST FAULTS AND DÉCOLLEMENT / Barry Voight
WEST INDIES ISLAND ARCS / Peter H. Mattson
CRYSTAL FORM AND STRUCTURE / Cecil J. Schneer
METEORITE CRATERS / G. J. H. McCall
AIR PHOTOGRAPHY AND COASTAL PROBLEMS / Mohamed T. El-Ashry
DIAGENESIS OF DEEP-SEA BIOGENIC SEDIMENTS / Gerrit J. van der Lingen
ANCIENT CONTINENTAL DEPOSITS / Franklyn B. Van Houten
DRAINAGE BASIN MORPHOLOGY / Stanley A. Schumm
SEA WATER: Cycles of the Major Elements / J. I. Drever

Benchmark Papers in Geology/41

A BENCHMARK® Books Series

DRAINAGE BASIN MORPHOLOGY

Edited by
STANLEY A. SCHUMM
Colorado State University

Dowden, Hutchinson & Ross, Inc.

STROUDSBURG, PENNSYLVANIA

Copyright © 1977 by **Dowden, Hutchinson & Ross, Inc.**
Benchmark Papers in Geology, Volume 41
Library of Congress Catalog Card Number: 77-7365
ISBN: 0-87933-284-0

All rights reserved. No part of this book covered by the copyrights hereon may be reproduced or transmitted in any form or by any means—graphic, electronic, or mechanical, including photocopying, recording, taping, or information storage and retrieval systems—without written permission of the publisher.

79 78 77   1 2 3 4 5
Manufactured in the United States of America.

LIBRARY OF CONGRESS CATALOGING IN PUBLICATION DATA
Main entry under title:
Drainage basin morphology.
  (Benchmark papers in geology ; 41)
  Includes indexes.
  1. Watersheds—Addresses, essays, lectures. 2. Landforms—Addresses, essays, lectures. I. Schumm, Stanley Alfred, 1927–
GB561.D7   551.4'8   77-7365
ISBN 0-87933-284-0

Exclusive Distributor: **Halsted Press**
A Division of John Wiley & Sons, Inc.
ISBN: 0-470-99223-9

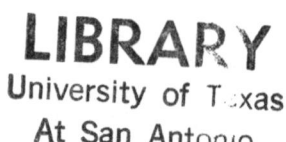

SERIES EDITOR'S FOREWORD

The philosophy behind the "Benchmark Papers in Geology" is one of collection, sifting, and rediffusion. Scientific literature today is so vast, so dispersed, and, in the case of old papers, so inaccessible for readers not in the immediate neighborhood of major libraries that much valuable information has been ignored by default. It has become just so difficult, or so time consuming, to search out the key papers in any basic area of research that one can hardly blame a busy man for skimping on some of his "homework."

This series of volumes has been devised, therefore, to make a practical contribution to this critical problem. The geologist, perhaps even more than any other scientist, often suffers from twin difficulties—isolation from central library resources and immensely diffused sources of material. New colleges and industrial libraries simply cannot afford to purchase complete runs of all the world's earth science literature. Specialists simply cannot locate reprints or copies of all their principal reference materials. So it is that we are now making a concerted effort to gather into single volumes the critical material needed to reconstruct the background of any and every major topic of our discipline.

We are interpreting "geology" in its broadest sense: the fundamental science of the planet Earth, its materials, its history, and its dynamics. Because of training and experience in "earthy" materials, we also take in astrogeology, the corresponding aspect of the planetary sciences. Besides the classical core disciplines such as mineralogy, petrology, structure, geomorphology, paleontology, and stratigraphy, we embrace the newer fields of geophysics and geochemistry, applied also to oceanography, geochronology, and paleoecology. We recognize the work of the mining geologists, the petroleum geologists, the hydrologists, the engineering and environmental geologists. Each specialist needs his working library. We are endeavoring to make his task a little easier.

Each volume in the series contains an Introduction prepared by a specialist (the volume editor)—a "state of the art" opening or a summary of the object and content of the volume. The articles, usually some twenty to fifty reproduced either in their entirety or in significant extracts, are selected in an attempt to cover the field, from the key papers of the last century to fairly recent work. Where the original works are in foreign

Series Editor's Foreword

languages, we have endeavored to locate or commission translations. Geologists, because of their global subject, are often acutely aware of the oneness of our world. The selections cannot, therefore, be restricted to any one country, and whenever possible an attempt is made to scan the world literature.

To each article, or group of kindred articles, some sort of "highlight commentary" is usually supplied by the volume editor. This commentary should serve to bring that article into historical perspective and to emphasize its particular role in the growth of the field. References, or citations, wherever possible, will be reproduced in their entirety—for by this means the observant reader can assess the background material available to that particular author, or, if he wishes, he, too, can double check the earlier sources.

A "benchmark," in surveyor's terminology, is an established point on the ground, recorded on our maps. It is usually anything that is a vantage point, from a modest hill to a mountain peak. From the historical viewpoint, these benchmarks are the bricks of our scientific edifice.

RHODES W. FAIRBRIDGE

PREFACE

It is the purpose of this collection of readings, the third in a trilogy relating to fluvial landforms (see *River Morphology*, S. A. Schumm and *Slope Morphology*, S. A. Schumm and M. P. Mosley, both in Benchmark Papers in Geology), to show that in spite of the morphologic complexity of drainage basins, they can be investigated quantitatively. In addition, their morphology can be explained and their erosional evolution described within the constraints imposed by geology, climate, soils, and hydrology.

The need for a fuller understanding of the dynamics of the fluvial system by geologists, geomorphologists, hydrologists, engineers, and planners is apparent, as environmental concerns increase. Although the components of a drainage basin may be modified by wind, glaciers, weathering, and subsurface solution, the most important process is running water. Therefore, the emphasis here is on the development, evolution, and morphology of drainage basins as a result of fluvial processes.

In view of the relatively recent development of drainage basin studies, many of the papers selected for this volume are not classic or benchmark papers in the normal sense, but they do inform the reader of progress in this geomorphic specialty and of potential research directions. The papers are grouped in five sections as follows: (1) traditional studies, (2) the Hortonian rejuvenation, (3) variables that influence the drainage basin, (4) erosional evolution of drainage basins, and (5) applications and directions of future research. It is hoped that this arrangement of the papers will lead the interested reader through the historical changes that have occurred in drainage basin geomorphology as well as to an appreciation of the magnitude of the research effort that remains.

STANLEY A. SCHUMM

CONTENTS

Series Editor's Foreword | v
Preface | vii
Contents by Author | xi
Introduction | 1

PART I: TRADITIONAL APPROACHES

Editor's Comments on Papers 1, 2, and 3 | 6

1. **DAVIS, W. M.:** The Development of Certain English Rivers | 9
   Geog. Jour. **5**:127-137, 142 (1895)

2. **DAVIS, W. M.:** The Geographical Cycle | 21
   Geog. Jour. **14**:481-504 (1899)

3. **ZERNITZ, E. R.:** Drainage Patterns and Their Significance | 45
   Jour. Geol. **40**:498-521 (1932)

PART II: HORTONIAN REJUVENATION

Editor's Comments on Papers 4 and 5 | 70

4. **HORTON, R. E.:** Erosional Development of Streams and Their Drainage Basins; Hydrophysical Approach to Quantitative Morphology | 73
   Geol. Soc. America Bull. **56**:275-370 (Mar. 1945)

5. **STRAHLER, A. N.:** Quantitative Analysis of Watershed Geomorphology | 169
   Am. Geophys. Union Trans. **38**(6):913-920 (1957)

PART III: DRAINAGE BASIN CONTROLS

Editor's Comments on Papers 6 Through 9 | 178

6. **CARLSTON, C. W.:** Drainage Density and Streamflow | 181
   U. S. Geol. Survey Prof. Paper 422-C:1-8 (1963)

7. **CHORLEY, R. J.:** Climate and Morphometry | 189
   Jour. Geol. **65**:628-638 (1957)

8. **LEOPOLD, L. B., and J. P. MILLER:** Ephemeral Streams—Hydraulic Factors and Their Relation to the Drainage Net | 200
   U. S. Geol. Survey Prof. Paper 282-A:16-24 (1956)

Contents

9   MELTON, M. A.: Correlation Structure of Morphometric Properties of Drainage Systems and Their Controlling Agents   210
Jour. Geol. **66**:442–460 (1958)

PART IV: DRAINAGE BASIN EVOLUTION

Editor's Comments on Papers 10 Through 13   230

10   JAGGAR, T. A., Jr.: Experiments Illustrating Erosion and Sedimentation   234
Harvard Univ. Mus. Comp. Zoology Bull. **49**:285–304 (Mar. 1908)

11   GLOCK, W. S.: The Development of Drainage Systems: A Synoptic View   261
Geog. Rev. **21**(3):475–482 (1931)

12   SCHUMM, S. A.: Evolution of Drainage Systems and Slopes in Badlands at Perth Amboy, New Jersey   269
Geol. Soc. America Bull. **67**:597–598, 599–622, 636–641, 645–646 (May 1956)

13   LEOPOLD, L. B., and W. B. LANGBEIN: The Concept of Entropy in Landscape Evolution   306
U. S. Geol. Survey Prof. Paper **500-A**:14–19 (1962)

PART V: APPLICATIONS AND DIRECTIONS

Editor's Comments on Papers 14 Through 17   314

14   HADLEY, R. F., and S. A. SCHUMM: Sediment Sources and Drainage Basin Characteristics in Upper Cheyenne River Basin   316
U. S. Geol. Survey Water-Supply Paper **1531-B**:169–176, 177 (1961)

15   MORISAWA, M. E.: Quantitative Geomorphology of Some Watersheds in the Appalachian Plateau   325
Geol. Soc. America Bull. **73**:1042–1045, 1028 (Sept. 1962)

16   HICKOK, R. B., R. V. KEPPEL, and B. R. RAFFERTY: Hydrograph Synthesis for Small Arid-Land Watersheds   330
Agricultural Eng. **40**(10):608–611 (1959)

17   SCHUMM, S. A.: Geomorphic Thresholds and Complex Response of Drainage Systems   335
Fluvial Geomorphology, M. E. Morisawa, ed., State University New York, Binghamton, 1973, pp. 299–310

Author Citation Index   347
Subject Index   351
About the Editor   353

CONTENTS BY AUTHOR

Carlston, C. W., 181
Chorley, R. J., 189
Davis, W. M., 9, 21
Glock, W. S., 261
Hadley, R. F., 316
Hickok, R. B., 330
Horton, R. E., 73
Jaggar, T. A., Jr., 234
Keppel, R. V., 330

Langbein, W. B., 306
Leopold, L. B., 200, 306
Melton, M. A., 210
Miller, J. P., 200
Morisawa, M. E., 325
Rafferty, B. R., 330
Schumm, S. A., 269, 316, 335
Strahler, A. N., 169
Zernitz, E. R., 45

DRAINAGE BASIN MORPHOLOGY

INTRODUCTION

The surface of the continents is a complex of rivers and slopes. Frequently, the earth scientist is so concerned with the morphology and dynamics of rivers and slopes that he fails to appreciate that they are but components of a larger unit, the drainage basin. A drainage basin may be only a few square meters in area or it may be a major river system such as the Nile, Amazon, or Mississippi. The drainage basin, regardless of its dimensions, is comprised of a drainage network of stream channels, the hills and mountains between the channels, and a drainage divide that is the boundary of the system. Obviously the drainage system is both a fundamental and complex unit of the landscape (Chorley, 1969).

In spite of the morphologic complexity inherent to the system and the complexity of its evolutionary development, which is imposed by climate, tectonic or eustatic change, there is order within the drainage basin as Playfair's Law makes clear (Horton, Paper 4, p. **78**). This order is important to the geologist who interprets the lithologic and structural variations of an area from its drainage patterns (Paper 3), to the hydrologist who relates drainage basin morphology to runoff and sediment yield and to other hydrologic aspects of a watershed (Papers 14, 15, and 16), and to the land manager who must predict drainage basin adjustment to man-induced changes of land use and climate (Paper 17).

Investigation of drainage basin morphology has followed two diverse courses. On the one hand geomorphologists have studied the inception and erosional evolution of drainage networks. Geologists, for the most part, have studied the morphology of the existing system in order to obtain clues to the structural and lithologic nature of the

Introduction

drainage basin and its hydrologic characteristics. Generally, stratigraphers and sedimentologists have not been concerned with drainage basins, but in fact, fluvial depositional systems can be significantly controlled by the evolutionary development and response of drainage basins to changing climatic and baselevel conditions. Therefore, it is necessary to understand both the evolutionary development of drainage basins and the influence of basin morphology on runoff, flood peaks, and sediment yield during every stage of their evolutionary development.

The geomorphic and hydrologic approaches were combined in a classic paper by R. E. Horton (Paper 4). The publication of this paper changed the emphasis of geomorphology and led to what has been referred to by at least one author as the Hortonian Revolution in geomorphology (Butzer, 1973; see also Salisbury, 1971). Horton's paper and a series of papers by A. N. Strahler (for reference, see Paper 5) introduced a period when the study of modern erosional processes became a major objective of geomorphologists throughout the world. The collection of basic data on modern erosional and depositional processes led in turn to a development of mathematical and theoretical models of drainage development. Unfortunately, basic data have not been collected sufficiently long enough for many of the theoretical models to be properly evaluated; hence, the emphasis in this volume will be on empirical relationships and applications based primarily on field and experimental studies.

Many of the field studies of drainage basins are restricted to relatively small areas in which the geology, climate, and vegetation are relatively uniform and erosion is measurable. In effect, students of drainage basins are handicapped by the numerous variables that influence drainage basin morphology, as well as by the slowness with which the drainage system changes through time; therefore, they have been forced to study small and simple drainage basins. In addition, because of their complexity, emphasis is usually placed on only one aspect of the drainage basin, hillslopes or channel morphology, for example, or on only one of the many variables that control drainage basin morphology.

Although many works have treated drainage basin morphology and evolution as part of a broader discussion of landforms (Leopold et al., 1964; Morisawa, 1968; Tricart, 1965; Scheidegger, 1970; Chorley and Kennedy, 1971; Haggett and Chorley, 1969), it is only very recently that a book has been published that is entirely devoted to this subject. Gregory and Walling's volume (1973) provides a comprehensive review of the morphology and dynamics of drainage basins, and it is strongly recommended.

REFERENCES

Butzer, K. W. 1973. Pluralism in geomorphology. *Assoc. Am. Geographers Proc.* 5:39-43.

Chorley, R. J. 1969. The drainage basin as the fundamental geomorphic unit. In *Water, Earth and Man*, ed. R. J. Chorley. London: Methuen and Co. pp. 77-99.

Chorley, R. J. and Kennedy, B. A. 1971. *Physical geography, a systems approach.* London: Prentice-Hall Internatl.

Gregory, K. J. and Walling, D. E. 1973. *Drainage basin form and process.* New York: John Wiley and Sons.

Haggett, P. and Chorley, R. J. 1969. *Network analysis in geography.* London: Edward Arnold.

Leopold, L. B.; Wolman, M. G; and Miller, J. P. 1964. *Fluvial processes in geomorphology.* San Francisco: W. H. Freeman.

Morisawa, M. 1968. *Streams, their dynamics and morphology.* New York: McGraw-Hill Book Co.

Salisbury, N. E. 1971. Threads of inquiry in quantitative geomorphology. In *Quantative geomorphology: Some aspects and applications*, ed. M. Morisawa, pp. 9-60. Publications in Geomorphology. Binghamton, N.Y.: SUNY.

Scheidegger, A. E. 1970. *Theoretical geomorphology*, 2nd edition. New York: Springer-Verlag.

Tricart, J. 1965. *Principles et methodes de la geómorphologie.* Paris: Masson et Cie.

Part I
TRADITIONAL APPROACHES

Editor's Comments on Papers 1, 2, and 3

1 **DAVIS**
Excerpts from *The Development of Certain English Rivers*

2 **DAVIS**
The Geographical Cycle

3 **ZERNITZ**
Drainage Patterns and Their Significance

A major objective of geomorphic research has been the description of the Tertiary and Quaternary erosional and depositional events that have produced the present landscape. The erosional or geographical cycle of W. M. Davis (Paper 2), which emphasized the erosional evolution of the landscape through time, provided a genetic classification of landforms for this purpose. In an earlier work Davis (Paper 1) classified and described the erosional development of rivers of eastern England. These are the first two papers in this volume because they illustrate the emphasis on classification and interpretation that was characteristic of early drainage basin and drainage pattern studies. Terminology that dominated discussion of river pattern morphology for decades is presented in these papers, and a sketch of the erosional cycle is introduced in Davis' 1895 paper prior to the fundamental paper of 1899. The excerpt from the 1895 paper is selected in preference to others that deal with the classification of stream channels (Davis, 1889, 1890) because of its brevity.

Davis dissects the drainage network into its components for purposes of discussion and explanation, and this is the technique used by Horton fifty years later in a quantitative analysis of drainage basin morphology (Paper 4).

The example of trellis drainage pattern evolution on a tilted coastal plain, as presented in Paper 1, is also typical of Appalachian Mountain drainage patterns, where much early geomorphic research was performed. In Paper 2, a comprehensive discussion of drainage basin evolution is presented with drainage patterns, slopes, and channel patterns receiving consideration. Furthermore, a discussion of interruptions of the ideal erosion cycle by baselevel fluctuation, climate change, and volcanic activity is presented. On page 25 the analogy to organic evolution is clearly stated. Darwin's influence on early geomorphic theory was, of course, highly significant. Note also the stress on geography. To Davis, geomorphology was a geographic discipline.

The qualitative description of streams, their assumed evolution, and their adjustment to geologic controls through time were characteristic of traditional geomorphology. Although Davis presented an idealized model of river pattern development and landform evolution, nevertheless, note the complexity of his scheme in comparison to later mathematical models of drainage system development (Paper 13). The depth of Davis' insights into the operation of geomorphic processes becomes apparent in his deduction that alluvium will be deposited in a river valley following rejuvenation and incision of the drainage system. This is clearly state in both Papers 1 and 2, and it is a very important concept that has been ignored until recently (Schumm, Paper 17 and 1976). For a full discussion of Davis' contributions see the definitive biography by Chorley et al. (1973).

The paper by Zernitz is a discussion of geologic effects on drainage patterns, and it logically follows the Davis 1898 paper by demonstrating the effect of structure on pattern. More recent papers have treated this subject in greater detail (Parvis, 1950; Howard, 1967), and most books on aerial photography interpretation include sections on this topic (Lueder, 1959; Miller and Miller, 1961). The application of drainage pattern interpretation to geologic mapping and especially to economic geology is very important. The characteristic drainage patterns reveal the effect of geologic controls on drainage systems. The Davis influence is apparent here, and there is no question that the qualitative descriptive approach utilized in this way has been of major significance in both scientific and economic terms.

These three papers are important contributions, and they are representative of pre-Horton drainage basin geomorphology. For additional information on the nature of this work the reader is directed to any geomorphology textbook, especially ones by Lobeck (1939) and Thornbury (1956).

REFERENCES

Chorley, R. J.; Beckinsale, R. P.; and Dunn, A. J. 1973. *The history of the study of landforms or the development of geomorphology*. The life and work of William Morris Davis, Vol. 2. London: Methuen and Co.

Davis, W. M. 1889. The rivers and valleys of Pennsylvania. *Natl. Geog. Mag.* 1:183-253.

———. 1890. The rivers of northern New Jersey with notes on the classification of rivers in general. *Natl. Geog. Mag.* 2:81-110.

Editor's Comments on Papers 1, 2, and 3

Howard, A. D. 1967. Drainage analysis in geologic interpretation: A summation. *Am. Assoc. Petroleum Geologists Bull.* **51**:2246-2259.

Lueder, D. R. 1959. *Aerial photographic interpretation.* New York: McGraw-Hill Book Co.

Lobeck, A. K. 1939. *Geomorphology.* New York: McGraw-Hill Book Co.

Miller, V. C. and Miller, C. F. 1961. *Photogeology.* New York: McGraw-Hill Book Co.

Parvis, M. 1950. Drainage pattern significance in air photo interpretation of soils and bedrocks, *Photogramm. Eng.* **16**:387-409.

Schumm, S. A. 1976. Episodic erosion: A modification of the geomorphic cycle. In *Theories of Landform Development,* eds. W. N. Melhorn and R. C. Flemal, pp. 69-85. Binghamton, N.Y.: SUNY.

Thornbury, W. D. 1956. *Principles of Geomorphology.* New York: John Wiley and and Sons.

THE DEVELOPMENT OF CERTAIN ENGLISH RIVERS.

By WILLIAM MORRIS DAVIS, Professor of Physical Geography, Harvard University.

Thesis.—The rivers of Eastern England have been developed in their present courses by the spontaneous growth of drainage lines on an original gently inclined plain, composed of sedimentary strata of varying resistance. In the course of this development, the land has been at least once worn down to a lowland of faint relief, and afterwards broadly uplifted, thus opening a second cycle of denudation, and reviving the rivers to new activities; and in the second cycle of denudation, the adjustment of streams to structures has been carried to a higher degree of perfection than it could have reached in the first cycle.

Outline.—1. Introduction. 2. Division of region. 3. Statement of the case. 4. General geological history of region. 5. Deductive scheme of river development.

6. The growth of subsequent streams. 7. Diagrams of river adjustments. 8. Conditions determining river adjustments. 9. Consequences of uplift of land-area. 10. Necessity of explicit statement of theory. 11. Evidence of uplift of land-area. 12. Marine or subaërial denudation. 13. Criteria for discrimination. 14. Subordinate conditions. 15. Consequences of theory confronted by facts. 16. The rivers of Eastern England. 17. Conclusion.

1. After much study of English maps and writings at home, I have been tempted to entertain certain theories regarding the development of English rivers, and to announce them in recent years to my college classes; but realizing the danger of theorizing at a distance, a part of the last summer has been given to an excursion over the ground, so as to test the theories by examination of the country and by conference with English students of the question. The experience has been most enjoyable, and in the hope of securing comment and discussion on a subject of so much interest in physical geography, the following statement of the problem and its results is submitted to those who may examine the question more deliberately.

2. Broadly speaking, England may be divided by the irregular lowland belt of New Red Sandstone into a region of older structures and more complicated forms on the west, and a region of younger structure and more simple forms on the east. The western region has three centres with various dependencies, all worthy of finer subdivision: the Lake District, with the great stretch of Carboniferous beds from Northumberland to Derbyshire; Wales, with the small isolated midland Carboniferous districts; and Cornwall. The eastern region, characterized by the escarpments of the Oölite and Chalk, with their uplands and dip slopes and Tertiary margins, is divided by the transverse depressions of the Humber and the Wash north of the London basin, and somewhat complicated by the uplift of the Weald in the south-east.

3. The rivers of the newer eastern region, and of the intermediate lowland, as well as some of the streams of the older western region, present certain geographical features of systematic arrangement that seem capable of explanation, because their arrangement falls into so close accord with the expectations concerning river growth that are deducible from the geological history of the region. In short, the rivers of to-day, in the mature stage of the present cycle of denudation, appear to be the revived and matured successors of a well-adjusted system of consequent and subsequent drainage inherited from an earlier and far-advanced cycle of denudation. Perhaps the best way of explaining what is meant by this condensed statement will be first to follow through an ideal but reasonable and pertinent system of river development, and afterwards to match its deductions with the facts presented by the existing rivers.

4. Imagine a well-varied mountainous region gradually wasting away by denudation, and as gradually sinking beneath the sea. Its windward

slopes will be well washed by sufficient rains; its leeward slopes will be drier, and, if the height of the mountains is great enough, the lowlands at the leeward base may possess saline lakes, around and within which the waste from the mountains will be deposited, with the remains of land animals and plants, or with forms representing fresh or brackish water. As the depression of the region continues, the mountains lose their former height both by wearing down and by sinking; the Piedmont lowlands may be submerged, and marine sediments will then bury the subaërial and lacustrine deposits, and encroach upon the fading mountains. The ridges will stand out as capes and promontories; the valleys will be drowned into estuaries or fiords; the streams will be shortened by the submergence of their lower courses, and often enough what was at first a single river-tree with many branches is now converted into several smaller independent rivers by the loss of the trunk stream in the drowned valley. Having frequently to speak of such dissevered rivers, which are very common in many parts of the world, I have fallen into the habit of calling them "betrunked;" the term being analogous to "beheaded," the two terms serving very well to suggest the opposite processes by which rivers are often affected.

The strata deposited during the submergence of the region will diminish in thickness off shore, and an increasing proportion of calcareous material may be expected as the neighbouring land-area decreases. The once lofty mountain summits, now worn down to a tolerably accordant height as the sea rises, may finally be quite submerged. Although the mountain range may have been denuded by thousands of feet during this long chapter of earth-history, and although the sediments furnished by so great a denudation may have accumulated to a thickness of thousands of feet in the encroaching sea, yet the littoral sea may not have been at most more than a hundred fathoms deep during the whole process; deposition may have so nearly balanced depression that shallow water was maintained during the whole period. Indeed, by introducing such episodes as halts in the depression, or even slight movements of elevation; by admitting climatic variations from dry to wet, from warm to cool, all manner of natural complications may be introduced in the series of strata by which the old mountains are finally buried.

5. Now reverse the process. Let gradual and intermittent elevation replace depression. Let the greater elevation be in the region of the old mountains. As the sea-bottom rises to form a new land, it will take the shape of a gently sloping plain. Streams will gather, guided by the slight inequalities of the constructional surface, and run down its faint incline to the sea, their length increasing as the shore recedes. If the uplift is uneven, with domes and troughs, the streams will assume an appropriate arrangement, in accordance with the special directions of the dip slopes. If certain residual summits of the old mountains were not completely drowned and buried at the time of greatest depression, the

short betrunked streams that ran from them will be the initial headwaters of the new drainage system on the rising coastal plain; but many entirely new streams will be formed on intermediate parts of the plain. The latter may be called simple consequent streams. The former will be old streams in their upper courses, surviving on the mountain heads from the former cycle of history, and now extended by the mouthward addition of consequent lower courses. Various independent old betrunked head-waters may in this way be engrafted on a single new consequent trunk stream. In describing existing river systems, it is certainly desirable to bear in mind the changes of this kind that the past may have played in bringing forth the present. The term "consequent" was introduced by Powell; "extended" has been used by Tarr in the sense here adopted.

Let there now be conceived a pause in the process of elevation after the uplift has reached a measure of 2000 or 3000 feet, and after the growing plain has attained a breadth of 100 or 200 miles. The consequent streams proceed to entrench themselves in the slanting plain, and in a geologically brief period, while they are yet young, they will cut their valleys down so close to baselevel that they cannot for the time being cut them any deeper; that is, the streams will, of their own accord, reduce their valley lines to such a grade that their capacity to do work shall be just equal to the work they have to do. When this condition is reached, the streams may be described as having attained a "profile of equilibrium;" or, more briefly, they may be said to be *graded*. It may be noted, in passing, that inasmuch as the work that a stream has to do is constantly varying, it must as constantly seek to assume new adjustments of grade. In the normal course of river events, undisturbed by outside interference, the change in the work is so slow that the desired adjustment of capacity to work is continually maintained. It may be that during the adolescence of river life, the work to be done is on the increase, on account of the increasingly rapid delivery of landwaste from the slopes of the growing valley branches; and in this case, part of the increase of waste must be laid down in the valley trough so as to steepen the grade, and thus enable the stream to gain capacity to carry the rest. Such a stream may be said to aggrade its valley, adopting a good term suggested by Salisbury; and in this way certain floodplains (but by no means all flood plains) may have originated. Aggrading of the valley line may often characterize the adolescence of a river's life; but later on, through maturity and old age, the work to be done decreases, and degrading is begun again, this time not to be interrupted. The longer the river works, the fainter is the grade that it adopts.

6. While the original consequent streams are thus at work cutting down, aggrading, and degrading their valley lines, lateral branches will be developed by headward erosion in greater or less number. In regions of tilted structure, with alternations of distinctly harder and softer

strata, the lateral branches will be developed along the strike of the softer masses; and inasmuch as such streams can be developed only after opportunity is offered to them by the deepening of the original consequent valleys, it seems appropriate to call them subsequent streams; the term "subsequent" having been used by Jukes in his important essay on certain valleys in the south of Ireland (*Quart. Jour. Geol. Soc.*, xviii., 1862), essentially in the meaning here given, although not as a technical term. The peculiarity of subsequent streams is, therefore, that they run along the strike of weak strata; while consequent streams run down the dip, crossing harder and softer strata alike. The stronger the dip, and the more marked the contrasts of harder and softer strata, the better the definition of the subsequent streams. Professor Green clearly recognizes streams of this kind, and gives examples and illustrations of them in his 'Geology' (pp. 425, 426), but he conceives them as beginning on a plain of marine denudation, not on the side of a consequent valley in an original constructional land surface. He calls them "longitudinal," thus not separating them from many longitudinal consequent streams, such as are found in the synclinal troughs of the Jura, or of which the Kennett-Thames may be taken as an example. He does not mention their relation to adjustments, considered below, and he seems to regard them as of exceptional or peculiar occurrence, for he refers to the subsequent streams of southern Ireland as "erratic" (p. 428), and calls the longitudinal and transverse drainage systems of the Oxford and the Weald districts "anomalous" (p. 429). It seems to me, on the contrary, that the drainage of these districts is singularly normal and systematic, as will be further shown below. Ramsay, Greenwood, Foster, Topley, Whitaker, and others, also explain examples of subsequent streams in various localities, but I believe in all cases they assume that a beginning was made on a plain of marine erosion, and the generality of the process by which subsequent streams are developed is hardly recognized.

In the case of the new coastal plain, imagined above, let it be supposed that the dip of the uplifted strata and the variations in their resistance is sufficient to give a fairly distinct guidance to the growth of subsequent branches of the consequent streams. A most interesting result follows from this supposition. The larger consequent streams will deepen their channels more quickly and to a lower grade than the smaller consequent streams. The lateral subsequent branches of the larger consequent streams will grow headward along the strike of the guiding weak strata more rapidly than the corresponding branches of the smaller consequent streams, and as a result the upper courses of the smaller consequent streams will be abstracted or diverted by the victorious subsequent streams, leaving the lower courses beheaded. Thus of all the original consequent streams, the smaller ones are naturally selected to be broken into two parts: an upper part, which is diverted by a subsequent stream into a larger consequent stream; and a lower part, which pursues its way alone to the sea. In this there arises a mature adjustment of streams to structures that is of great significance in river history.

7. Figs. 1 and 2 will, perhaps, make the scheme of adjustment more apparent. The straight consequent streams, a to i (Fig. 1), follow the direct courses suggested by the constructional surface, and flow immediately to the constructional shore-line. They differ in volume, c being the largest, and h the next, while f is the smallest. All of them go to work at once, cutting down their channels to the grade adapted to their volume and their load, and sending out subsequent branches wherever the faster wasting of softer strata on the valley slopes may determine,

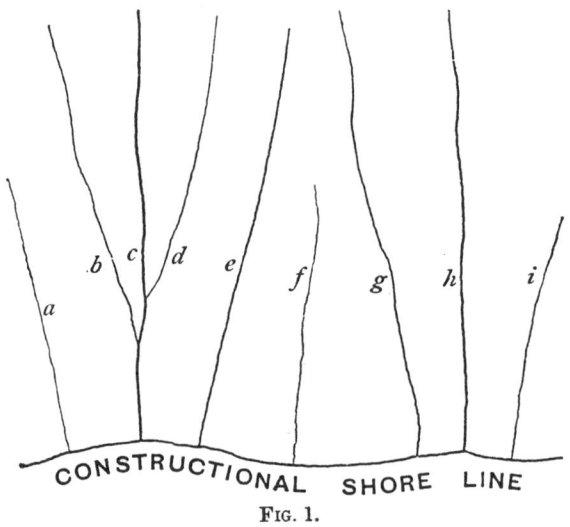

Fig. 1.

and at the same time the slowly retreating edges of the more resistant strata take the form of inland-facing escarpments, of which more is said below. Stream a is tapped in its upper course by m', a subsequent branch of b; the upper part of a is then diverted to swell the volume of b, while the lower beheaded part of a is reduced correspondingly. A pretty consequence of this change may sometimes be seen in the obstruction of the beheaded stream by the waste brought down by rivulets on its valley slopes; such being essentially the case of the lakes of the Engadine, at the head of the river Inn, as described by Heim. Moreover, for a short time after the capture is made, the diverted and diverting streams will flow in a local gorge, just above and below the "elbow of capture," as might be illustrated by special examples. In the same way, the upper part of b is tapped by n'; e is divided into three parts, being tapped by m'' and n''. It is manifest, that of the three captures made by n'', the first was d, and the last was g. At a still later time, even the head of h may be diverted by the further extension of n''. The mature arrangement of the streams, therefore, differs systematically from their arrangement when only their initial consequent courses existed, and any region which offers an example of so specialized a drainage system may be accepted as one of the most trustworthy witnesses to the theory of the uniformity of geological processes. No cataclysmic struggles could account for so particular a relation between internal structure and superficial drainage, as must necessarily result from the patient processes of inorganic natural selection here outlined.

Mr. Jukes-Brown has, of all English writers known to me, made the most explicit reference to the spontaneous rearrangement of river courses by the process here described ("On the Relative Ages of Certain River Valleys in Lincolnshire," *Quart. Jour. Geol. Soc.*, 1883, pp. 596–610); but he, like the other writers already referred to, assumes a plain of marine denudation as the beginning of the river history. His recognition of the general applicability of adjustments to explain river courses will appear from the following extract: "I believe that the principles above enunciated and exemplified will explain the courses of certain other English rivers, and purpose recurring to the subject in a future paper" (p. 610). So far as I know, the future paper here referred to has not yet been prepared.

Mr. Cadell has described certain valleys in the southern part of the Highlands of Scotland as rearranged by the spontaneous action of the streams themselves (*Scot. Geog. Mag.*, ii., 1886, pp. 337–347); but he ascribes the changes to the action of the rivers from up-stream downwards, and this does not seem admissible. All such changes are caused, not by the downward action of the stream that is to be diverted, but by the headward or backward action of some other stream, which for some reason has an advantage over the first. The change is always passive so far as the changed stream is concerned. Messrs. Foster and Topley at a much earlier date approached the subject of adjustments in their essay "On the Superficial Deposits of the Valley of the Medway" (*Quart. Jour. Geol. Soc.*, 1865, pp. 443–474), but the systematic relations of the various parts of an adjusted drainage system were not then fully worked out. It is chiefly from the writings of Gilbert, Heim, Löwl, and Philippson that I have gained the ideas here presented.

8. The degree of adjustment of mature streams varies with many determining conditions. Those which favour a high degree of adjustment are: 1. Considerable diversity in the size of the initial consequent streams; the old-land streams that are extended across the new coastal plain by the addition of consequent lower courses generally having an advantage on this account. 2. Considerable altitude of the land-mass exposed to erosion, thus allowing the larger consequent streams to cut valleys of a significantly greater depth during their adolescence than is

allowed to the smaller consequent streams. 3. Considerable diversity of resistance in the strata that are cut through by the streams; a very resistant stratum serving to increase the contrast of depth in the adolescent valleys of the larger and smaller streams, and a very weak stratum beneath the resistant stratum serving to guide the rapid headward growth of subsequent branch streams. 4. A significant amount of inclination in the strata, so as to give distinct guidance to the growth of the subsequent branches. When all these variable elements conspire to favour an adjustment of streams to structures, the systematic arrangement of land features becomes very striking during the mature stage of river life. The denuded and retreating margin of each resistant stratum takes the form of an inland facing escarpment: its dip-slope towards the shore is more or less stripped of any weaker strata that may have

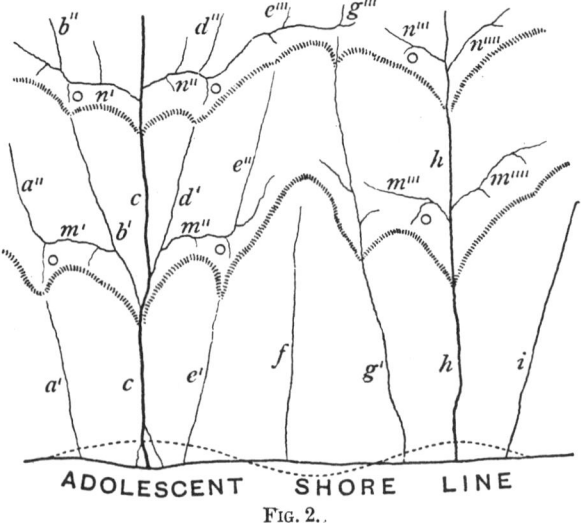

Fig. 2.

originally covered it; its escarpment face sheds short, back-flowing streams into the longitudinal subsequent valley that has been developed along the weak underlying stratum: and, even at the risk of multiplying terms unduly, I would suggest that these streams be called *obsequent*,[*] as their direction is opposed to that of the initial consequent streams. Several of them are indicated by the letter o in Fig. 2. In cases of recent capture, obsequent streams are wanting, as at the elbow where g''' turns into n''; in cases of more remote capture, the obsequent streams acquire all the drainage on the inland slope of the retreating escarpment. The crest of the escarpment turns inland and reaches greater height about midway between the consequent streams, but curves seaward and

---

[*] In earlier writings, I have used the term "inverted" for streams of this kind; but, as its meaning is equivocal, it should be abandoned.

loses height on approaching the transverse valleys that are kept open by the successful consequent streams. The notches in the escarpment, cut by unsuccessful consequent streams before their upper part was diverted, are of varying depth; generally being least perceptible nearest the largest successful or persistent consequent streams, for here the diversion was early accomplished, and most perceptible farthest from the persistent streams, for there the diversion was longest delayed and the notch was cut deepest before it was abandoned.

When the uplifted coastal plain contains more than one resistant stratum, there will be a corresponding increase in the number of inland facing escarpments, of longitudinal subsequent streams, and of adjustments, until a most beautiful complication is attained. The arrangement of the various parts will be remarkably systematic, hence when an explorer sees what seems to be any one of these parts, he should always look for its associated fellows, and in accordance with the perfection of their display he may easily and briefly describe the geographical features of the region that he traverses.

The best examples of the adjustment of streams to structures will be found in and shortly after that stage of the development of a region when its relief is strongest and most varied; that is, during maturity and advancing older age. But as time passes on, and even the harder rocks are worn down lower and lower, the streams meander somewhat freely from side to side, and depart to a greater or less degree from the close adjustments of maturity; yet we may seldom expect to find a region worn down so low in its extreme old age, that the streams shall have lost all traces of the adjustments that they must have once possessed. It is truly possible to imagine a plain of denudation so well completed that it has lost all perceptible relief; but in actual geographical experience, no such plains are discovered. Some residual relief occurs in all known examples of this kind, and, in order to emphasize this general truth, it seems to me advisable to call such almost-denuded surfaces, *peneplains*. The river systems of peneplains are in the condition of fading adjustment. A significant peculiarity of relief in a peneplain produced by the far-advanced denudation of a coastal region such as has been described above, is found in the increasing straightness of its fading escarpments.

9. Now, if it is not too fatiguing to pursue deductive considerations still further, let a second cycle of erosion be introduced by a new uplift of the peneplain that was produced in the first cycle. Thereupon a new strip of coastal plain will be added on the seaward margin of the peneplain; the old rivers of the first cycle will be extended across the new plain to the new shore; and (unless the slope of the new plain is less than the grade of the extended rivers, which is very unlikely) all the old rivers will be revived into new activity. From wandering idly on the old low peneplain, they will at once proceed to incise the newly uplifted peneplain. But in two significant respects, the features of the second cycle

differ from those of the first. In the first place, at the beginning of the initial cycle there were no subsequent streams; all the drainage was consequent. At the beginning of the second cycle, a considerable share of the drainage may be along revived subsequent streams; and with this opportunity so early afforded, the adjustments of the second cycle may exceed those of the first. Consequently, the adjustments in the maturity of a second cycle, following the older age of a first cycle, may reach a high degree of perfection. In the second place, the crest-lines of the escarpments or ridges in the second cycle will for some time retain the evenly bevelled form to which they were reduced at the close of the first cycle. When a region presents these two special features together, it can hardly be doubted that two cycles of subaërial denudation have been more or less completely passed through in its geographical development. A type of such an example of composite topography, that is, of a region whose features are referable to the work of two geographical cycles, is given in Fig. 3. The added strip of new coastal plain lies between the old and the new shore-lines. In their extension across it, the streams c and e, g and h, are engrafted in pairs. The escarpments, having been reduced to faint linear mounds at the close of the first cycle, now run straight along the strike of the controlling strata. The larger streams in particular possess strongly meandering courses, whereby they have departed more or less from their original lines. The subsequent streams have been much advanced, partly in the first cycle, partly in the second cycle; and they have shifted their courses a little down the dip of the weak strata that they follow. The identification of the several parts, as e', e'', e''', of a once continuous consequent stream becomes difficult and doubtful. The importance of the obsequent streams has somewhat increased. Supposing stream h to have been doubly beheaded by m'' and n'' in the second cycle, and therefore at a date after it had learned to meander in large curves, its lower course, h', will now be found "wriggling" in diminished volume along the large curves that it once followed smoothly. The stream c, the master of the region, with increasing volume from its successive captures, will fill its curves well, and will even extend their radius as it cuts down to a new grade in the second cycle.

10. Although this is an overlong introduction to the remainder of the article, it is absolutely essential to the appreciation of the points that I desire to make; for unless the expectations and consequences of a reasonable theory are well worked out, there are no definite mental conceptions to be confronted by the facts. The deductions must be made as carefully as the inductions, the two classes of knowledge being held carefully apart until their comparison is undertaken. Then just in accordance with the degree to which the deductions and inductions match one another will confidence be allowed or compelled in the theory by which the deductions were reached. As a matter of practical

experience, it is seldom that every deduction will find its mate among the inductions, and *vice versâ*. There will nearly always be some fact of observation not accounted for by theory; some expectation of theory not discovered among the facts; and no part of careful investigation is more exhilarating than the search thus excited in quest of the missing mates. Finally, when the pairing off is essentially complete, conviction grows spontaneously, and the will has little to do with the belief.

The theory here outlined seems reasonable, because it appeals to no extraordinary processes, and because every one of its requirements in the earlier or later stages of its advance may be compared with the facts of some actual district, where the inferred conditions really exist. The once lofty mountains in Wales and northward, near whose leeward base there were interior saline lakes, may be compared with the now lofty mountains of California and Oregon, which withhold the rainfall from the arid plains to the east; indeed, under no other conditions can a dry climate be produced near the eastern side of a great ocean and in so high a latitude as 50°. The relation of simple and extended consequent streams on a new coastal plain may be well studied in the rivers of North Carolina. The progressive development of escarpments and of adjusted rivers can be followed in various stages on different parts of the coastal plain of our Atlantic and Gulf states; or again in north-eastern France, where the interaction of certain rivers is most beautifully shown in the neighbourhood of Châlons-sur-Marne. The even crest-lines of ridges and escarpments in the maturity of a second cycle following the old age of a first, are wonderfully displayed in the Appalachians of Pennsylvania, where the adjustment of the rivers to the structures is also most admirable. The processes of river development are not newly invented, but are such as have been worked out by Jukes, Heim, Löwl, Philipson, and others in Europe; by Powell, Gilbert, and others in America.

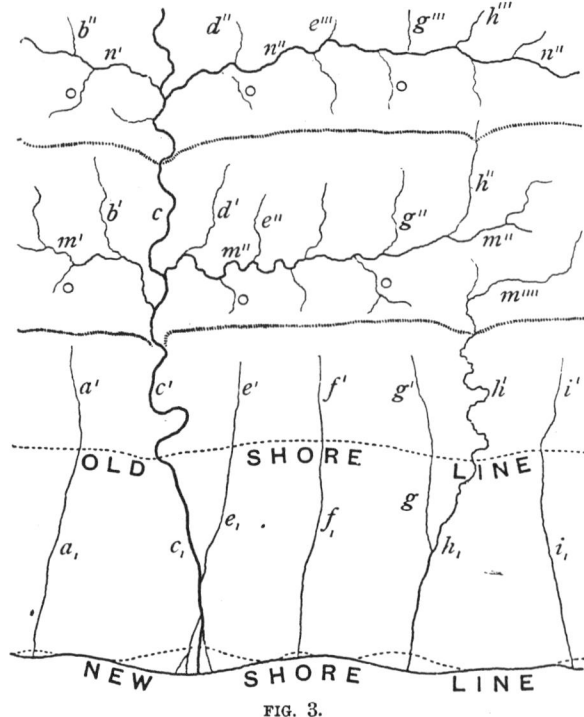

FIG. 3.

[*Editor's Note:* The remainder of this paper, which describes an application of the above scheme of drainage development to the interpretation of some English rivers, has been deleted.]

THE GEOGRAPHICAL CYCLE.

By WILLIAM M. DAVIS, Professor of Physical Geography in Harvar University.

THE GENETIC CLASSIFICATION OF LAND-FORMS.—All the varied forms of the lands are dependent upon—or, as the mathematician would say, are functions of—three variable quantities, which may be called structure, process, and time. In the beginning, when the forces of deformation and uplift determine the structure and attitude of a region, the form of its surface is in sympathy with its internal arrangement, and its height depends on the amount of uplift that it has suffered. If its rocks were unchangeable under the attack of external processes, its surface would remain unaltered until the forces of deformation and uplift acted again;

and in this case structure would be alone in control of form. But no rocks are unchangeable; even the most resistant yield under the attack of the atmosphere, and their waste creeps and washes downhill as long as any hills remain; hence all forms, however high and however resistant, must be laid low, and thus destructive process gains rank equal to that of structure in determining the shape of a land-mass. Process cannot, however, complete its work instantly, and the amount of change from initial form is therefore a function of time. Time thus completes the trio of geographical controls, and is, of the three, the one of most frequent application and of most practical value in geographical description.

Structure is the foundation of all geographical classifications in which the trio of controls is recognized. The Alleghany plateau is a unit, a "region," because all through its great extent it is composed of widespread horizontal rock-layers. The Swiss Jura and the Pennsylvanian Appalachians are units, for they consist of corrugated strata. The Laurentian highlands of Canada are essentially a unit, for they consist of greatly disturbed crystalline rocks. These geographical units have, however, no such simplicity as mathematical units; each one has a certain variety. The strata of plateaus are not strictly horizontal, for they slant or roll gently, now this way, now that. The corrugations of the Jura or of the Appalachians are not all alike; they might, indeed, be more truly described as all different, yet they preserve their essential features with much constancy. The disordered rocks of the Laurentian highlands have so excessively complicated a structure as at present to defy description, unless item by item; yet, in spite of the free variations from a single structural pattern, it is legitimate and useful to look in a broad way at such a region, and to regard it as a structural unit. The forces by which structures and attitudes have been determined do not come within the scope of geographical inquiry, but the structures acquired by the action of these forces serve as the essential basis for the genetic classification of geographical forms. For the purpose of this article, it will suffice to recognize two great structural groups: first, the group of horizontal structures, including plains, plateaus, and their derivatives, for which no single name has been suggested; second, the group of disordered structures, including mountains and their derivatives, likewise without a single name. The second group may be more elaborately subdivided than the first.

The destructive processes are of great variety—the chemical action of air and water, and the mechanical action of wind, heat, and cold, of rain and snow, rivers and glaciers, waves and currents. But as most of the land surface of the Earth is acted on chiefly by weather changes and running water, these will be treated as forming a normal group of destructive processes; while the wind of arid deserts and the ice of frigid deserts will be considered as climatic modifications of the norm, and set apart for particular discussion; and a special chapter will be needed to explain the action of waves and currents on the shore-lines at the edge of the lands. The various processes by which destructive work is done are in their turn geographical features, and many of them are well recognized as such, as rivers, falls, and glaciers; but they are too commonly considered by geographers apart from the work that they do, this phase of their study being, for some unsatisfactory reason, given over to physical geology. There should be no such separation of agency and work in physical geography, although it is profitable to give separate consideration to the active agent and to the inert mass on which it works.

TIME AS AN ELEMENT IN GEOGRAPHICAL TERMINOLOGY.—The amount of change caused by destructive processes increases with the passage of time, but neither the amount nor the rate of change is a simple function of time. The amount of change is limited, in the first place, by the altitude of a region above the sea; for, however long the time, the normal destructive forces cannot wear a land surface below this ultimate baselevel of their action; and glacial and marine forces cannot wear down a land-mass indefinitely beneath sea-level. The rate of change under normal processes, which alone will be considered for the present, is at the very first relatively moderate; it then advances rather rapidly to a maximum, and next slowly decreases to an indefinitely postponed minimum.

Evidently a longer period must be required for the complete denudation of a resistant than of a weak land-mass, but no measure in terms of years or centuries can now be given to the period needed for the effective wearing down of highlands to featureless lowlands. All historic time is hardly more than a negligible fraction of so vast a duration. The best that can be done at present is to give a convenient name to this unmeasured part of eternity, and for this purpose nothing seems more appropriate than a "*geographical cycle.*" When it is possible to establish a ratio between geographical and geological units, there will probably be found an approach to equality between the duration of an average cycle and that of Cretaceous or Tertiary time, as has been indicated by the studies of several geomorphologists.

"THEORETICAL" GEOGRAPHY.—It is evident that a scheme of geographical classification that is founded on structure, process, and time, must be deductive in a high degree. This is intentionally and avowedly the case in the present instance. As a consequence, the scheme gains a very "theoretical" flavour that is not relished by some geographers, whose work implies that geography, unlike all other sciences, should be developed by the use of only certain ones of the mental faculties, chiefly observation, description, and generalization. But nothing seems to me clearer than that geography has already suffered too long from the disuse of imagination, invention, deduction, and the various other mental faculties that contribute towards the attainment of a well-tested explanation. It is like walking on one foot, or looking with one eye, to exclude from geography the "theoretical" half of the brain-power, which other sciences call upon as well as the "practical" half. Indeed, it is only as a result of misunderstanding that an antipathy is implied between theory and practice, for in geography, as in all sound scientific work, the two advance most amiably and effectively together. Surely the fullest development of geography will not be reached until all the mental faculties that are in any way pertinent to its cultivation are well trained and exercised in geographical investigation.

All this may be stated in another way. One of the most effective aids to the appreciation of a subject is a correct explanation of the facts that it presents. Understanding thus comes to aid the memory. But a genetic classification of geographical forms is, in effect, an explanation of them; hence such a classification must be helpful to the travelling, studying, or teaching geographer, provided only that it is a true and natural classification. True and natural a genetic classification may certainly be, for the time is past when even geographers can look on the forms of lands as "ready made." Indeed, geographical definitions and descriptions are untrue and unnatural just so far as they give the impression that the forms of the lands are of unknown origin, not susceptible of rational explanation. From the very beginning of geography in the lower schools, the pupils should be possessed with the belief that geographical forms have meaning, and that the meaning or origin of so many forms is already so well assured that there is every reason to think that the meaning of all the others will be discovered in due time. The explorer of the Earth should be as fully convinced of this principle, and as well prepared to apply it, as the explorer of the sky is to carry physical principles to the furthest reach of his telescope, his spectroscope, and his camera. The preparation of route-maps and the determination of latitude, longitude, and altitude for the more important points is only the beginning of exploration, which has no end till all the facts of observation are carried forward to explanation.

It is important, however, to insist that the geographer needs to know the meaning, the explanation, the origin, of the forms that he looks at, simply because of the aid thus received when he attempts to observe and describe the forms carefully. It is necessary clearly to recognize this principle, and constantly to bear it in mind, if we would avoid the error of confounding the objects of geographical and geological study. The latter examines the changes of the past for their own sake, inasmuch as geology is concerned with the history of the Earth; the former examines the changes of the past only so far as they serve to illuminate the present, for geography is concerned essentially with the Earth as it now exists. Structure is a pertinent element of geographical study when, as nearly always, it influences form; no one would to-day attempt to describe the Weald without some reference to the resistant chalk layers that determine its rimming hills. Process is equally pertinent to our subject, for it has everywhere been influential in determining form to a greater or less degree, and it is everywhere in operation to-day. It is truly curious to find geographical text-books which accept the movement of winds, currents, and rivers as part of their responsibility, and yet which leave the weathering of the lands and the movement of land-waste entirely out of consideration. Time is certainly an important geographical element, for where the forces of uplift or deformation have lately (as the Earth views time) initiated a cycle of change, the destructive processes can have accomplished but little work, and the land-form is "young;" where more time has elapsed, the surface will have been more thoroughly carved, and the form thus becomes "mature;" and where so much time has passed that the originally uplifted surface is worn down to a lowland of small relief, standing but little above sea-level, the form deserves to be called "old." A whole series of forms must be in this way evolved in the life-history of a single region, and all the forms of such a series, however unlike they may seem at first sight, should be associated under the element of time, as merely expressing the different stages of development of a single structure. The larva, the pupa, and the imago of an insect; or the acorn, the full-grown oak, and the fallen old trunk, are no more naturally associated as representing the different phases in the life-history of a single organic species, than are the young mountain block, the maturely carved mountain-peaks and valleys, and the old mountain peneplain, as representing the different stages in the life-history of a single geographic group. Like land-forms, the agencies that work upon them change their behaviour and their appearance with the passage of time. A young land-form has young streams of torrential activity, while an old form would have old streams of deliberate or even of feeble current, as will be more fully set forth below.

THE IDEAL GEOGRAPHICAL CYCLE.—The sequence in the developmental changes of land-forms is, in its own way, as systematic as the sequence of changes found in the more evident development of organic forms. Indeed, it is chiefly for this reason that the study of the origin of land-forms—or geomorphogeny, as some call it—becomes a practical aid, helpful to the geographer at every turn. This will be made clearer by the specific consideration of an ideal case, and here a graphic form of expression will be found of assistance.

The base-line, $a\omega$, of Fig. 1 represents the passage of time, while verticals above the base-line measure altitude above sea-level. At the epoch 1, let a region of whatever structure and form be uplifted, B representing the average altitude of its higher parts, and A that of its lower parts; thus AB measuring its average initial relief. The surface rocks are attacked by the weather. Rain falls on the weathered surface, and washes some of the loosened waste down the initial slopes to the trough-lines where two converging slopes meet; there the streams are formed, flowing in directions consequent upon the descent of the trough-lines. The machinery of the destructive processes is thus put in motion, and the destructive development of the region is begun. The larger rivers, whose channels initially had an altitude, A, quickly deepen their valleys, and at the epoch 2 have reduced their main channels to a moderate altitude, represented by C. The higher parts of the inter-stream uplands, acted on only by the weather without the concentration of water in streams, waste away much more slowly, and at epoch 2 are reduced in height only to D. The relief of the surface has thus been increased from AB to CD. The main rivers then deepen their channels very slowly for the rest of their life, as shown by the curve CEGJ; and the wasting of the uplands, much dissected by branch streams, comes to be more rapid than the deepening of the main valleys, as shown by comparing the curves DFHK and CEGJ. The period 3-4 is the time of the most rapid consumption of the uplands, and

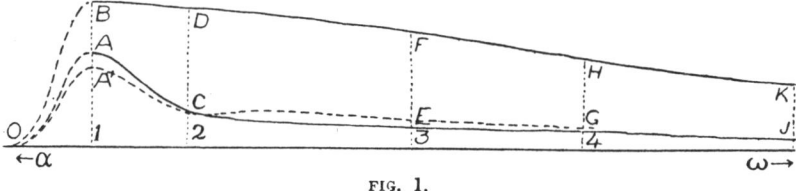

FIG. 1.

thus stands in strong contrast with the period 1-2, when there was the most rapid deepening of the main valleys. In the earlier period, the relief was rapidly increasing in value, as steep-sided valleys were cut beneath the initial troughs. Through the period 2-3 the maximum value of relief is reached, and the variety of form is greatly increased by the headward growth of side valleys. During the period 3-4 relief is decreasing faster than at any other time, and the slope of the valley sides is becoming much gentler than before; but these changes advance much more slowly than those of the first period. From epoch 4 onward the remaining relief is gradually reduced to smaller and smaller measures, and the slopes become fainter and fainter, so that some time after the latest stage of the diagram the region is only a rolling lowland, whatever may have been its original height. So slowly do the later changes advance, that the reduction of the reduced relief JK to half of its value might well require as much time as all that which has already elapsed; and from the gentle slopes that would then remain, the further removal of waste must indeed be exceedingly slow. The frequency of torrential floods and of landslides in young and in mature mountains, in contrast to the quiescence of the sluggish streams and the slow movement of the soil on lowlands of denudation, suffices to show that rate of denudation is a matter of strictly geographical as well as of geological interest.

It follows from this brief analysis that a geographical cycle may be subdivided into parts of unequal duration, each one of which will be characterized by the strength and variety of relief, and by the rate of change, as well as by the amount of change that has been accomplished since the initiation of the cycle. There will be a brief youth of rapidly increasing relief, a maturity of strongest relief and greatest variety of form, a transition period of most rapidly yet slowly decreasing relief, and an indefinitely long old age of faint relief, on which further changes are exceedingly slow. There are, of course, no breaks between these subdivisions or stages; each one merges into its successor, yet each one is in the main distinctly characterized by features found at no other time.

THE DEVELOPMENT OF CONSEQUENT STREAMS.—The preceding section gives only the barest outline of the systematic sequence of changes that run their course through a geographical cycle. The outline must be at once gone over, in order to fill in the more important details. In the first place, it should not be implied, as was done in Fig. 1, that the forces of uplift or deformation act so rapidly that no destructive changes occur during their operation. A more probable relation at the opening of a cycle of change places the beginning of uplift at O (Fig. 1), and its end at 1. The divergence of the curves OB and OA then implies that certain parts of the disturbed region were uplifted more than others, and that, from a surface of no relief at sea-level at epoch O, an upland having AB relief would be produced at epoch 1. But even during uplift, the streams that gather in the troughs as soon as they are defined do some work, and hence young valleys are already incised in the trough-bottoms when epoch 1 is reached, as shown by the curve OA'. The uplands also waste more or less during the period of disturbance, and hence no absolutely unchanged initial surface should be found, even for some time anterior to epoch 1. Instead of looking for initial divides separating initial slopes that descend to initial troughs followed by initial streams, such as were implied in Fig. 1 at the epoch of instantaneous uplift, we must always expect to find some greater or less advance in the sequence of developmental changes, even in the youngest known land-forms. "Initial" is therefore a term adapted to ideal rather than to actual cases, in treating which the term "sequential" and its derivatives will be found more appropriate. All the changes which directly follow the guidance of the ideal initial forms may be called consequent; thus a young form would possess consequent divides, separating consequent slopes which descend to consequent valleys; the initial troughs being changed to consequent valleys in so far as their form is modified by the action of the consequent drainage.

THE GRADE OF VALLEY FLOORS.—The larger rivers soon—in terms of the cycle—deepen their main valleys, so that their channels are but little above the baselevel of the region; but the valley floor cannot be reduced to the absolute baselevel, because the river must slope down to its mouth at the sea-shore. The altitude of any point on a well-matured valley floor must therefore depend on river-slope and distance from mouth. Distance from mouth may here be treated as a constant, although a fuller statement would consider its increase in consequence of delta-growth. River-slope cannot be less, as engineers know very well, than a certain minimum that is determined by volume and by quantity and texture of detritus or load. Volume may be temporarily taken as a constant, although it may easily be shown to suffer important changes during the progress of a normal cycle. Load is small at the beginning, and rapidly increases in quantity and coarseness during youth, when the region is entrenched by steep-sided valleys; it continues to increase in quantity, but probably not in coarseness, during early maturity, when ramifying valleys are growing by headward erosion, and are thus increasing the area of wasting slopes; but after full maturity, load continually decreases in quantity and in coarseness of texture; and during old age, the small load that is carried must be of very fine texture or else must go off in solution. Let us now consider how the minimum slope of a main river will be determined.

In order to free the problem from unnecessary complications, let it be supposed that the young consequent rivers have at first slopes that are steep enough to make them all more than competent to carry the load that is washed into them from the wasting surface on either side, and hence competent to entrench themselves beneath the floor of the initial troughs,—this being the condition tacitly postulated in Fig. 1, although it evidently departs from those cases in which deformation produces basins where lakes must form and where deposition (negative denudation) must take place, and also from those cases in which a main-trough stream of moderate slope is, even in its youth, over-supplied with detritus by active side streams that descend steep and long wasting surfaces; but all these more involved cases may be set aside for the present.

If a young consequent river be followed from end to end, it may be imagined as everywhere deepening its valley, unless at the very mouth. Valley-deepening will go on most rapidly at some point, probably nearer head than mouth. Above this point the river will find its slope increased; below, decreased. Let the part up-stream from the point of most rapid deepening be called the headwaters; and the part down-stream, the lower course or trunk. In consequence of the changes thus systematically brought about, the lower course of the river will find its slope and velocity decreasing, and its load increasing; that is, its ability to do work is becoming less, while the work that it has to do is becoming greater. The original excess of ability over work will thus in time be corrected, and when an equality of these two quantities is brought about, the river is *graded*, this being a simple form of expression, suggested by Gilbert, to replace the more cumbersome phrases that are required by the use of " profile of equilibrium " of French engineers. When the graded condition is reached, alteration of slope can take place only as volume and load change their relation; and changes of this kind are very slow.

In a land-mass of homogeneous texture, the graded condition of a river would be (in such cases as are above considered) first attained at the mouth, and would then advance retrogressively up-stream. When the trunk streams are graded, early maturity is reached; when the smaller headwaters and side streams are also graded, maturity is far advanced; and when even the wet-weather rills are graded, old age is attained. In a land-mass of heterogeneous texture, the rivers will be divided into sections by the belts of weaker and stronger rocks that they traverse; each section of weaker rocks will in due time be graded with reference to the section of harder rock next down-stream, and thus the river will come to consist of alternating quiet reaches and hurried falls or rapids. The less resistant of the harder rocks will be slowly worn down to grade with respect to the more resistant ones that are further down stream; thus the rapids will decrease in number, and only those on the very strongest rocks will long survive. Even these must vanish in time, and the graded condition will then be extended from mouth to head. The slope that is adopted when grade is assumed varies inversely with the volume; hence rivers retain steep headwaters long after their lower course is worn down almost level; but in old age, even the headwaters must have a gentle declivity and moderate velocity, free from all torrential features. The so-called "normal river," with torrential headwaters and well-graded middle and lower course, is therefore simply a maturely developed river. A young river may normally have falls even in its lower course, and an old river must be free from rapid movement even near its head.

If an initial consequent stream is for any reason incompetent to carry away the load that is washed into it, it cannot degrade its channel, but must aggrade instead (to use an excellent term suggested by Salisbury). Such a river then lays down the coarser part of the offered load, thus forming a broadening flood-land, building up its valley floor, and steepening its slope until it gains sufficient velocity to do the required work. In this case the graded condition is reached by filling up the initial trough instead of by cutting it down. Where basins occur, consequent lakes rise in them to the level of the outlet at the lowest point of the rim. As the outlet is cut down, it forms a sinking local baselevel with respect to which the basin is aggraded; and as the lake is thus destroyed, it forms a sinking baselevel with respect to which the tributary streams grade their valleys; but, as in the case of falls and rapids, the local baselevels of outlet and lake are temporary, and lose their control when the main drainage lines are graded with respect to absolute baselevel in early or late maturity.

THE DEVELOPMENT OF RIVER BRANCHES. — Several classes of side streams may be recognized. Some of them are defined by slight initial depressions in the side slopes of the main river-troughs: these form lateral or secondary consequents, branching from a main consequent; they generally run in the direction of the dip of the strata. Others are developed by headward erosion under the guidance of weak substructures that have been laid bare on the valley walls of the consequent streams: they follow the strike of the strata, and are entirely regardless of the form of the initial land surface; they may be called subsequent, this term having been used by Jukes in describing the development of such streams. Still others grow here and there, to all appearance by accident, seemingly independent of systematic guidance; they are common in horizontal or massive structures. While waiting to learn just what their control may be, their independence of apparent control may be indicated by calling them "insequent." Additional classes of streams are well known, but cannot be described here for lack of space.

RELATION OF RIVER ABILITY AND LOAD. — As the dissection of a land-mass proceeds with the fuller development of its consequent, subsequent, and insequent streams, the area of steep valley sides greatly increases from youth into early and full maturity. The waste that is delivered by the side branches to the main stream comes chiefly from the valley sides, and hence its quantity increases with the increase of strong dissection, reaching a maximum when the formation of new branch streams ceases, or when the decrease in the slope of the wasting valley sides comes to balance their increase of area. It is interesting to note in this connection the consequences that follow from two contrasted relations of the date for the maximum discharge of waste and of that for the grading of the trunk streams. If the first is not later than the second, the graded rivers will slowly assume gentler slopes as their load lessens; but as the change in the discharge of waste is almost infinitesimal compared to the amount discharged at any one time, the rivers will essentially preserve their graded condition in spite of the minute excess of ability over work. On the other hand, if the maximum of load is not reached until after the first attainment of the graded condition by the trunk rivers, then the valley floors will be aggraded by the deposition of a part of the increasing load, and thus a steeper slope and a greater velocity will be gained whereby the remainder of the increase can be borne along. The bottom of the V-shaped valley, previously carved, is thus slowly filled with a gravelly flood-plain, which continues to rise until the epoch of the maximum load is reached, after which the slow degradation above stated is entered upon. Early maturity may therefore witness a slight shallowing of the main valleys, instead of the slight deepening (indicated by the dotted line CE in Fig. 1); but late maturity and all old age will be normally occupied by the slow continuation of valley erosion that was so vigorously begun during youth.

THE DEVELOPMENT OF DIVIDES.—There is no more beautiful process to be found in the systematic advance of a geographical cycle than the definition, subdivision, and rearrangement of the divides (water-partings) by which the major and minor drainage basins are separated. The forces of crustal upheaval and deformation act in a much broader way than the processes of land-sculpture; hence at the opening of a cycle one would expect to find a moderate number of large river-basins, somewhat indefinitely separated on the flat crests of broad swells or arches of land surface, or occasionally more sharply limited by the raised edge of faulted blocks. The action of the lateral consequent streams alone would, during youth and early maturity, sharpen all the vague initial divides into well-defined consequent divides, and the further action of insequent and subsequent streams would split up many consequent drainage slopes into subordinate drainage basins, separated by sub-divides either insequent or subsequent. Just as the subsequent valleys are eroded by their gnawing streams along weak structural belts, so the subsequent divides or ridges stand up where maintained by strong structural belts. However imperfect the division of drainage areas and the discharge of rainfall may have been in early youth, both are well developed by the time full maturity is reached. Indeed, the more prompt discharge of rainfall that may be expected to result from the development of an elaborate system of subdivides and of slopes from divides to streams should cause an increased percentage of run-off; and it is possible that the increase of river-volume thus brought about from youth to maturity may more or less fully counteract the tendency of increase in river load to cause aggradation. But, on the other hand, as soon as the uplands begin to lose height, the rainfall must decrease; for it is well known that the obstruction to wind-movement caused by highlands is an effective cause of precipitation. While it is a gross exaggeration to maintain that the quaternary Alpine glaciers caused their own destruction by reducing the height of the mountains on which their snows were gathered, it is perfectly logical to deduce a decrease of precipitation as an accompaniment of loss of height from the youth to the old age of a land-mass. Thus many factors must be considered before the life-history of a river can be fully analyzed.

The growth of subsequent streams and drainage areas must be at the expense of the original consequent streams and consequent drainage areas. All changes of this kind are promoted by the occurrence of inclined instead of horizontal rock-layers, and hence are of common occurrence in mountainous regions, but rare in strictly horizontal plains. The changes are also favoured by the occurrence of strong contrasts in the resistance of adjacent strata. In consequence of the migration of divides thus caused, many streams come to follow valleys that are worn down along belts of weak strata, while the divides come to occupy the ridges that stand up along the belts of stronger strata; in other words, the simple consequent drainage of youth is modified by the development of subsequent drainage lines, so as to bring about an *increasing adjustment of streams to structures*, than which nothing is more characteristic of the mature stage of the geographical cycle. Not only so: adjustments of this kind form one of the strongest, even if one of the latest, proofs of the erosion of valleys by the streams that occupy them, and of the long continued action in the past of the slow processes of weathering and washing that are in operation to-day.

There is nothing more significant of the advance in geographical development than the changes thus brought about. The processes here involved are too complicated to be now presented in detail, but they may be briefly illustrated by taking the drainage of a denuded arch, suggested

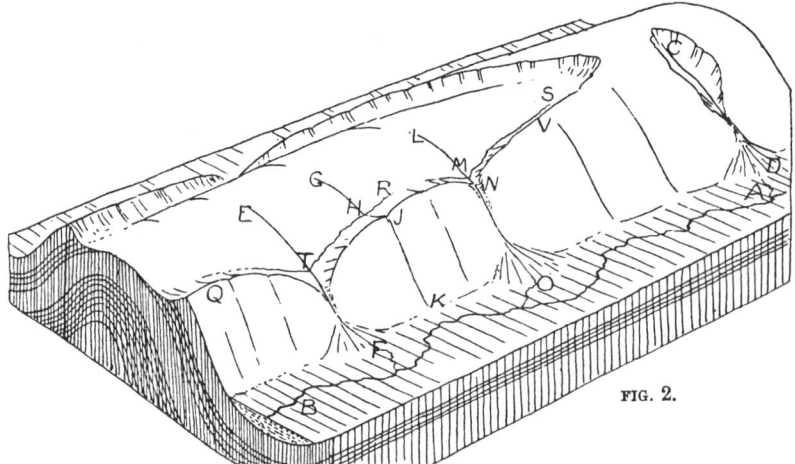

FIG. 2.

by the Jura mountains, as a type example. AB, Fig. 2, is a main longitudinal consequent stream following a trough whose floor has been somewhat aggraded by the waste actively supplied by the lateral consequents, CD, LO, EF, etc. At an earlier stage of denudation, before the hard outer layer was worn away from the crown of the mountain arch, all the lateral consequents headed at the line of the mountain crest. But, guided by a weak under-stratum, subsequent streams, TR, MS, have been developed as the branches of certain lateral consequents, EF, LO, and thus the hard outer layer has been undermined and partly removed, and many small lateral consequents have been beheaded. To-day, many of the laterals, like JK, have their source on the crest of the lateral ridge VJQ, and the headwaters, such as GH, that once belonged to them, are now diverted by the subsequent streams to swell the volume of the more successful laterals, like EF. Similar changes having taken place on the further slope of the mountain arch, we now find the original consequent divide of the arch-crest supplemented by the subsequent divides formed by the lateral ridges. A number of short streams, like JH, belonging to a class not mentioned above, run down the inner face of the lateral ridges to a subsequent stream, RT. These short streams have a direction opposite to that of the original consequents, and may therefore be called obsequents. As denudation progresses, the edge of the lateral ridge will be worn further from the arch-crest; in other words, the subsequent divide will migrate towards the main valley, and thus a greater length will be gained by the diverted consequent headwaters, GH, and a greater volume by the subsequents, SM and RT. During these changes the inequality that must naturally prevail between adjacent successful consequents, EF and LO, will eventually allow the subsequent branch, RT, of the larger consequent, EF, to capture the headwaters, LM and SM, of the smaller consequent, LO. In late maturity the headwaters of so many lateral consequents may be diverted to swell the volume of EF, that the main longitudinal consequent above the point F may be reduced to relatively small volume.

THE DEVELOPMENT OF RIVER MEANDERS.—It has been thus far implied that rivers cut their channels vertically downward, but this is far from being the whole truth. Every turn in the course of a young consequent stream causes the stronger currents to press toward the outer bank, and each irregular, or, perhaps, subangular bend is thus rounded out to a comparatively smooth curve. The river therefore tends to depart from its irregular initial path (background block of Fig. 3) towards a serpentine course, in which it swings to right and left over a broader belt than at first. As the river cuts downwards and outwards at the same time, the valley-slopes become unsymmetrical (middle block of Fig. 3), being steeper on the side toward which the current is urged by centrifugal force. The steeper valley side thus gains the form of a half-amphitheatre, into which the gentler sloping side enters as a spur of the opposite uplands. When the graded condition is attained by the stream, downward cutting practically ceases, but outward cutting continues; a normal flood-plain is then formed as the channel is withdrawn from the gently sloping side of the valley (foreground block of Fig. 3). Flood-plains of this kind are easily distinguished in their early stages from those already mentioned (formed by aggrading the flat courses of incompetent young rivers, or by aggrading the graded valleys of overloaded rivers in early maturity); for these occur in detached lunate areas, first on one side, then on the other side of the stream, and always systematically placed at the foot of the gentler sloping spurs. But, as time passes, the river impinges on the up-stream side, and withdraws from the down-stream side of every spur, and thus the spurs are gradually consumed; they are first sharpened, so as better to observe their name; they are next reduced to short cusps; then they are worn back to blunt salients; and finally, they are entirely consumed, and the river wanders freely on its open flood-plain, occasionally swinging against the valley side, now here, now there. By this time the curves of youth are changed into systematic meanders, of radius appropriate to river volume; and, for all the rest of an undisturbed life, the river persists in the habit of serpentine flow. The less the slope of the flood-plain becomes in advancing old age, the larger the arc of each meander,

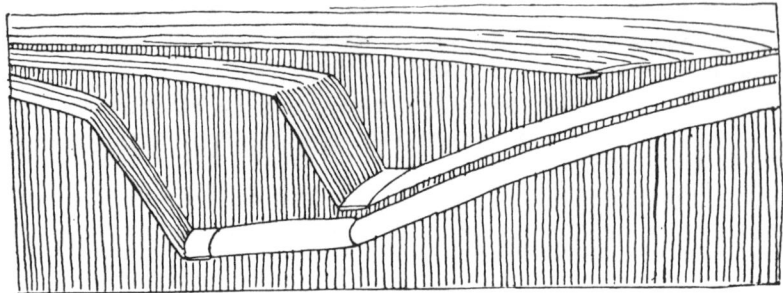

FIG. 3.

and hence the longer the course of the river from any point to its mouth. Increase of length from this cause must tend to diminish fall, and thus to render the river less competent than it was before; and the result of this tendency will be to retard the already slow process by which a gently sloping flood-plain is degraded so as to approach coincidence with a level surface; but it is not likely that old rivers often remain undisturbed long enough for the full realization of these theoretical conditions.

The migration of divides must now and then result in a sudden increase in the volume of one river and in a correspondingly sudden decrease of another. After such changes, accommodation to the changed volume must be made in the meanders of each river affected. The one that is increased will call for enlarged dimensions; it will usually adopt a gentler slope, thus terracing its flood-plain, and demand a greater freedom of swinging, thus widening its valley. The one that is decreased will have to be satisfied with smaller dimensions; it will wander aimlessly in relatively minute meanders on its flood-plain, and from increase of length, as well as from loss of volume, it will become incompetent to transport the load brought in by the side streams, and thus its flood-plain must be aggraded. There are beautiful examples known of both these peculiar conditions.

THE DEVELOPMENT OF GRADED VALLEY SIDES.—When the migration of divides ceases in late maturity, and the valley floors of the adjusted streams are well graded, even far toward the headwaters, there is still to be completed another and perhaps even more remarkable sequence of systematic changes than any yet described: this is the development of graded waste slopes on the valley sides. It is briefly stated that valleys are eroded by their rivers; yet there is a vast amount of work performed in the erosion of valleys in which rivers have no part. It is true that rivers deepen the valleys in the youth, and widen the valley floors during the maturity and old age of a cycle, and that they carry to the sea the waste denuded from the land; it is this work of transportation to the sea that is peculiarly the function of rivers; but the material to be transported is supplied chiefly by the action of the weather on the steeper consequent slopes and on the valley sides. The transportation of the weathered material from its source to the stream in the valley bottom is the work of various slow-acting processes, such as the surface wash of rain, the action of ground water, changes of temperature, freezing and thawing, chemical disintegration and hydration, the growth of plant-roots, the activities of burrowing animals. All these cause the weathered rock waste to wash and creep slowly downhill, and in the motion thus ensuing there is much that is analogous to the flow of a river. Indeed, when considered in a very broad and general way, a river is seen to be a moving mixture of water and waste in variable proportions, but mostly water; while a creeping sheet of hillside waste is a moving mixture of waste and water in variable proportions, but mostly waste. Although the river and the hillside waste-sheet do not resemble each other at first sight, they are only the extreme members of a continuous series; and when this generalization is appreciated, one may fairly extend the "river" all over its basin, and up to its very divides. Ordinarily treated, the river is like the veins of a leaf; broadly viewed, it is like the entire leaf. The verity of this comparison may be more fully accepted when the analogy, indeed, the homology, of waste-sheets and water-streams is set forth.

In the first place, a waste-sheet moves fastest at the surface and slowest at the bottom, like a water-stream. A graded waste-sheet may be defined in the very terms applicable to a graded water-stream; it is one in which the ability of the transporting forces to do work is equal to the work that they have to do. This is the condition that obtains on those evenly slanting, waste-covered mountain-sides which have been reduced to a slope that engineers call "the angle of repose," because of the apparently stationary condition of the creeping waste, but that should be called, from the physiographic standpoint, "the angle of first-developed grade." The rocky cliffs and ledges that often surmount graded slopes are not yet graded; waste is removed from them faster than it is supplied by local weathering and by creeping from still higher slopes, and hence the cliffs and ledges are left almost bare;

they correspond to falls and rapids in water-streams, where the current is so rapid that its cross-section is much reduced. A hollow on an initial slope will be filled to the angle of grade by waste from above; the waste will accumulate until it reaches the lowest point on the rim of the hollow, and then outflow of waste will balance inflow; and here is the evident homologue of a lake.

In the second place, it will be understood, from what has already been said, that rivers normally grade their valleys retrogressively from the mouth headwards, and that small side streams may not be graded till long after the trunk river is graded. So with waste-sheets; they normally begin to establish a graded condition at their base, and then extend it up the slope of the valley side whose waste they "drain." When rock-masses of various resistance are exposed on the valley side, each one of the weaker is graded with reference to the stronger one next downhill; and the less resistant of the stronger ones are graded with reference to the more resistant (or with reference to the base of the hill): this is perfectly comparable to the development of graded stretches and to the extinction of falls and rapids in rivers. Ledges remain ungraded on ridge-crests and on the convex front of hill spurs long after the graded condition is reached in the channels of wet-weather streams in the ravines between the spurs; this corresponds nicely with the slower attainment of grade in small side streams than in large trunk rivers. But as late maturity passes into old age, even the ledges on ridge-crests and spur-fronts disappear, all being concealed in a universal sheet of slowly creeping waste. From any point on such a surface a graded slope leads the waste down to the streams. At any point the agencies of removal are just able to cope with the waste that is there weathered *plus* that which comes from further uphill. This wonderful condition is reached in certain well-denuded mountains, now subdued from their mature vigour to the rounded profiles of incipient old age. When the full meaning of their graded form is apprehended, it constitutes one of the strongest possible arguments for the sculpture of the lands by the slow processes of weathering, long continued. To look upon a landscape of this kind without any recognition of the labour expended in producing it, or of the extraordinary adjustments of streams to structures, and of waste to weather, is like visiting Rome in the ignorant belief that the Romans of to-day have had no ancestors.

Just as graded rivers slowly degrade their courses after the period of maximum load is past, so graded waste-sheets adopt gentler and gentler slopes when the upper ledges are consumed and coarse waste is no longer plentifully shed to the valley sides below. A changing adjustment of a most delicate kind is here discovered. When the graded slopes are first developed, they are steep, and the waste that covers them is coarse and of moderate thickness; here the strong agencies of removal have all they can do to dispose of the plentiful supply of coarse waste from the strong ledges above, and the no less plentiful supply of waste that is weathered from the weaker rocks beneath the thin cover of detritus. In a more advanced stage of the cycle, the graded slopes are moderate, and the waste that covers them is of finer texture and greater depth than before; here the weakened agencies of removal are favoured by the slower weathering of the rocks beneath the thickened waste cover, and by the greater refinement (reduction to finer texture) of the loose waste during its slow journey. In old age, when all the slopes are very gentle, the agencies of waste-removal must everywhere be weak, and their equality with the processes of waste-supply can be maintained only by the reduction of the latter to very low values. The waste-sheet then assumes a great thickness—even 50 or 100 feet—so that the progress of weathering is almost *nil*; at the same time, the surface waste is reduced to extremely fine texture, so that some of its particles may be moved even on faint slopes. Hence the occurrence of deep soils is an essential feature of old age, just as the occurrence of bare ledges is of youth. The relationships here obtaining are as significant as those which led Playfair to his famous statement concerning the origin of valleys by the rivers that drain them.

OLD AGE.—Maturity is past and old age is fully entered upon when the hilltops and the hillsides, as well as the valley floors, are graded. No new features are now developed, and those that have been earlier developed are weakened or even lost. The search for weak structures and the establishment of valleys along them has already been thoroughly carried out; now the larger streams meander freely in open valleys and begin to wander away from the adjustments of maturity. The active streams of the time of greatest relief now lose their headmost branches, for the rainfall is lessened by the destruction of the highlands, and the run-off of the rain water is retarded by the flat slopes and deep soils. The landscape is slowly tamed from its earlier strength, and presents only a succession of gently rolling swells alternating with shallow valleys, a surface everywhere open to occupation. As time passes, the relief becomes less and less; whatever the uplifts of youth, whatever the disorder and hardness of the rocks, an almost featureless plain (a peneplain) showing little sympathy with structure, and controlled only by a close approach to baselevel, must characterize the penultimate stage of the uninterrupted cycle; and the ultimate stage would be a plain without relief.

Some observers have doubted whether even the penultimate stage of a cycle is ever reached, so frequently do movements in the Earth's crust cause changes in its position with respect to baselevel. But, on the other hand, there are certain regions of greatly disordered structure, whose small relief and deep soils cannot be explained without supposing them to have, in effect, passed through all the stages above described—and doubtless many more, if the whole truth were told—before reaching the penultimate, whose features they verify. In spite of the great disturbances that such regions have suffered in past geological periods, they have afterwards stood still so long, so patiently, as to be worn down to pene-plains over large areas, only here and there showing residual reliefs where the most resistant rocks still stand up above the general level. Thus verification is found for the penultimate as well as for many earlier stages of the ideal cycle. Indeed, although the scheme of the cycle is here presented only in theoretical form, the progress of developmental changes through the cycle has been tested over and over again for many structures and for various stages; and on recognizing the numerous accordances that are discovered when the consequences of theory are confronted with the facts of observation, one must feel a growing belief in the verity and value of the theory that leads to results so satisfactory.

It is necessary to repeat what has already been said as to the practical application of the principles of the geographical cycle. Its value to the geographer is not simply in giving explanation to landforms; its greater value is in enabling him to see what he looks at, and to say what he sees. His standards of comparison, by which the unknown are likened to the known, are greatly increased over the short list included in the terminology of his school-days. Significant features are consciously sought for; exploration becomes more systematic and less haphazard. "A hilly region" of the unprepared traveller becomes (if such it really be) "a maturely dissected upland" in the language of the better prepared traveller; and the reader of travels at home gains greatly by the change. "A hilly region" brings no definite picture before the mental eyes. "A maturely dissected upland" suggests a systematic association of well-defined features; all the streams at grade, except the small headwaters; the larger rivers already meandering on flood-plained valley floors; the upper branches ramifying among spurs and hills, whose flanks show a good beginning of graded slopes; the most resistant rocks still cropping out in ungraded ledges, whose arrangement suggests the structure of the region. The practical value of this kind of theoretical study seems to me so great that, among various lines of work that may be encouraged by the Councils of the great Geographical Societies, I believe there is none that would bring larger reward than the encouragement of some such method as is here outlined for the systematic investigation of land-forms.

Some geographers urge that it is dangerous to use the theoretical or explanatory terminology involved in the practical application of the principles of the geographical cycle; mistakes may be made, and harm would thus be done. There are various sufficient answers to this objection. A very practical answer is that suggested by Penck, to the effect that a threefold terminology should be devised—one set of terms being purely empirical, as "high," "low," "cliff," "gorge," "lake," "island;" another set being based on structural relations, as "monoclinal ridge," "transverse valley," "lava-capped mesa;" and the third being reserved for explanatory relations, as "mature dissection," "adjusted drainage," "graded slopes." Another answer is that the explanatory terminology is not really a novelty, but only an attempt to give a complete and systematic expansion to a rather timid beginning already made; a sand-dune is not simply a hillock of sand, but a hillock heaped by the wind; a delta is not simply a plain at a river mouth, but a plain formed by river action; a volcano is not simply a mountain of somewhat conical form, but a mountain formed by eruption. It is chiefly a matter of experience and temperament where a geographer ceases to apply terms of this kind. But little more than half a century ago, the erosion of valleys by rivers was either doubted or not thought of by the practical geographer; to-day, the mature adjustment of rivers to structures is in the same position; and here is the third, and to my mind the most important, answer to those conservatives who would maintain an empirical position for geography, instead of pressing forward toward the rational and explanatory geography of the future. It cannot be doubted, in view of what has already been learned to-day, that an essentially explanatory treatment must in the next century be generally adopted in all branches of geographical study; it is full time that an energetic beginning should be made towards so desirable an end.

INTERRUPTIONS OF THE IDEAL CYCLE.—One of the first objections that might be raised against a terminology based on the sequence of changes through the ideal uninterrupted cycle, is that such a terminology can have little practical application on an Earth whose crust has the habit of rising and sinking frequently during the passage of geological time. To this it may be answered, that if the scheme of the geographical cycle were so rigid as to be incapable of accommodating itself to the actual condition of the Earth's crust, it would certainly have to be abandoned as a theoretical abstraction; but such is by no means the case. Having traced the normal sequence of events through an ideal cycle, our next duty is to consider the effects of any and all kinds of movements of the land-mass with respect to its baselevel. Such movements must be imagined as small or great, simple or complex, rare or frequent, gradual or rapid, early or late. Whatever their character, they will be called "interruptions," because they determine a more or less complete break in processes previously in operation, by beginning a new series of processes with respect to the new baselevel. Whenever interruptions occur, the pre-existent conditions that they interrupt can be understood only after having analyzed them in accordance with the principles of the cycle, and herein lies one of the most practical applications of what at first seems remotely theoretical. A land-mass, uplifted to a greater altitude than it had before, is at once more intensely attacked by the denuding processes in the new cycle thus initiated; but the forms on which the new attack is made can only be understood by considering what had been accomplished in the preceding cycle previous to its interruption. It will be possible here to consider only one or two specific examples from among the multitude of interruptions that may be imagined.

Let it be supposed that a maturely dissected land-mass is evenly uplifted 500 feet above its former position. All the graded streams are hereby revived to new activities, and proceed to entrench their valley floors in order to develop graded courses with respect to the new baselevel. The larger streams first show the effect of the change; the smaller streams follow suit as rapidly as possible. Falls reappear for a time in the river-channels, and then are again worn away. Adjustments of streams to structures are carried further in the second effort of the new cycle than was possible in the single effort of the previous cycle. Graded hillsides are undercut; the waste washes and creeps down from them, leaving a long even slope of bare rock; the rocky slope is hacked into an uneven face by the weather, until at last a new graded slope is developed. Cliffs that had been extinguished on graded hillsides in the previous cycle are thus for a time brought to life again, like the falls in the rivers, only to disappear in the late maturity of the new cycle.

The combination of topographic features belonging to two cycles may be called "composite topography," and many examples could be cited in illustration of this interesting association. In every case, description is made concise and effective by employing a terminology derived from the scheme of the cycle. For example, Normandy is an uplifted peneplain, hardly yet in the mature stage of its new cycle; thus stated, explanation is concisely given to the meandering course of the rather narrow valley of the Seine, for this river has carried forward into the early stages of the new cycle the habit of swinging in strong meanders that it had learned in the later stages of the former cycle.

If the uplift of a dissected region be accompanied by a gentle tilting, then all the water-streams and waste-streams whose slope is increased will be revived to new activity; while all those whose slope is decreased will become less active. The divides will migrate into the basins of the less active streams, and the revived streams will gain length and drainage area. If the uplift be in the form of an arch, some of the weaker streams whose course is across the axis of the arch may be, as it were, "broken in half;" a reversed direction of flow may be thus given to one part of the broken stream; but the stronger rivers may still persevere across the rising arch in spite of its uplift, cutting down their channels fast enough to maintain their direction of flow unchanged; and such rivers are known as "antecedent."

The changes introduced by an interruption involving depression are easily deduced. Among their most interesting features is the invasion of the lower valley floors by the sea, thus "drowning" the valleys to a certain depth, and converting them into bays. Movements that tend to produce trough-like depressions across the course of a river usually give birth to a lake of water or waste in the depressed part of the river valley. In mountain ranges frequent and various interruptions occur during the long period of deformation; the Alps show so many recent interruptions that a student there would find little use for the ideal cycle; but in mountain regions of ancient deformation, the disturbing forces seem to have become almost extinct, and there the ideal cycle is almost realized. Central France gives good illustration of this principle. It is manifest that one might imagine an endless number of possible combinations among the several factors of structure, stage of development at time of interruption, character of interruption, and time since interruption; but space cannot be here given to their further consideration.

ACCIDENTAL DEPARTURES FROM THE IDEAL CYCLE.—Besides the interruptions that involve movements of a land-mass with respect to baselevel, there are two other classes of departure from the normal or ideal cycle that do not necessarily involve any such movements: these are changes of climate and volcanic eruptions, both of which occur so arbitrarily as to place and time that they may be called "accidents." Changes of climate may vary from the normal towards the frigid or the arid, each change causing significant departures from normal geographical development. If a reverse change of climate brings back more normal conditions, the effects of the abnormal "accident" may last for some small part of a cycle's duration before they are obliterated. It is here that features of glacial origin belong, so common in northwestern Europe and north-eastern America. Judging by the present analysis of glacial and interglacial epochs during quaternary time, or of humid and arid epochs in the Great Salt Lake region, it must be concluded that accidental changes may occur over and over again within a single cycle.

In brief illustration of the combined interruptions and accidents, it may be said that southern New England is an old mountain region, which had been reduced to a pretty good peneplain when further denudation was interrupted by a slanting uplift, with gentle descent to the south-east; that in the cycle thus introduced the tilted peneplain was denuded to a sub-mature or late mature stage (according to the strength or weakness of its rocks); and that the maturely dissected region was then glaciated and slightly depressed so recently that little change has happened since. An instructive picture of the region may be conceived from this brief description.

Many volcanic eruptions produce forms so large that they deserve to be treated as new structural regions; but when viewed in a more general way, a great number of eruptions, if not the greater number, produce forms of small dimensions compared to those of the structures on which they are superposed: the volcanoes of central France are good instances of this relation. Thus considered, volcanoes and lava-flows are so arbitrarily placed in time and space that their classification under the head of "accidents" is warranted. Still further ground for this classification is found when the effects of a volcanic eruption on the pre-existent processes of land-sculpture are examined. A valley may be blockaded by a growing cone and its lava-flows; lakes may form in the up-stream portion of such a valley, even if it be mature or old. If the blockade be low, the lake will overflow to one side of the barrier, and thus the river will be locally displaced from its former course, however well adjusted to a weak structure that course may have been. If the blockade be higher than some points on the headwater divides, the lake will overflow " backwards," and the upper part of the river system will become tributary to an adjacent system. The river must cut a gorge across the divide, however hard the rocks are there; thus systematic adjustments to structure are seriously interfered with, and accidental relations are introduced. The form of the volcanic cone and the sprawling flow of its lava-streams are quite out of accord with the forms that characterize the surrounding region. The cone arbitrarily forms a mountain, even though the subjacent rocks may be weak; the lava-flows aggrade valleys that should be degraded. During the dissection of the cone, a process that is systematic enough if considered for itself alone, a radial arrangement of spurs and ravines will be developed; in long future time the streams of such ravines may cut down through the volcanic structures, and thus superpose themselves most curiously on the underlying structures. The lava-flows, being usually more resistant than the rocks of the district that they invade, gain a local relief as the adjoining surface is lowered by denudation; thus an inversion of topography is brought about, and a " table-mountain " comes to stand where formerly there had been the valley that guided the original course of the lava-flow. The table-mountain may be quite isolated from its volcanic source, where the cone is by this time reduced to a knob or "butte." But although these various considerations seem to me to warrant the classification of volcanic forms as " accidental," in contrast to the systematic forms with which they are usually associated, great importance should not be attached to this method of arrangement; it should be given up as soon as a more truthful or more convenient classification is introduced.

THE FORMS ASSUMED BY LAND WASTE.—An extension of the subject treated in the section on Graded Valley Sides, would lead to a general discussion of the forms assumed by the waste of the land on the way to the sea; one of the most interesting and profitable topics for investigation that has come under my notice. Geographers are well accustomed to giving due consideration to the forms assumed by the water-drainage of the land on the way to the sea, and a good terminology is already in use for naming them; but much less consideration is given to the forms assumed by the waste that slowly moves from the land to the sea. They are seldom presented in their true relations; many of them have no generally accepted names—for example, the long slopes of waste that reach forward from the mountains into the desert basins of Persia; forms as common as alluvial fans are unmentioned in all but the most recent school-books; and such features as till plains, moraines, and drumlins are usually given over to the geologist, as if the geographer had nothing to do with them! There can be no question of the great importance of waste-forms to the geographer, but it is not possible here to enter into their consideration. Suffice it to say that waste-forms constitute a geographical group which, like water-forms, stand quite apart from such groups as mountains and plateaus. The latter are forms of structure, and should be classified according to the arrangement of their rocks, and to their age or stage of development. The former are forms of process, and should be classified according to the processes involved, and to the stage that they have reached. The application of this general principle gives much assistance in the description of actual landscapes.

Lack of space prevents due consideration here of the development of shore-lines, a subject not less interesting, suggestive, and helpful than the development of inland forms; but I shall hope to return on some later occasion to a discussion of shore features, when it may be found that much of the terminology already introduced is again applicable. In closing this article, I must revert, if even for a third time, to the practical side of the theoretical cycle, with its interruptions and accidents. It cannot be too carefully borne in mind that the explanation of the origin of land-forms is not for its own sake added to the study of geography, but for the sake of the aid that explanation gives to the observation and description of existing geographical features. The sequence of forms developed through the cycle is not an abstraction that one leaves at home when he goes abroad; it is literally a *vade-mecum* of the most serviceable kind. During the current year that I am spending in Europe, the scheme and the terminology of the cycle have been of the greatest assistance in my studies. Application of both scheme and terminology is found equally well in the minute and infantile coastal plains that border certain stretches of the Scotch shore-line in consequence of the slight post-glacial elevation of the land, and in the broad and aged central plateau of France, where the young valleys of to-day result from the uplift of the region, and the revival of its rivers after they had submaturely dissected a pre-existent peneplain. The adjustments of streams to structures brought about by the interaction of the waxing Severn and the waning Thames, prove to be even more striking than when I first noticed them in 1894.* The large ancient delta of the Var, between

---

* See *Geographical Journal*, 1895; and *Proceedings Geologists' Association*, 1899.

Nice and Cannes, now uplifted more than 200 metres, and maturely dissected, must come to be the type example of this class of forms. The Italian Riviera, west of Genoa, may be concisely described as a region of subdued mountains that has been partly submerged and that is now approaching maturity of shore-line features in the cycle thus initiated: one may picture, from this brief statement, the mountain spurs with well-graded slopes, limited by a very irregular shore-line when first depressed, but now fronting in a comparatively simple shore-line of cliffed headlands and filled bays. The peninsula of Sorrento, on its northern side, once resembled the Riviera, but it has now been elevated 50 metres, and its uplifted bay-plains have cliffed fronts. The lower Tiber, whose mature valley floor is now somewhat wider than its meander belt, is consequent upon a volcanic accident, for it follows the trough between the slopes of the Bracciano volcanic centre on the north-west, and the Alban centre on the south-east; further up-stream, as far as Orvieto, the river, as a rule, follows a trough between the Apennines and the three volcanic centres of Bolsena, Vico, and Bracciano. The Lepini mountains, a maturely carved block of moderately deformed Cretaceous limestones south of the Alban volcanic group, has along a part of its north-eastern base a very young fault cliff, by which the graded slopes of the spurs and ravines are abruptly cut off; the fault cliff is easily recognized from the train on the line between Rome and Naples.

Botanists and zoologists know very well that a trained observer can easily recognize and describe many small items of form that pass without notice from the untrained observer. It is the same in geography, and the only question is—How can the desired training be secured? Of the many methods of geographical training, I believe that, as far as the forms of the land are concerned, no method can equal the value of one in which explanation is made an essential feature along with observation, for there is no other in which so many mental faculties are exercised.

DRAINAGE PATTERNS AND THEIR SIGNIFICANCE[1]

EMILIE R. ZERNITZ

Columbia University

ABSTRACT

A study of the patterns assumed by drainage lines suggests the existence of types additional to those commonly recognized. A more detailed classification of drainage patterns is herewith presented, in the belief that clearer conceptions regarding the various types will increase their usefulness in interpreting structural controls of drainage evolution.

INTRODUCTION

The patterns which streams form are determined by inequalities of surface slope and inequalities of rock resistance. This being true, it is evident that drainage patterns may reflect original slope and original structure or the successive episodes by which the surface has been modified, including uplift, depression, tilting, warping, folding, faulting, and jointing, as well as deposition by the sea, glaciers, volcanoes, winds, and rivers. A single drainage pattern may be the result of one or of several of these factors. Moreover, as streams are long lived, comprising among physiographic features "some of the oldest survivors or surviving remnants and also some of the youngest developments in response to earth movements," they may embody a long record of the geologic history of a region. As Bailey Willis has well said:

> Those [streams] of an older generation often continue to exist in sections as parts of a younger system, by which they have been captured and dismembered. The direction of flow and the angles in the course thus register older and younger controls which were inherent in the structure of the rocks or due to disturbing earth movements.[2]

Since drainage patterns are the reflection of so many factors, it is evident that they are of very real significance. They form one of the most immediate approaches to an understanding of geologic structure.

---

[1] Based on an essay prepared under the direction of Professor Douglas Johnson, and submitted in partial fulfilment of requirements for the degree of Master of Arts in Columbia University. The writer is indebted to Professor Johnson for access to his unpublished manuscript notes on drainage patterns.

[2] *Geologic Structures* (New York: McGraw-Hill, 1929), p. 310.

This article is based on a somewhat detailed study of stream patterns with a view to clarifying the fundamental principles of each type. It is believed that clearer conceptions of the different patterns will increase their usefulness as preliminary diagnostic criteria in the interpretation of land forms and geologic structure.

DENDRITIC DRAINAGE

The dendritic drainage pattern (Fig. 1) is characterized by irregular branching in all directions with the tributaries joining the main stream at all angles. The streams are insequent in origin. In the ideal case there are no true consequent streams, for such take their courses in consequence of some obvious initial slope of the land, and just in proportion as slope becomes the controlling factor does the pattern depart from the true dendritic type. Nor can a true dendritic pattern develop where there are marked zones of weakness which determine valley growth by subsequent streams. Only insequent drainage can develop a perfect dendritic pattern. It may happen that some of the tributaries are by chance parallel, but such are mere coincidences and have no significance in the classification of the drainage as a whole.

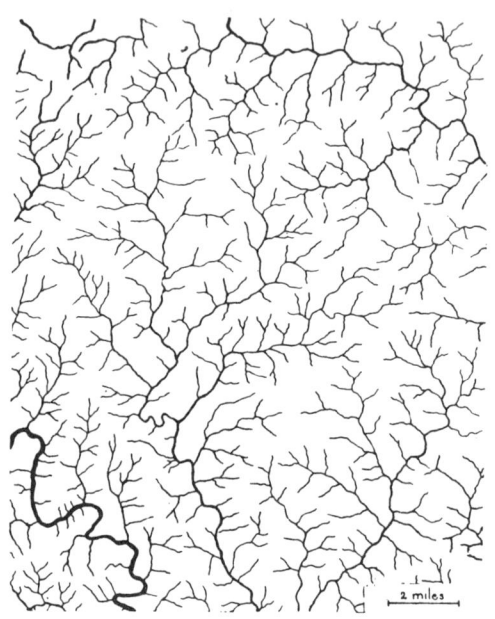

FIG. 1.—Dendritic drainage, northwestern part of Palmyra (Va.) quadrangle. Scale 1:125,000.

The pattern is called "dendritic" because it branches like a tree. This does not mean that the term "dendritic" should be applied to any branching drainage system, for practically all streams branch.

The dendritic pattern may be compared to the branching of an apple tree, while branching that resembles a pine tree or a Lombardy poplar is not dendritic. These latter indicate pronounced structural or slope control, which is lacking in the true dendritic pattern.

Dendritic drainage will develop where rocks offer uniform resistance in a horizontal direction. Such conditions are found in the flat-lying beds of plains and plateaus, and in massive crystalline rocks. The Holden (W.Va.) quadrangle[1] shows typical dendritic drainage in the Allegheny Plateau, while the McCormick (Ga.–S.C.) quadrangle illustrates dendritic drainage in the southern part of the Atlantic coastal plain. The northern third of the Cucamonga (Calif.) and the Mount Washington (N.H.) quadrangles illustrate this drainage on massive crystallines.

Rocks differing in composition but of equal resistance may occur in regions which have suffered intense metamorphism. Original differences in rock hardness tend to be obliterated by metamorphic action, and on such rocks there may develop a pattern that is essentially dendritic. The Mount Monadnock (N.H.) and Chesterfield (Mass.) sheets have been cited as illustrations,[2] but glaciation and the presence of infolded belts of limestone in this and adjacent sections of the highly metamorphosed New England area have led to the development of a drainage pattern that differs so much from the true dendritic as hardly to justify classifying it under that heading. This type of drainage will be later discussed as the complex pattern.

Since dendritic drainage develops as a result of several structural conditions, the structure in any given case cannot be as readily inferred as it can from other drainage patterns. It might suggest horizontal sedimentary rocks, superposed drainage on folded sedimentary rocks of equal resistance, massive igneous rocks, or complex metamorphics. In general, dendritic drainage implies a lack of marked structural control.

An early use of the term "dendritic" occurs in I. C. Russell's

---

[1] All quadrangles cited in the text refer to topographic maps of the U.S. Geol. Surv. unless otherwise indicated.

[2] Dake and Brown, *Interpretation of Topographic and Geologic Maps* (New York: McGraw-Hill, 1925), pp. 127–28.

Rivers of North America.[1] But the distinctive type of drainage that the term represents was recognized many years before and associated with branching like a tree. In 1882 C. E. Dutton[2] wrote: "Every canyon wall throughout its trunk, branches, and twigs. . . . ." The idea of impartial branching in all directions he expresses thus: "an intricate network, like the fibres of a leaf"; while the drainage system is described as "a minutely ramified plexus of ravines."

The term does not appear to have come into general use in 1895, for Bailey Willis in discussing drainage patterns in the Appalachians at that time referred to this type as one where "the streams flow diversely over rocks which lie in horizontal beds."[3] A year later William Morris Davis wrote: "Regions of essentially horizontal structure normally have wandering streams; no systematic arrangement of drainage is here to be expected."[4] That the concept of dendritic drainage was well understood, however, is shown by the foregoing as well as the following quotation from Davis written in 1895: "Uplifted platforms have therefore, when they rise above the sea, and for some time thereafter, a drainage system that is essentially indifferent to the rock structure."[5] Hobbs,[6] as late as 1899–1900, instead of using the term "dendritic" speaks of "twiglike branches" that are normally produced in homogeneous rocks. It thus appears that the concept of dendritic drainage was widely recognized before the term itself came into general use.

TRELLIS DRAINAGE

The essential characteristic of trellis drainage (Fig. 2) is the presence of secondary tributaries parallel to the master-stream or

[1] New York: G. P. Putnam, 1898), p. 204. "The student should study also the well-developed dendritic drainage in Lebanon Valley."

[2] "The Tertiary History of the Grand Canyon District," *U.S. Geol. Surv. Mon. 2* (1882), pp. 6, 62, 63.

[3] "The Northern Appalachians," *Nat. Geog. Soc. Mon. 1* (1895), p. 186.

[4] "Plains of Marine and Subaerial Denudation," *Bull. Geol. Soc. Amer.*, Vol. VII (1896), p. 398. Reprinted in *Geog. Essays* (Boston: Ginn, 1909), p. 346.

[5] "Physical Geography of Southern New England," *Nat. Geog. Soc. Mon. 1* (1895), p. 277.

[6] W. H. Hobbs, "The Newark System of the Pomperaug Valley, Connecticut," *U.S. Geol. Surv. Ann. Rept. 21*, Part III (1899–1900), p. 145.

other stream into which the primary tributaries enter. These secondary tributaries are usually conspicuously elongated and approximately at right angles to the streams into which they flow. The term "trellis" should not be applied where but one set of tributaries joins a master-stream at right angles,[1] for such may represent the beginning of dendritic or of rectangular drainage; whereas the trellis pattern implies a lattice effect which the elongated parallel secondary tributaries furnish.

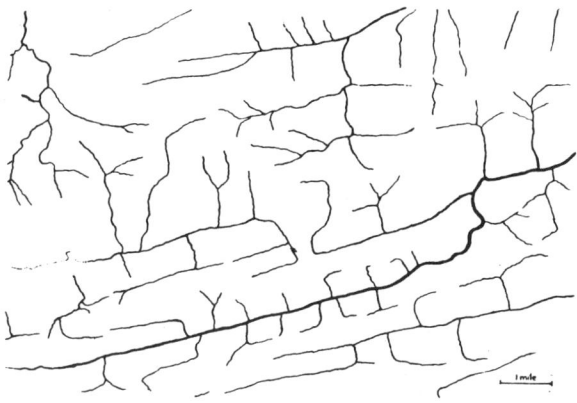

Fig. 2.—Trellis drainage, southern part of Williamsport (Pa.) quadrangle. Scale 1:62,500.

Drainage wholly consequent in origin may assume the trellis pattern as the result of glaciation. In the northeastern part of Geneva (N.Y.) quadrangle the character of the topography is transitional between typical drumlins and fluted ground moraine. Because of the linear alignment of those depositional features, a trellis pattern that is consequent in origin has developed. Consequent trellis drainage is unusual, however. As a rule, trellis drainage is made up mainly of subsequent streams connected by resequents or obsequents, and develops where the edges of formations varying in resistance outcrop in parallel belts. It is therefore characteristic of mountains of folded or tilted strata, and of maturely dissected belted coastal plains. The ridge and valley section of the Appalachians furnishes abundant illustrations of trellis drainage that is largely

[1] Douglas Johnson, unpublished notes on drainage patterns.

subsequent in origin. The Williamsport (Pa.) and Monterey (Va.–W.Va.) quadrangles show particularly fine examples.

Sometimes the tributaries in the trellis pattern do not enter the main stream at right angles. This may happen where they are affected by the general declivity of the land toward the ocean, which, although usually slight, may be sufficient to cause the streams to flow somewhat obliquely. The thousands of minor inequalities which occur in rocks cause sinuosities which are an added factor in preventing right-angled joinings. The Hancock (W.Va.–Pa.) quadrangle shows good trellis drainage where many of the joinings are not right angled.

Bailey Willis in 1895 described the drainage in the Appalachians and referred to it as the "trellis" or "grapevine" system. He wrote: "This arrangement of parallel brooks, which swell the volume of a creek generally flowing at right angles to their courses, resembles a vine from whose central stem branches are trained on a trellis. It is sometimes called the trellis or grapevine system."[1] This is the earliest use of the term in print which the writer has been able to find. It was in common usage, however, in discussion during the preceding decade.[2]

RECTANGULAR DRAINAGE

The rectangular pattern (Fig. 3) is characterized by right-angled bends in both the main stream and its tributaries. It differs from the trellis pattern in that it is more irregular; there is not such perfect parallelism of side streams; these latter are not necessarily as conspicuously elongated; and secondary tributaries need not be present. Structural control is prominent, as the pattern is directly conditioned by the right-angled jointing or faulting of rocks.

Approximate uniformity of fracturing at right angles is characteristic of certain joint systems, and when these determine the pattern of streams, good rectangular drainage results. The evenly spaced, sharp, right-angled turns of Ausable Chasm in northeastern New York, and the occurrence of tributary ravines at approximately uniform distances, is an illustration in point, for here the drainage pat-

---

[1] *Op. cit.*, p. 187.
[2] Bailey Willis, personal communication.

tern has apparently been determined by joints in the Potsdam sandstone.

The adjustment of drainage in the eastern and southern Adirondacks to lines of weakness developed by faulting was first noted by Kemp, who regarded the valleys as chiefly due to faults. He states: "In many cases these faults afforded a start for lines of drainage which have now worn out the valleys to broad reaches and have masked their origin. The old scarps at present are rounded and worn down."[1] The Elizabethtown (N.Y.) quadrangle exhibits excellent rectangular drainage, somewhat modified by glaciation, owing apparently to two systems of faults intersecting at right angles.

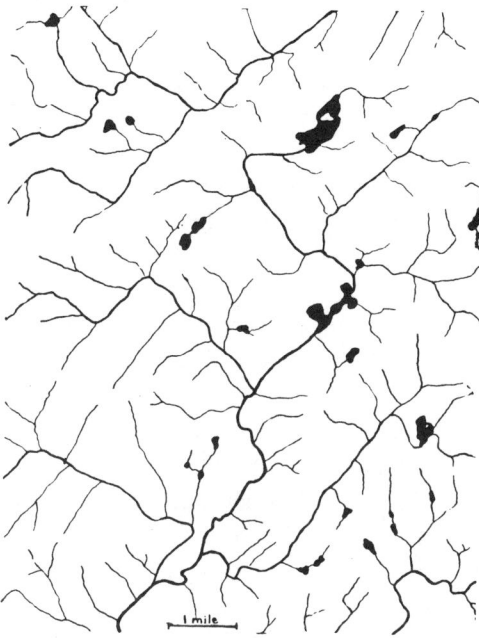

FIG. 3.—Rectangular drainage, southwestern part of Elizabethtown (N.Y.) quadrangle. Scale 1:62,500.

The rectangular drainage of the Adirondacks was first called "trellised,"[2] and was considered as being of the same general pattern as the Appalachian drainage. The sole criterion of the trellis pattern recognized at that time appears to have been the existence of rectangular tributaries, for Davis writes: "When the headwater streams captured the drainage .... they developed subsequent rectangular branches growing like a well trained grapevine";[3] while

[1] J. F. Kemp, "Preliminary Report on the Geology of Essex County," *New York State Mus. Ann. Rept. 47* (1894), pp. 632, 634.

[2] A. P. Brigham, "Note on Trellised Drainage in the Adirondacks," *Amer. Geologist*, Vol. XXI (1898), p. 220.

[3] "Rivers and Valleys of Pennsylvania," *Nat. Geog. Mag.*, Vol. I (1889), p. 248; reprinted in *Geog. Essays*, p. 479.

in his "Rivers of Northern New Jersey"[1] he refers to the rectangular courses of the streams that cross the Watchung Mountains and which are part of a trellis pattern. The present classification stresses elongated secondary tributaries parallel to the main stream as the determining factor in differentiating trellis from rectangular drainage, inasmuch as rectangular joinings apply to both patterns.

Rectangular drainage is conspicuously developed along the Norwegian coast, although submergence partially obscures the pattern. The Bergen, Björnör, and Dönna topographic sheets published by the Norwegian Topographic Survey show the rectangular pattern in great detail. Professor Th. Kjerulf, former head of the geological survey of Norway, from a study of the topographic maps came to the conclusion, according to Hobbs,[2] that the bounding lines of the valleys, lakes, and fjords of Norway were almost universally along fault and joint planes. They are, however, not always rectangular. Brögger's field studies in Southern Norway[3] confirmed this generalization for the particular areas studied. He found that the influence of faults and joints upon the formation of the valleys was most profound; in fact, almost every valley and every depression corresponded to a line of dislocation.

The gorge of the Zambesi (Fig. 4), below the famous Victoria Falls, is strikingly rectangular in pattern. A major factor in determining the river's course has been the fractures in the basaltic plateau through which the river flows. The mile-long transverse chasm into which the waters of the falls descend has been scooped out along one of the vertical fractures.[4] These fractures are independent of the columnar jointing characteristic of basalt. The tributaries which drain the basaltic country also make right-angled joinings, owing probably to east and west planes of fracture.[5]

[1] "The Rivers of Northern New Jersey with Notes on the Classification of Rivers in General," *Nat. Geog. Mag.*, Vol. II (1890), p. 99; reprinted in *Geog. Essays*, p. 504.

[2] *Op. cit.*, p. 149.

[3] W. C. Brögger, "Spaltenverwerfungen in der Gegend Langesund-Skien," *Nyt Magazin for Naturvidens kaberne*, Vol. XXVIII (1884), pp. 253–419; "Über die Bildungsgeschichte des Kristianiafjords," *ibid.*, Vol. XXX (1886), pp. 99–231.

[4] G. W. Lamplugh, "Geology of the Zambesi Basin," *Quart. Jour. Geol. Soc. London*, Vol. LXIII (1907), pp. 187–92.

[5] G. W. Lamplugh, "The Gorge of the Zambesi," *Geog. Jour.*, Vol. XXXI (1908), pp. 287–303.

RADIAL DRAINAGE

In radial drainage (Fig. 5) the streams radiate from a central area, like the spokes of a wheel. The consequent drainage of dome mountains is radial, except where an antecedent stream maintains its course across the site of uplift.[1] On more or less circular monadnocks rude radial drainage develops as these residuals assume eminence

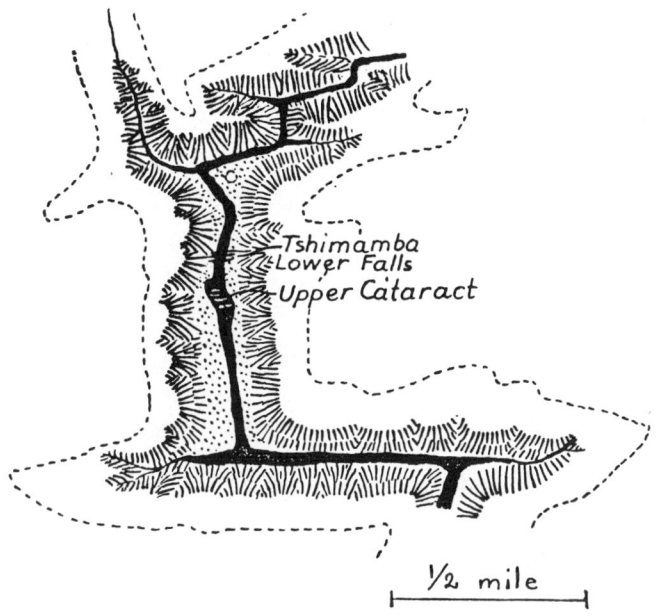

FIG. 4.—Section of gorge of Zambesi below Victoria Falls. (After Lamplugh.) Approximate scale: 1 mile = 2 inches.

through the degradation of the surrounding country. It is the typical consequent drainage pattern of volcanoes, persisting throughout all stages of their life-history. Volcanoes furnish the most perfect examples of this type of drainage pattern, owing to the marked symmetry of form which usually characterizes them and to the conical nature of their internal structures.

All the streams do not necessarily flow away from each other in normal radial pattern. Individual streams may, owing to irregularities in the initial slope of dome or volcano, or to other causes, flow

[1] Dake and Brown, *op. cit.*, p. 134.

for parts of their courses obliquely toward each other; and they may even join. A new lava flow may spread across several valleys and cause those parts of the drainage to unite. As erosion advances some of the streams may become tributaries of their more aggressive neighbors through capture. Gullies may develop on certain valley walls, then grow headward up the slope, and therefore nearly parallel to the main stream. These are all parts of the normal radial pattern.

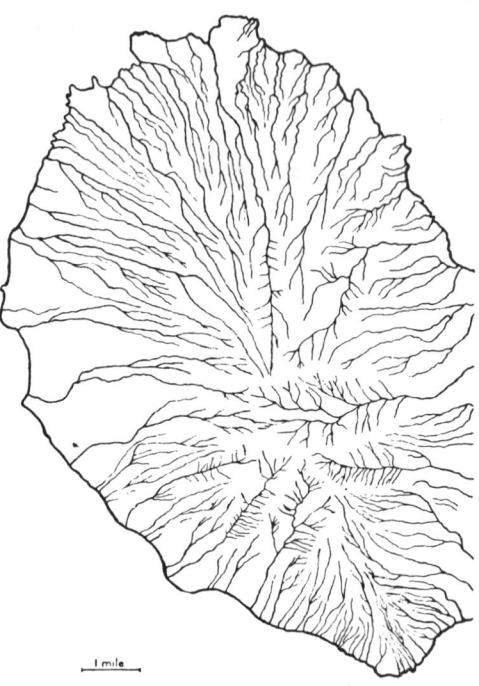

FIG. 5.—Radial drainage, Lahaina (Hawaii) quadrangle. Scale 1:62,500.

Numerous examples of radial drainage on unbreached domes are given in Jaggar's report on the Black Hills.[1] Radial drainage on a monadnock is well shown on the Mount Monadnock (N.H.) quadrangle. Shasta Special (Calif.), Mount Hood (Ore.-Wash.) and Mount Rainier National Park (Wash.) quadrangles, as well as many of the Hawaiian topographic sheets, show excellent radial drainage on volcanoes. Numerous beautiful illustrations are found on maps of the Dutch East Indies.

ANNULAR DRAINAGE

Annular drainage (Fig. 6), as the name implies, is ringlike in pattern. It is subsequent in origin and associated with maturely dissected dome or basin structures.

During the initial stage of dissection of a dome mountain, the

[1] T. A. Jaggar, Jr., "The Laccoliths of the Black Hills," *U.S. Geol. Surv. Ann. Rept.* 21, Part III (1899–1900), pp. 163–303.

streams are consequent in origin and radial in pattern. As erosion advances nonresistant layers are exposed along which subsequent tributaries develop. As these grow in length they reach successive original consequent streams that extend radially down the slopes. By capturing the upper portions of these they increase their own volume and reduce that of the consequent streams. The more fortunate subsequent streams will, furthermore, increase their length at

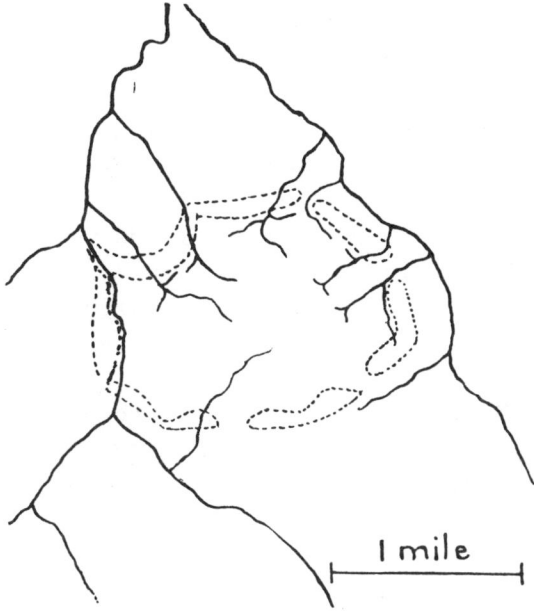

FIG. 6.—Beginnings of annular drainage, Elk Horn Peak, Black Hills. (From *U.S. Geol. Surv. Ann. Rept. 21*, Part III, Pl. XLI, p. 270.)

the expense of less fortunate subsequents as they extend their valleys headward in the weak rock which encircles the dome. Holding their own, in this struggle for survival, will be a few consequent masterstreams that have dug channels deeper than those of the piratical subsequents. Thus there results a series of subsequent streams which tend to assume a circular or annular pattern and a few trunk streams which are consequent in origin and radial in pattern.

The annular, like the trellis pattern, is an excellent illustration of the increasing influence of structure over slope as drainage approaches maturity. Slope alone controls the initial courses of streams; structure and slope, the adjusted courses of maturity. The annular valleys carved out by subsequent streams on domes and basins will be separated from one another by rimming hogbacks, the outcrops of harder strata.

The annular pattern develops most perfectly where erosion of the dome exposes rimming sedimentary strata of greatly varying degrees of hardness. When a thick layer of weak rock is revealed in the central area of the dome, a lowland is there formed. On such a dome the annular pattern may be very poorly or not at all developed, as witness the drainage of Nashville Basin, and that of Little Buffalo and Spring Creek basins on the Meeteetse (Wyo.) quadrangle.

On that excellent example of dome structure, the Black Hills, the most conspicuous member of the annular drainage is the well-known Red Valley, called the "Race Track" by the Indians. It forms a continuous depression encircling the dome and has been etched out of gypsiferous sandy red clays. Minor domes due to local igneous intrusions occur in the Black Hills region. On these, more or less complete annular drainage has developed in many instances. A vivid description of the highly symmetrical and well-developed annular valleys on Inyankara Mountain, one of these minor domes, is quoted by Jaggar from an earlier report and is well worth reading.[1] The domes of the Henry Mountains in Utah, Turkey Mountain on the Watrous (N.M.) quadrangle, and the Wealdan dome in Southeastern England have more or less well-developed annular drainage.

In describing the drainage of the Henry Mountains in 1877, Gilbert showed a clear understanding of initial radial drainage on dome structures when he wrote: "The waterways .... formed by the first rain .... radiated from the crest in all directions,"[2] but he gave no clear account of annular drainage. Jaggar, in his report on the Black Hills[3] in 1899, used the term "annular valleys" in discussing the erosion of dome mountains.

[1] *Ibid.*, pp. 249–50, 270–71.

[2] G. K. Gilbert, "Geology of the Henry Mountains," *U.S. Geol. and Geog. Surv. of Rocky Mt. Reg.* (1877), p. 145.

[3] *Op. cit.*, pp. 270–74.

PARALLEL DRAINAGE

The drainage pattern is called "parallel" (Fig. 7) when the streams over a considerable area or in a number of successive cases flow nearly parallel to one another. Parallel drainage implies either a pronounced regional slope, or a slope control by parallel topographic features such as glacially remodeled surfaces of the drumloidal or fluted ground moraine type, or control by parallel folded or faulted structures.

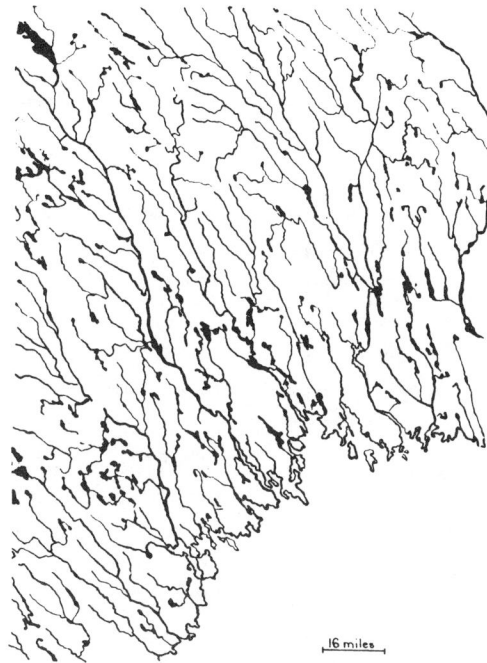

Fig. 7.—Parallel drainage, Northwestern Finland. (Section of Map 3, *Atlas of Finland* [Geog. Soc. of Finland, 1925].) Scale 1:1,000,000.

On newly emerged tilted coastal or lake plains, on the steep slopes of glacial troughs, on the walls of young stream valleys, on lake or sea cliffs, and on fault scarps the stream courses would be controlled by the slope. Such streams would be consequent in origin and parallel in pattern. The glaciated tilted coastal plain in the northwestern part of Finland which borders the Gulf of Bothnia has fairly good parallel drainage, owing to a combination of tilting and glaciation.

On the northern two-thirds of the Highwood (Ill.) quadrangle there is parallel drainage on a slightly dissected lake plain. The streams consequent upon the glacially remodeled slopes of Seneca Lake basin on the Watkins Glen (N.Y.) quadrangle, and many of the streams on the Cloud Peak (Mont.) sheet are illustrations of parallel drainage on the slopes of glacial troughs. On the Kotsima-Kuskulana sheet (Alaska map, No. 601C) are excellent examples of the same feature. The parallel tributaries involved in the radial drainage pattern on certain volcanoes, as, for example, those of the Lahaina (Hawaii) sheet, have been discussed under radial drainage. Examples of parallel streams on the sides of young valleys can be found on any number of quadrangles representing mature or submature mountains. As dissection advances, the short parallel streams on the face of fault scarps may become parallel canyons. Such are well shown on the western side of the Wasatch Mountains and in some of the Basin ranges.

Numerous parallel streams, subsequent in origin, occur in the folded Appalachians. These are, however, part of the more comprehensive trellis pattern. The parallelism in trellis drainage is in large measure restricted to a parallel relationship between tributaries. In the simple parallel pattern no such relationship is necessary; the parallelism may be between unconnected streams or it may exist between tributaries.

Parallel streams may develop along parallel faults, the streams etching out subsequent valleys along fault zones; or faulting may cause parallel "rift valleys" in which the consequent streams would be parallel.

In the dendritic pattern it is frequently possible to pick out numbers of streams or parts of streams which happen to be parallel. But the significant features of the drainage as a whole, and not the accidental details of selected parts, determine the classification of the pattern. When a stream intrenches its valley rapidly, parallel consequent streams may develop on the valley walls; and when these grow by headward erosion back into the plateau, the lower courses of the elongated streams will be parallel, while the upper courses will be dendritic. The term "parallel drainage" is not commonly applied to the restricted lower portions of such streams, the drainage being classified as a whole under the term "dendritic." This statement applies also to streams originating as parallel gullies on the face of a sea cliff or lake cliff but immediately growing in a variety of directions by headward erosion.

An excellent example of the influence of steep slope in producing parallel gullying is found on the basaltic cliffs of the Hawaiian Islands where the constant trickling of water down the steep faces of the cliffs has produced remarkably parallel even-spaced depressions which give a fluted appearance to the cliff walls.[1] Jaggar in his drainage experiments observed the importance of uniformity of slope in producing parallelism. He writes: "Lateral tributaries have a tendency to parallelism and rhythmic spacing according as the general slope, initial sheet flood and valley slopes were uniform."[2]

MODIFIED TYPES OF DENDRITIC DRAINAGE

THE PINNATE, SUBDENDRITIC, AND ANASTOMOTIC PATTERNS

In the foregoing pages the six major types of drainage patterns at present commonly recognized have been discussed. These, however, do not include all patterns which streams more or less commonly form. There are additional types which deviate in varying degrees from the established classification.

The greatest diversity occurs in modifications of typical dendritic drainage. We find patterns which resemble the dendritic but which should not be classified as such because the major streams are consequent in origin and show definite slope control. Take, for example, the drainage on the high plains in western South Dakota, western Kansas, and eastern Colorado.[3] It is obvious from the general west-east direction of the major streams and of many of the tributaries that here the slope of the plains largely determines the general direction of the stream courses. The more or less parallel and rhythmical arrangement of the lesser tributaries is doubtless due to uniformity of slope on the sides of major valleys in a region characterized by homogeneity of rock resistance. Acute-angled joinings predominate, indicating a pronounced slope influence. These acute-angled joinings with the rather evenly spaced and parallel tributaries form a pattern so much like that of a feather that it might appropriately be called "pinnate." The Phillipsburg (Kan.) quadrangle illustrates this type of drainage on a larger scale while Figure 8 is an example of pinnate

[1] H. S. Palmer, "Lapies in Hawaiian Basalts," *Geog. Rev.*, Vol. XVII (1927), pp. 627–31.

[2] "Experiments Illustrating Erosion and Sedimentation," *Harvard Coll. Bull. Mus. Comp. Zoöl.*, Vol. XLIX (1908), p. 295.

[3] *Rand McNally Commercial Atlas of America* (55th ed., 1924), pp. 351, 369, 437. Scale: approximately 16 miles = 1 inch.

drainage on a still larger scale. It is a prominent type of drainage in Roumania[1] and Southwestern Persia.[2]

Sometimes the tributaries are more nearly parallel to the main streams than in pinnate drainage and the pattern resembles the Lombardy poplar type of branching. Since in this case parallelism is the dominant feature, this pattern is classified as "subparallel." It is found in certain sections of the high plains, for example, in eastern Colorado.

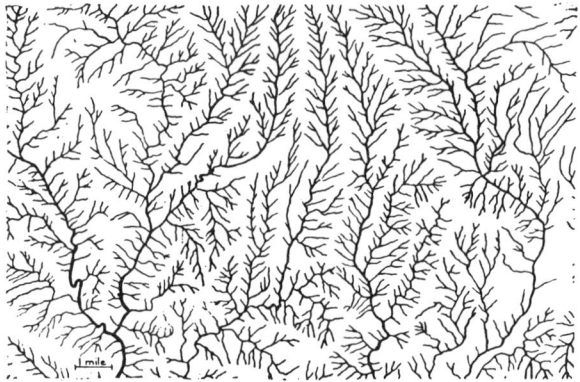

FIG. 8.—Pinnate drainage. (*Sumatra's Westkust geologische Kaart van een gedeelte van Het Gouvernement in 7 Bladen*, ed. R. D. M. Verbeek [1875–79], Blad 5: Solok.) Scale 1:100,000.

On the Columbus (Ohio) quadrangle the main streams are consequent upon the southward-sloping glacial till plain. As the tributaries of these consequent streams extend horizontally, they encounter rock of homogeneous composition which exerts too slight a structural control to be detected. As dissection advances the slope control of these minor tributaries becomes less pronounced and the drainage tends toward a dendritic pattern; nevertheless, the consequent origin of the major streams continues to be manifest in the north-south direction of their courses. This type of drainage is here designated as "subdendritic" (Fig. 9). It is a rather common variation of typical dendritic drainage.

[1] *Stieler Hand-Atlas* (Gotha: Justus Perthes, 1926–27), pp. 56, 57.

[2] Bushire (Asia—North H39) sheet, *International Map of World* (Roy. Geog. Soc., 1918), 1:1,000,000.

Flood-plain and delta drainage assumes a characteristic pattern. The tortuous windings of interlocking channels, sloughs, bayous, and oxbow lakes form a network pattern that can be best described as "anastomotic" (Fig. 10).[1] The Mississippi flood plain and delta furnish examples. It is the characteristic drainage over considerable areas of the Hungarian plain and Danubian delta, over the plains of the Indus and Irrawaddy, to cite but a few illustrations. It differs from a braided stream in that it includes features not shown by the latter, as can be seen by comparing the Memphis (Tenn.-Ark.) quadrangle, which shows a section of Mississippi flood-plain drainage, with the Paxton (Neb.) quadrangle upon which the braided Platte River is represented.

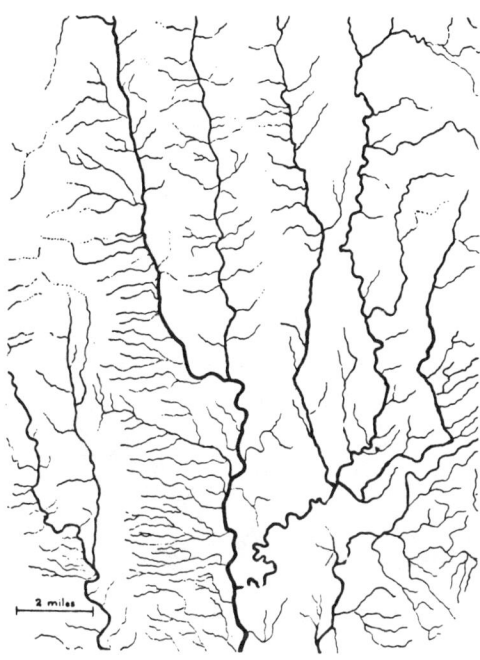

FIG. 9.—Subdendritic drainage, Columbus (Ohio) quadrangle. Scale 1:125,000.

On very flat, youthful coastal plains a network of interlocking streams, swamps, and lakes may occur, as in parts of Florida. The pattern, however, differs from that developed on flood plains. These two varieties of the anastomotic pattern are phases in the development of dendritic drainage, usually in restricted areas.

A MODIFIED TYPE OF TRELLIS DRAINAGE

THE FAULT TRELLIS PATTERN

Trellis drainage may develop where there is a series of parallel faults. These will act as planes of weakness along which erosion

[1] *Anastomotic:* having a network of lines, branches, streams, or the like (Webster's *New International Dictionary*).

readily acts. As a result of faulting, hard and soft layers may be brought into juxtaposition, when lowlands will be developed on the softer layers; or parallel fault valleys may be formed on tilted or down-dropped blocks. Such drainage has been described as "fault trellis" and the San Mateo and Priest Valley (Calif.) quadrangle cited as examples.[1] The spacing between the parallel subsequents is, however, wider than in the typical trellis pattern and the intervening blocks have dendritic drainage. Inasmuch as dendritic drainage is as prominent as fault trellis on these two maps, the drainage as a whole might be better described as complex.

A MODIFIED TYPE OF RECTANGULAR DRAINAGE

THE ANGULATE PATTERN

All fault or joint-controlled drainage patterns are not necessarily rectangular. Intersecting fault systems might cause an endless variety of stream patterns, depending upon the complexity of the fracturing. Hobbs, who has particularly investigated the influence of faults and joints upon stream courses, attributes to them a far more important rôle than do most students of the subject. He believes the modern school of physiographers ascribes too little importance to structural planes as a factor in determining the position and orientation of water courses.[2] His view is that joints are capillary openings in which water accumulates, and by freezing and by chemical action brings about disintegration. This will produce softer layers on each side of the joint which are easily eroded. Deviations

Fig. 10.—Anastomotic drainage. (Section of Apatin und Erdut [Zone 23, Col. XX], *Specialkarte der österreichisch-ungarischen Monarchie* [1905].) Scale 1:75,000.

[1] Dake and Brown, *op. cit.*, pp. 129, 191.
[2] "The River System of Connecticut," *Jour. Geol.*, Vol. IX (1901), pp. 469–85.

from a straight course are believed to be often due to other joint planes crossing the first.[1]

Hobbs, from his researches in the Pomperaug Valley in Connecticut, reached the decision that the faults in the Newark system of rocks controlled the stream patterns in their vicinity; and since the stream courses in the adjacent crystallines were strikingly like those in the Newark, he assumed that the fracture system could be inferred from the drainage, and was therefore the same as that which prevailed in the Newark rocks. The abnormal branching of headwater streams he considered as evidence of several intersecting fault systems. In regard to the rivers about Manhattan Island,[2] he advanced the theory that their channels were largely determined by lines of jointing and displacement. He saw a correspondence between drainage lines and joints in the Driftless Area of Wisconsin,[3] and believed that the striking rectilinear extension and peculiar pattern assumed by the valleys in the Finger Lake district of New York could be explained by structural fractures. His contentions have not, however, been generally accepted.

One of the earliest students of the influence of joints and faults upon drainage was Daubrée,[4] who believed that the development of many river systems was controlled by faults and joints. He explained the parallel arrangement of the main streams and the network of smaller streams in Northern France as due to the existence of a number of sets of intersecting joint planes in the Cretaceous beds of that area, and illustrated his theory with several plates which show the supposed correspondence between the stream courses and the conjectured joints. This region has been cited by Hobbs as an example of joint-controlled drainage.[5] Johnson[6] ex-

[1] E. C. Harder, "The Joint System in the Rocks of Southwestern Wisconsin and Its Relation to the Drainage Network," *Bull. Univ. Wis.*, No. 138, "Sci. Ser.," Vol. III, No. 5 (1906), pp. 207–46.

[2] "Origin of the Channels Surrounding Manhattan Island, New York," *Geol. Soc. Amer. Bull. 16* (1905), pp. 151–82.

[3] "Examples of Joint Controlled Drainage from Wisconsin and New York," *Jour. Geol.*, Vol. XIII (1903), p. 365.

[4] A. Daubrée, *Géologie expérimentale* (Paris: Dunod, 1879), pp. 300–373.

[5] "The Newark System in the Pomperaug Valley, Connecticut," *U.S. Geol. Surv. Ann. Rept. 21*, Part III (1899–1900), p. 146.

[6] Douglas Johnson, *Battlefields of the World War* (London: Oxford University Press, 1921), p. 238.

plains the east-west trend of these particular rivers and their valleys as due to the southward dip of the rocks and to a series of shallow folds that are too faint to be noticeable to the eye but "sufficient to affect the direction of running water and hence the lines of greatest and least erosion."

The influence of joints and fractures upon drainage patterns is likely to be greatly obscured or reduced by the manner in which rocks disintegrate, each rock weathering after its own fashion according-ing to its mineralogical composition and state of aggregation.[1] On the other hand, Jaggar's experiments in producing miniature drainage patterns in the laboratory led him to the conclusion that river patterns are too frequently attributed to the influence of joints and faults.[2]

FIG. 11.—Angulate drainage, northwestern part of Huron-Ottawa territory, Ontario, Canada, 1896. Scale: 5 miles = 1 inch.

Figure 11 illustrates a modification of the rectangular pattern. Parallelism due to faults and joints exists, but the joinings form acute or obtuse angles and not right angles. To call such drainage "rectangular" would be a misnomer, yet it is sufficiently like the latter to warrant a somewhat similar name, hence the term "angulate" is suggested. This drainage pattern is found in the Timiskaming and Nipissing areas of Canada and in parts of Norway.

A MODIFIED TYPE OF RADIAL DRAINAGE
THE CENTRIPETAL OR RADIAL INWARD PATTERN

"Centripetal" or radial inward drainage occurs on the inner slopes of craters and calderas, and where rimming ridges surround basins or valleys as on structural basins or breached domes. Like radial outward drainage, the centripetal drainage of craters and

[1] Geikie, *Earth Sculpture* (London: John Murray, 1898), p. 160.
[2] *Op. cit.*, p. 286.

calderas is consequent in origin. The map of Coon Butte (Fig. 12) is a good example of this pattern, which also is shown on the Sierraville (Calif.) quadrangle. Often the centripetal streams are too insignificant to appear on a map, and one gets the impression that this pattern occurs less frequently than is actually the case. Excellent examples are reported from the Java volcanoes. Davis[1] early used the term "centripetal" in describing radial inward drainage. The streams flowing inward down the slopes of the Tarim Basin in Central Asia are an example of this pattern.[2] Centripetal drainage on structural basins may be consequent or resequent, that on breached domes is obsequent.

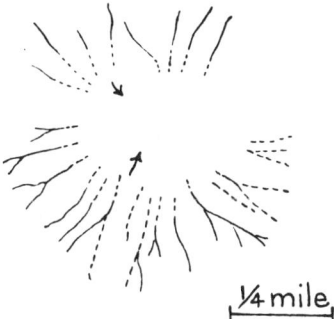

FIG. 12.—Centripetal drainage, Coon Butte, Ariz. Scale 1:30,000.

MODIFIED TYPES OF PARALLEL DRAINAGE

SUBPARALLEL AND COLINEAR DRAINAGE

A drainage pattern which has the streams oriented in a similar direction, but which lacks the regularity of the parallel pattern, may be designated as "subparallel." On the Ramapo (N.Y.-N.J.) quadrangle the streams on the eastern part of the map flow in a general north-south direction, owing to the structure of the beveled Triassic rocks. Figure 13 represents a portion of this drainage. In various parts of New England the schistosity of the rocks, faulting, and complex folding are factors contributing to the development of a subparallel arrangement of many of the streams. This is well shown on a section of the Brookfield (Mass.-Conn.) quadrangle.

Subparallelism between streams may be due to slope control. As previously stated, in some cases this gives a Lombardy-poplar pattern to the drainage. In regions of high relief, steep slopes of major valleys frequently give a subparallel if not a parallel pattern to the tributaries. Sometimes the drainage of a glaciated region assumes a

---

[1] "Rivers and Valleys of Pennsylvania," *Nat. Geog. Mag.*, Vol. I (1889), p. 249. Reprinted in *Geog. Essays*, p. 479.

[2] *Stielers Hand-Atlas*, p. 69.

subparallel pattern, owing to the linear arrangement of the glacial deposits. The north-south trend of the drumlins on the Weedsport (N.Y.) sheet has caused a subparallel alignment of the streams.

An interesting pattern is found in the region west of Budapest, Hungary, and in the Yang-Yang and Louga areas, Senegal, West Africa, which might be called "colinear" (Fig. 14), since it consists of a succession of streams extending along the same straight line. West of Budapest the streams flow in very straight lines, disap-

Fig. 13.—Subparallel drainage, eastern part of Ramapo (N.Y.–N.J.) quadrangle. Scale 1:62,500.

Fig. 14.—Colinear drainage, Senegal, West Africa. (Eastern part of Louga sheet [D28, XX], *Carte régulière* [Afrique Occid'le Française].) Scale 1:200,000.

pear and emerge again farther on in the same straight line. This is a region of loess and sand, and the wind forms furrows in the easily blown material. These furrows direct the courses of the streams. The permeable nature of the loess and sand causes the streams to disappear where the water table becomes low and to reappear where it again reaches the surface. In the Senegal area many of the streams are intermittent, as the area is a desert; but they are unusually straight and markedly colinear in pattern. The sand, as in the Budapest area, is easily blown into furrows or elongated dunes between which streams flow after infrequent but heavy rain storms. The uniform composition of the sandy surface and the torrential character of the temporary streams are additional factors in causing the very straight courses. Colinear drainage is usually part of a larger parallel pattern, but its peculiar character justifies treating it separately.

IRREGULAR AND COMPLEX DRAINAGE

Glaciated regions frequently exhibit a distinctive irregular drainage pattern. In such areas a lack of relationship between the stream courses and the higher lands is usually apparent. Streams may exist without valleys. Abandoned valleys may point to stream diversion. Numerous lakes and swamps may testify to the undeveloped character of the drainage. Glacial deposits may more or less completely obscure the hard rock surface. In a word, the former surface has been glacially remodeled, the preglacial drainage effaced, and a new drainage system which bears no relation to the underlying structure has been developed on the new topography. Such conditions exist where the retreat of the ice has been relatively recent, and thus are found in many areas covered by the last or Wisconsin ice sheet.

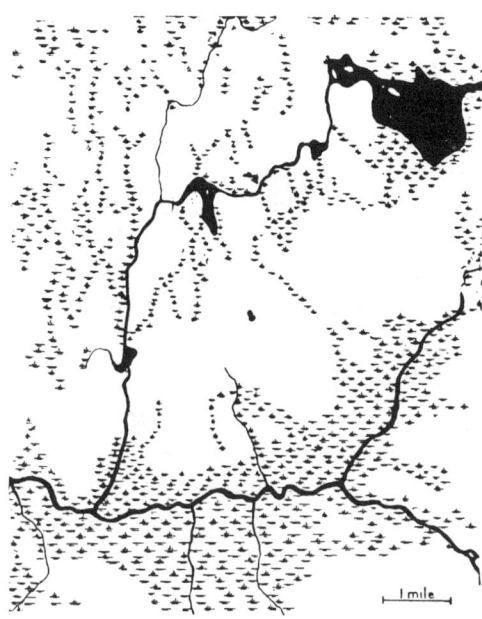

FIG. 15.—Glacially disturbed drainage, section of northern part of Whitewater (Wis.) quadrangle. Scale 1:62,500.

Such terms as "deranged" or "glacially superimposed" might be used for this "glacially disturbed" drainage pattern (Fig. 15), but the latter term is less cumbersome than "glacially superimposed" and suggests the type better than does "deranged." Its wide distribution justifies a separate classification despite the fact that it is only a transitional phase in the development of drainage. The Whitewater (Wis.) and St. Croix Dalles (Wis.-Minn.) quadrangles show familiar examples of this type.

In regions of complex structure the drainage is usually "complex" in pattern. Sections of such drainage may represent various patterns, depending upon the structural control. For example, on the Guilford (Conn.) quadrangle, there is fairly good rectangular drainage in the west-central part, subparallel in the southern, and dendritic between these two. A structural control furthermore seems to account for the alignment of some of the tributaries in the central and eastern parts. Complexity of pattern is the outstanding characteristic here and determines its classification.

The importance of drainage patterns exceeds the attention thus far devoted to them. They have been fittingly called "the key to a landscape." It is hoped that the present analysis and classification will lead to a better understanding of the various types and thus increase their usefulness to the student of geomorphology. It should, of course, be understood that inferences regarding the geology made from drainage patterns are to be regarded merely as tentative until corroborated by further information.

Part II
HORTONIAN REJUVENATION

Editor's Comments on Papers 4 and 5

4   HORTON
    Erosional Development of Streams and Their Drainage Basins; Hydrophysical Approach to Quantitative Morphology

5   STRAHLER
    Quantiative Analysis of Watershed Geomorphology

Robert E. Horton was an outstanding hydrologist who realized that the hydrologic characteristics of drainage systems could not be separated from their morphologic character (Horton, 1932). In his 1945 paper he laid the basis for the development of quantitative geomorphology. In contrast to the qualitative description of drainage patterns, Horton, Langbein (1947), and Strahler (Paper 5) developed quantitative methods of landform description that could be used to establish empirical relations between landform characteristics and the controlling variables of climate, hydrology, vegetation, and soil properties. Earlier but neglected work by Gravelius (1914) pioneered the quantitative approach to drainage basin description.

No collection of papers on drainage basins would be complete without Horton's 1945 paper, and perhaps the most important impact of his paper was philosophical. We have seen in Zernitz' presentation (Paper 3) that there is a category of drainage pattern named dendritic or insequent. This is supposedly a randomly developed drainage pattern without geologic controls. It is this type of dendritic pattern that Horton studied in detail. He demonstrated by dissecting the patterns, through classification of streams by order number, that "laws" of drainage development could be expressed. These laws indicate that the drainage network develops and is controlled by the interrelations between the forces exerted by water moving over the surface of a drainage basin and the resistance to erosion of the material composing the drainage basin. Horton's statement of laws of drainage composition and his introduction of concepts such as the length of overland flow, stream frequency, and drainage density, led directly to subsequent vigorous quantitative efforts to understand drainage basin morphology and evolution.

Horton's long paper is reprinted in its entirety in this section, although it is composed of three distinct parts. The first part (p. **79-104**) is concerned with quantitative geomorphology and is appropriate to this section of the book. The second part (pp. **104-129**), which represents Horton's infiltration theory of surface runoff and slope erosion, should appear in section three (Drainage Basin Controls), but it is best not to dismember this important paper. The third part of the paper (**129-167**) presents Horton's model of drainage system development and it should appear in section four (Drainage Basin Evolution). A discussion of the second and third parts of the Horton paper will appear later where appropriate.

Following the direction indicated by Horton, Strahler in a series of inovative papers laid the basis for a truly quantitative geomorphology. For reasons of space it is not possible to reprint all of these papers and instead a short review of his work is included as the fifth selection. A more detailed treatment appears elsewhere (Strahler, 1964). Strahler modified the technique of assigning order numbers to channels and further refinements were made by Shreve (1966).

The efforts of Horton and Strahler led to a resurgence of geomorphology on an international scale (Avena et al., 1967; Rzhanitsyn, 1960; Christofoletti, 1969; Ghose and Pandy, 1963). Strahler's use of statistics further improved the quality and reliability of geomorphic investigations, and it laid the basis for the application of geomorphology to hydrology, land management, and erosion control.

Following the pioneering work of Horton and Strahler, which involved the study of natural drainage basins, a new approach to drainage basin studies was taken by Shreve (1966, 1967, 1969) and others that involved the theoretical analysis and the mathematical modeling of drainage networks. No attempt will be made here to consider this extensive literature, and the reader is referred to Scheidegger (1970), Smart (1972), Howard (1971) and Werrity (1972) for further information.

REFERENCES

Avena, G. C.; Giuliana, G; and Lupia-Palmieri, E. 1967. Sullavalutazione quantitativa della gerarchizzazione ed evoluzione dei reticoli fluvial. *Geol. Soc. Italy. Bull.* **86**:781-796.

Christofoletti, A. 1969. Análise morfométrica das bacias hidrográficas. *Noticia Geomorfólogica Univ. Católica de Campinas (Brasil)* **9**:35-64.

Ghose, B. and Pandy, S. 1963. Quantitative geomorphology of drainage basins. *Indian Soc. Soil Sci. Jour.* **11**:259-274.

Gravelius, H. 1914. *Flusskunde*. Berlin: G. J. Göshen.

Horton, R. E. 1932. Drainage basin characteristics. *Am. Geophys. Union Trans.* **13**: 350-361.

Howard, A. D. 1971. Problems of interpretation of simulation models of geologic processes. In *Quantitative geomorphology: Some aspects and applications,* ed. M. Morisawa, pp. 61-82. Publications in Geomorphology. Binghamton, N.Y.: SUNY.

Langbein, W. B. 1947. Topographic characteristics of drainage basins. *U.S. Geol. Survey Water-Supply Paper 968-C,* pp. 125-155.

Rzhanitsyn, N. A. 1960. *Morphological and hydrological regularities of the structure of the river net.* Leningrad: Gidrometeoizdat. Translated by D. B. Krimgold and published by U.S. Dept. Agriculture, Agricultural Research Services (1963).

Scheidegger, A. E. 1970. *Theoretical geomorphology,* 2nd edition. New York: Springer-Verlag.

Shreve, R. L. 1966. Statistical law of stream numbers. *Jour. Geol.* 74:17-37.

———. 1967. Infinite topologically random channel networks. *Jour. Geol.* 75:178-186.

———. 1969. Stream lengths and basin areas in topologically random channel networks. *Jour. Geol.* 77:397-414.

Smart, J. S. 1972. Channel networks. *Advances in Hydroscience* 8:305-346.

Strahler, A. N. 1964. Quantitative geomorphology of drainage basins and channel networks: In *Handbook of applied hydrology,* ed. V. T. Chow, Section 4-II, pp. 4-39-4-76. New York: McGraw-Hill Book Co.

Werritty, A. 1972. The topology of stream networks. In *Spatial analyses in geomorphology,* ed. R. J. Chorley, pp. 167-196. London: Methuen and Co.

4

Reprinted from *Geol. Soc. America Bull.* 56:275-370 (Mar. 1945)

EROSIONAL DEVELOPMENT OF STREAMS AND THEIR DRAINAGE BASINS; HYDROPHYSICAL APPROACH TO QUANTITATIVE MORPHOLOGY

BY ROBERT E. HORTON

CONTENTS

	Page
Abstract	277
Acknowledgments	279
List of symbols used	279
Playfair's law	280
Quantitative physiographic factors	281
General considerations	281
Stream orders	281
Drainage density	283
Length of overland flow	284
Stream frequency	285
Composition of drainage net	285
Laws of drainage composition	286
Total length of streams of a given order	291
Channel-storage capacity	292
General equation of composition of stream systems	292
Relation of size of drainage area to stream order	293
Law of stream slopes	295
Determination of physiographic factors for drainage basins	295
Relation of geologic structures to drainage composition	300
Infiltration theory of surface runoff	306
General statement	306
Infiltration-capacity	307
Overland or sheet flow	309
Law of overland flow	309
Index of turbulence	312
Types of overland flow	312
Rain-wave trains	313
Profile of overland flow	314
Surface erosion by overland flow	315
Soil-erosion processes	315
Resistance to erosion	317
Eroding force	319
Critical length x_c—Belt of no erosion	320
Erosion rate	324
Total erosion and erosion depth	324
Relation of erosion to slope length	325
Rain intensity and erosion	326
Transportation and sedimentation	329

Origin and development of stream systems and their valleys by aqueous erosion............. 331
    Rill channels and rilled surface... 331
    Origin of rill channels... 332
    Cross-grading and micropiracy.. 333
    Hydrophysical basis of geometric-series laws of stream numbers and stream lengths...... 339
        General statement... 339
        First stage.. 339
        Second stage.. 339
        Subsequent stages.. 341
    Adventitious streams.. 341
    Stream development with progressively increasing land-exposure competition.......... 342
    End point of stream development.. 346
    Stream-entrance angles.. 349
    Drainage patterns.. 350
    Asymmetrical drainage patterns... 352
    Perched or sidehill streams.. 352
    Rejuvenated streams; epicycles of erosion..................................... 352
Drainage-basin topography... 355
    Marginal belt of no erosion; gradation of divides................................ 355
    Interfluve hills and plateaus... 360
    Concordant stream and valley junctions....................................... 362
    Stream-valley gradation.. 363
    Typical ovoid forms of drainage-basins.. 365
    Development of large drainage-basins... 366
    Davis stream-erosion cycle.. 366
References cited.. 369

ILLUSTRATIONS

Figure	Page
1. Well-drained basin..	282
2. Flat sandy area, poorly drained...	283
3. Bifurcation or relation of stream order to number of streams in different drainage basins.....	288
4. Relation of stream lengths to stream order in different drainage basins.................	289
5. Diagram of factor $\rho^e - 1/\rho - 1$.....................................	294
6. Law of stream slopes..	295
7. Drainage net, upper Hiwassee River......................................	297
8. Graphical determination of stream characteristics...........................	299
9. Drainage patterns of Laurel Fork and Glady Fork, Cheat River drainage basin...........	301
10. Stream numbers and stream lengths.....................................	303
11. Drainage patterns...	304
12. Relation of surface-runoff intensity (q_s) to average depth of surface detention..........	311
13. Half section of a small drainage basin...................................	314
14. Half profile of a valley slope...	316
15. Horton slope function for surface erosion.................................	321
16. Gradient and degree of erosion...	322
17. Relation of erosion to slope length......................................	326
18. Relation of initial infiltration-capacity to erosion...........................	327
19. Erosion of sodded area initiated by the breaking down of grass cover in intense rains......	328
20. Successive stages of rill-channel development..............................	333
21. Portion of Moon Mountain, Arizona-California, quadrangle, U. S. G. S..............	334
22. Development of a valley by cross-grading.................................	336
23. Successive stages of rill obliteration.....................................	337
24. Development of a stream from a rill system by cross-grading....................	338

25. Development of a drainage net in a stream basin	340
26. Beginning of erosion on newly exposed land	342
27. Development of first pair of tributaries on new stream system	343
28. Lines of flow after cross-grading of first pair of tributary areas	344
29. Development of lower pairs of main tributaries	345
30. Final development of two adjacent drainage basins on newly exposed land	346
31. End point of stream development	347
32. End point of a definite stream channel	347
33. Drainage basin of Pennypack Creek	348
34. Centripetal drainage pattern	351
35. Perched or hillside stream	353
36. Belts of no erosion	356
37. Belt of no erosion at a cross divide	358
38. Topography of an interior cross divide	360
39. Origin of ungraded or partially graded interfluve hills and plateaus	362
40. Gradation of stream valley	363

ABSTRACT

The composition of the stream system of a drainage basin can be expressed quantitatively in terms of stream order, drainage density, bifurcation ratio, and stream-length ratio.

Stream orders are so chosen that the fingertip or unbranched tributaries are of the 1st order; streams which receive 1st order tributaries, but these only, are of the 2d order; third order streams receive 2d or 1st and 2d order tributaries, and so on, until, finally, the main stream is of the highest order and characterizes the order of the drainage basin.

Two fundamental laws connect the numbers and lengths of streams of different orders in a drainage basin:

(1) The law of stream numbers. This expresses the relation between the number of streams of a given order and the stream order in terms of an inverse geometric series, of which the bifurcation ratio r_b is the base.

(2) The law of stream lengths expresses the average length of streams of a given order in terms of stream order, average length of streams of the 1st order, and the stream-length ratio. This law takes the form of a direct geometric series. These two laws extend Playfair's law and give it a quantitative meaning.

The infiltration theory of surface runoff is based on two fundamental concepts:

(1) There is a maximum or limiting rate at which the soil, when in a given condition, can absorb rain as it falls. This is the infiltration-capacity. It is a volume per unit of time.

(2) When runoff takes place from any soil surface, there is a definite functional relation between the depth of surface detention δ_s, or the quantity of water accumulated on the soil surface, and the rate q_s of surface runoff or channel inflow.

For a given terrain there is a minimum length x_c of overland flow required to produce sufficient runoff volume to initiate erosion. The critical length x_c depends on surface slope, runoff intensity, infiltration-capacity, and resistivity of the soil to erosion. This is the most important single factor involved in erosion phenomena and, in particular, in connection with the development of stream systems and their drainage basins by aqueous erosion.

The erosive force and the rate at which erosion can take place at a distance x from the watershed line is directly proportional to the runoff intensity, in inches per hour, the distance x, a function of the slope angle, and a proportionality factor K_e, which represents the quantity of material which can be torn loose and eroded per unit of time and surface area, with unit runoff intensity, slope, and terrain.

The rate of erosion is the quantity of material actually removed from the soil surface per unit of time and area, and this may be governed by either the transporting power of overland flow or the actual rate of erosion, whichever is smaller. If the quantity of material torn loose and carried in suspension in overland flow exceeds the quantity which can be transported, deposition or sedimentation on the soil surface will take place.

On newly exposed terrain, resulting, for example, from the recession of a coast line, sheet erosion occurs first where the distance from the watershed line to the coast line first exceeds the critical length x_c, and sheet erosion spreads laterally as the width of the exposed terrain increases. Erosion of such a newly exposed plane surface initially develops a series of shallow, close-spaced, shoestring gullies or rill channels. The rills flow parallel with or are consequent on the original slope. As a result of various causes, the divides between adjacent rill channels are broken down locally, and the flow in the shallower rill channels more remote from the initial rill is diverted into deeper rills more closely adjacent thereto, and a new system of rill channels is developed having a direction of flow at an angle to the initial rill channels and producing a resultant slope toward the initial rill. This is called cross-grading.

With progressive exposure of new terrain, streams develop first at points where the length of overland flow first exceeds the critical length x_c, and streams starting at these points generally become the primary or highest-order streams of the ultimate drainage basins. The development of a rilled surface on each side of the main stream, followed by cross-grading, creates lateral slopes toward the main stream, and on these slopes tributary streams develop, usually one on either side, at points where the length of overland flow in the new resultant slope direction first exceeds the critical length x_c.

Cross-grading and recross-grading of a given portion of the area will continue, accompanied in each case by the development of a new order of tributary streams, until finally the length of overland flow within the remaining areas is everywhere less than the critical length x_c. These processes fully account for the geometric-series laws of stream numbers and stream lengths.

A belt of no erosion exists around the margin of each drainage basin and interior subarea while the development of the stream system is in progress, and this belt of no erosion finally covers the entire area when the stream development becomes complete.

The development of interior divides between subordinate streams takes place as the result of competitive erosion, and such divides, as well as the exterior divide surrounding the drainage basin, are generally sinuous in plan and profile as a result of competitive erosion on the two sides of the divide, with the general result that isolated hills commonly occur along divides, particularly on cross divides, at their junctions with longitudinal divides. These interfluve hills are not uneroded areas, as their summits had been subjected to more or less repeated cross-grading previous to the development of the divide on which they are located.

With increased exposure of terrain weaker streams may be absorbed by the stronger, larger streams by competitive erosion, and the drainage basin grows in width at the same time that it increases in length. There is, however, always a triangular area of direct drainage to the coast line intermediate between any two major streams, with the result that the final form of a drainage basin is usually ovoid or pear-shaped.

The drainage basins of the first-order tributaries are the last developed on a given area, and such streams often have steep-sided, V-shaped, incised channels adjoined by belts of no erosion.

The end point of stream development occurs when the tributary subareas have been so completely subdivided by successive orders of stream development that there nowhere remains a length of overland flow exceeding the critical length x_c. Stream channels may, however, continue to develop to some extent through headward erosion, but stream channels do not, in general, extend to the watershed line.

Valley and stream development occur together and are closely related. At a given cross section the valley cannot grade below the stream, and the valley supplies the runoff and sediment which together determine the valley and stream profiles. As a result of cross-grading antecedent to the development of new tributaries, the tributaries and their valleys are concordant with the parent stream and valley at the time the new streams are formed and remain concordant thereafter.

Valley cross sections, when grading is complete, and except for first-order tributaries, are generally S-shaped on each side of the stream, with a point of contraflexure on the upper portion of the slope, and downslope from this point the final form is determined by a combination of factors, including erosion rate, transporting power, and the relative frequencies of occurrence of storms and runoff of different intensities. The longitudinal profile of a valley along the stream bank and the cross section of the valley are closely related, and both are related to the resultant slope at a given location.

Many areas on which meager stream development has taken place, and which are commonly classified as youthful, are really mature, because the end point of stream development and erosion for existing conditions has already been reached.

When the end point of stream and valley gradation has arrived in a given drainage basin, the remaining surface is usually concave upward, more or less remembling a segment of a parabaloid, ribbed by cross and longitudinal divides and containing interfluve hills and plateaus. This is called a "graded" surface, and it is suggested that the term "peneplain" is not appropriate, since this surface is neither a plane nor nearly a plane, nor does it approach a plane as an ultimate limiting form.

The hydrophysical concepts applied to stream and valley development account for observed phenomena from the time of exposure of the terrain. Details of these phenomena of stream and valley development on a given area may be modified by geologic structures and subsequent geologic changes, as well as local variations of infiltration-capacity and resistance to erosion.

In this paper stream development and drainage-basin topography are considered wholly from the viewpoint of the operation of hydrophysical processes. In connection with the Davis erosion cycle the same subject is treated largely with reference to the effects of antecedent geologic conditions and subsequent geologic changes. The two views bear much the same relation as two pictures of the same object taken in different lights, and one supplements the other. The Davis erosion cycle is, in effect, usually assumed to begin after the development of at least a partial stream system; the hydrophysical concept carries stream development back to the original newly exposed surface.

ACKNOWLEDGMENTS

The author is indebted to Dr. Howard A. Meyerhoff for many helpful suggestions and criticisms. Grateful acknowledgment is also given Dr. Alfred C. Lane, who, more than 40 years ago, gave the author both the incentive and an opportunity to begin the study of drainage basins with respect to possible interrelations of their hydraulic, hydrologic, hydrophysical, and geologic features.

LIST OF SYMBOLS USED

A = area of drainage basin in square miles.
α = slope angle.
c = distance from stream tip to watershed line.
δ = depth of sheet flow in inches at the stream margin or at the foot of a slope length l_o.
δ_a = average depth of surface detention or overland flow, in inches, on a unit strip of length l_o.
δ_x = depth of sheet flow in inches at a distance x from the crest of the slope or watershed line.
D_d = drainage density or average length of streams per unit of area.
e = energy expended by frictional resistance on soil surface, ft.-lbs. per sq. ft. per sec.
E_a = average erosion over a given strip of unit width and length l_o per unit of time.
e_r = erosion rate or quantity of material, preferably expressed in terms of depth of solid material, removed per hour by sheet erosion.
E_t = total erosion = total solid material removed from a given strip of unit length per unit of time.
f = infiltration-capacity at a given time t from the beginning of rain, inches per hour.
f_c = minimum infiltration-capacity for a given terrain.
f_0 = initial infiltration-capacity at beginning of rain.
F_1 = erosive force of overland flow, lbs. per sq. ft.
F_0 = tractive force of overland flow, lbs. per sq. ft. of surface.
F_s = stream frequency or number of streams per unit area.
i = rain intensity—usually inches per hour.
I = index of turbulence or percentage of the area covered by sheet flow on which the flow is turbulent.
K_a = coefficient in the runoff equation where δ_a is used instead of δ as the depth of sheet flow.
k_e = proportionality factor required to convert the rate of performance of work in sheet erosion into equivalent quantity of material removed per unit of time.
K_f = a proportionality factor which determines the time t_c required for infiltration-capacity to be reduced from its initial value f_0 to its constant value f_c.
K_l = corresponding coefficient (to K_s) in the equation for laminar overland flow.
K_s = constant or proportionality factor in equation expressing runoff intensity in terms of depth δ of overland flow.
l_a = average length of streams of order o.
l_g = maximum length of overland flow on a given area.
l_o = length of overland flow or length of flow over the ground surface before the runoff becomes concentrated in definite stream channels.
l_1, l_2 etc. = average lengths of streams of 1st and 2d orders, etc.
L' = extended stream length measured along stream from outlet and extended to watershed line.
L_0 = total length of tributaries of order o.
M = exponent in the equation: $q_s = K_s \delta^M$, expressing the runoff intensity in terms of depth of sheet flow along the stream margin.
n = surface roughness factor, as in the Manning formula.
N_o = number of streams of a given order in a drainage basin.
N_s = total number of streams in a drainage basin.
N_1, N_2 etc. = total number of streams of 1st, 2d orders, etc.

o = order of a given stream.
q_s = surface-runoff intensity—usually inches per hour.
q_1 = runoff intensity in cubic feet per second from a unit strip 1 foot wide and with a slope length l_o.
ρ = stream length ratio/bifurcation ratio = r_l/r_b.
r_b = bifurcation ratio or ratio of the average number of branchings or bifurcations of streams of a given order to that of streams of the next lower order. It is usually constant for all orders of streams in a given basin.
r_l = stream length ratio or ratio of average length of streams of a given order to that of streams of the next lower order.
r'_l = stream length ratio using extended stream lengths.
r_s = ratio of channel slope to ground slope for a given stream or in a given drainage basin, = s_c/s_g.
R_i = initial surface resistance to sheet erosion, lbs. per sq. ft.
R_1 = subsurface resistance to sheet erosion, lbs. per sq. ft., or resistance of a lower surface or horizon of the soil to erosion after the surface layer of resistance R_i is removed.
σ = supply rate = $i - f$.
s = order of main stream in a given drainage basin.
s_c = channel slope.
s_g = resultant slope of ground surface of area tributary to a given parent stream.
S = surface slope = fall/horizontal length.
t_e = duration of rainfall excess or time during which rain intensity exceeds infiltration-capacity.
v = mean velocity of overland flow, feet per second.
v_x = mean velocity of overland flow at the distance x from the watershed line.
V_d = depth of depression storage on a given area, inches.
w_1 = weight of runoff, including solids in suspension, lbs. per cu. ft.
w_b = width of marginal belt of no erosion = $x_c \sin A$, where A is the angle between the direction of the divide and the direction of overland flow at the divide.
x_c = critical length of overland flow or distance from the watershed line, measured in the direction of overland flow, within which sheet erosion does not occur.

The customary procedure of expressing rainfall, infiltration, and runoff in inches depth per hour, velocities, channel lengths, and discharge rates in foot-second units, and drainage areas in square miles, has been followed, with appropriate interconversion factors in the formulas.

PLAYFAIR'S LAW

More than a century ago Playfair (*in* Tarr and Martin, 1914, p. 177) stated:

"Every river appears to consist of a main trunk, fed from a variety of branches, each running in a valley proportioned to its size, and all of them together forming a system of vallies, communicating with one another, and having such a nice adjustment of their declivities that none of them join the principal valley either on too high or too low a level."

This has often been interpreted as meaning merely that tributary streams and their valleys enter the main streams and their valleys concordantly. A careful reading of Playfair's law implies that he envisaged a great deal more than merely the concordance of stream and valley junctions. He speaks of the "nice adjustment" of the entire system of valleys and states that each branch runs in a valley "proportioned to its size."

Playfair did pioneer work based on ocular observations. There were available to him neither the results of measurements nor the hydrophysical laws necessary to their quantitative interpretation. It appears that the time has now come when such a quantitative interpretation can be undertaken.

QUANTITATIVE PHYSIOGRAPHIC FACTORS

GENERAL CONSIDERATIONS

In spite of the general renaissance of science in the present century, physiography as related in particular to the development of land forms by erosional and gradational processes still remains largely qualitative. Stream basins and their drainage basins are described as "youthful," "mature," "old," "poorly drained," or "well drained," without specific information as to how, how much, or why. This is probably the result largely of lack of adequate tools with which to work, and these tools must be of two kinds: measuring tools and operating tools.

One purpose of this paper is to describe two sets of tools which permit an attack on the problems of the development of land forms, particularly drainage basins and their stream nets, along quantitative lines.

An effort will be made to show how the problem of erosional morphology may be approached quantitatively, and even in this respect only the effects of surface runoff will be considered in detail. Drainage-basin development by ground-water erosion, highly important as it is, will not be considered, and the discussion of drainage development by surface runoff will mainly be confined to processes occurring outside of stream channels. The equally important phase of the subject, channel development—including such problems as those of the growth of channel dimensions with increase of size of drainage basin, stream profiles, and stream bends—will not be considered in detail.

STREAM ORDERS

In continental Europe attempts have been made to classify stream systems on the basis of branching or bifurcation. In this system of stream orders, the largest, most branched, main or stem stream is usually designated as of order 1 and smaller tributary streams of increasingly higher orders (Gravelius, 1914). The smallest unbranched fingertip tributaries are given the highest order, and, although these streams are similar in characteristics in different drainage basins, they are designated as of different orders.

Feeling that the main or stem stream should be of the highest order, and that unbranched fingertip tributaries should always be designated by the same ordinal, the author has used a system of stream orders which is the inverse of the European system. In this system, unbranched fingertip tributaries are always designated as of order 1, tributaries or streams of the 2d order receive branches or tributaries of the 1st order, but these only; a 3d order stream must receive one or more tributaries of the 2d order but may also receive 1st order tributaries. A 4th order stream receives branches of the 3d and usually also of lower orders, and so on. Using this system the order of the main stream is the highest.

To determine which is the parent and which the tributary stream upstream from the last bifurcation, the following rules may be used:

(1) Starting below the junction, extend the parent stream upstream from the bifurcation in the same direction. The stream joining the parent stream at

the greatest angle is of the lower order. Exceptions may occur where geologic controls have affected the stream courses.

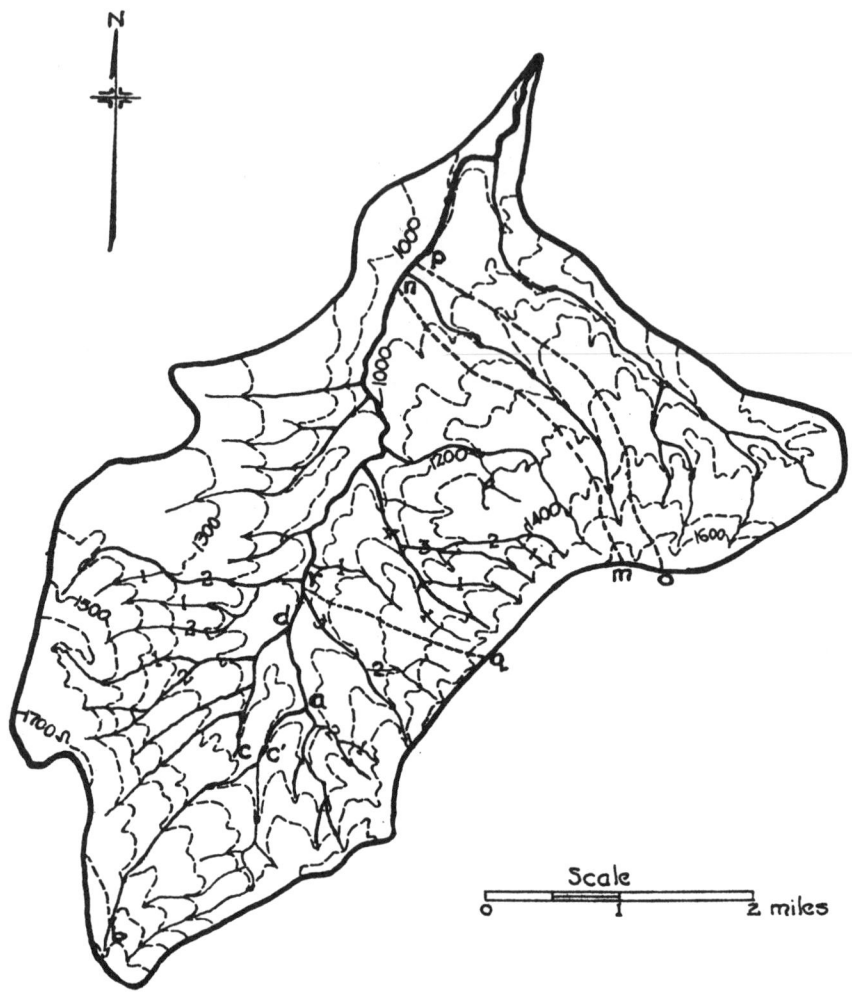

FIGURE 1.—*Well-drained basin*
(Cherry Creek, N. Y., quad., U. S. G. S.)

(2) If both streams are at about the same angle to the parent stream at the junction, the shorter is usually taken as of the lower order.

On Figure 1 several streams are numbered 1, and these are 1st order tributaries. Streams numbered 2 are of the 2nd order throughout their length, both below and above the junctions of their 1st order tributaries. The main stream is apparently $ac'b$ although it joins ad at nearly a right angle. It is probable that the original course of the stream was dcb, but the portion above dc was diverted by headwater erosion into stream ac'. The well-drained basin (Fig. 1) is of the 5th order, while the poorly drained basin (Fig. 2) is of the 2d order. Stream order therefore affords a simple quantitative basis for comparison of the degree of development in the drainage nets of basins of comparable size. Its usefulness as a basis for such comparisons is limited by the fact that, other things equal, the order of a drainage basin or its stream system generally increases with size of the drainage area.

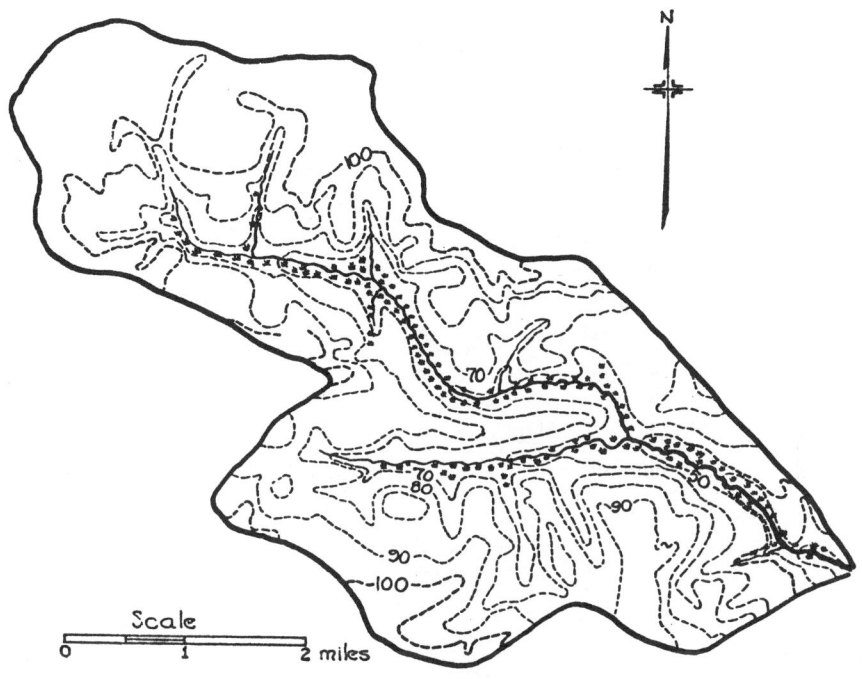

FIGURE 2.—*Flat sandy area, poorly drained*
(Bridgeton, N. J., quad. U. S. G. S.)

DRAINAGE DENSITY

Figures 1 and 2 show two small drainage basins, both on the same scale; one well drained, the other poorly drained. These terms, well drained and poorly drained, while in common use in textbooks on physiography, are purely qualitative, and something better is needed to characterize the degree of drainage development within a basin. The simplest and most convenient tool for this purpose is drainage density or average length of streams within the basin per unit of area (Horton, 1932). Expressed as an equation

$$\text{Drainage density, } D_d = \frac{\Sigma L}{A} \qquad (1)$$

where ΣL is the total length of streams and A is the area, both in units of the same system. The poorly drained basin has a drainage density 2.74, the well-drained one, 0.73, or one fourth as great.

For accuracy, drainage density must, if measured directly from maps, be determined from maps on a sufficiently large scale to show all permanent natural stream channels, as do the U. S. Geological Survey topographic maps. On these maps perennial streams are usually shown by solid blue lines, intermittent streams by dotted blue lines. Both should be included. If only perennial streams were included, a drainage basin containing only intermittent streams would, in accordance with equation (1), have zero drainage density, although it may have a considerable degree of basin development. Most of the work of valley and stream development by running water is performed during floods. Intermittent and ephemeral streams carry flood waters, hence should be included in determining drainage density. Most streams which are perennial in their lower reaches or throughout most of their courses have an intermittent or ephemeral reach or both, near their headwaters, where the stream channel has not cut down to the water table. These reaches should also be included in drainage-density determinations.

In textbooks on physiography, differences of drainage density are commonly attributed to differences of rainfall or relief, and these differences in drainage density are largely used to characterize physiographic age in the sense used by Davis (Davis, 1909; Wooldridge and Morgan, 1937). In the poorly drained area (Fig. 2) the mean annual rainfall is about 30 per cent greater than in the well-drained area (Fig. 1). Therefore some other factor or factors are far more important than either rainfall or relief in determining drainage density. These other factors are infiltration-capacity of the soil or terrain and initial resistivity of the terrain to erosion.

LENGTH OF OVERLAND FLOW

The term "length of overland flow," designated l_o, is used to describe the length of flow of water over the ground before it becomes concentrated in definite stream channels. To a large degree length of overland flow is synonymous with length of sheet flow as quite commonly used. The distinction between overland flow and channel flow is not so vague or uncertain as might at first appear. Overland flow is sustained by a relatively thin layer of surface detention. This disappears quickly—often in a few minutes—through absorption by the soil or infiltration after rain ends. Surface detention and surface runoff may, in fact, end before rain ends if, as is often the case, there is at the end of the storm an interval of residual rainfall having an intensity less than the infiltration-capacity. Channel flow is sustained by accumulated channel storage. This drains out slowly and lasts for hours or even days after channel inflow from surface runoff ends.

In addition to its obvious value in various ways in characterizing the degree of development of a drainage net within a basin, drainage density is particularly useful because of the fact that the average length of overland flow l_o is in most cases approximately half the average distance between the stream channels and hence is approximately equal to half the reciprocal of the drainage density, or

$$l_o = \frac{1}{2D_d} \quad (2)$$

Later it will be shown that length of overland flow is one of the most important independent variables affecting both the hydrologic and physiographic development of drainage basins.

In this paper it is frequently assumed for purposes of convenience that the average length of overland flow is sensibly equal to the reciprocal of twice the drainage density. From considerations of the geometry of streams and their drainage areas the author has shown (Horton, 1932) that the average length of overland flow is given by the equation

$$l_o = \frac{1}{2D_d \sqrt{1 - \left(\frac{s_c}{s_g}\right)^2}} \qquad (3)$$

where s_c is the channel or stream slope and s_g the average ground slope in the area.

Values of the correction factor or of the ratio $l_o/2D_d$ for different values of the slope ratio s_c/s_g are as follows:

s_c/s_g =	0.9	0.8	0.7	0.6	0.5	0.4	0.3	0.2	0.1
$l_o/2D_d$ =	1.86	1.67	1.40	1.25	1.15	1.09	1.05	1.02	1.005

The ground slope or resultant slope of the area tributary to a stream on either side is necessarily always greater than the channel slope since the ground surface has a component of slope parallel with and of the same order of magnitude as that of the stream, and in addition it has a component of slope at right angles to the stream.

Table 4 shows the average channel and ground slopes of streams in the Delaware River and some other drainage basins, derived from topographic maps. The channel slope in these instances is commonly from half to one fourth the ground slope. Often on an area which as a whole is nearly horizontal, but the surface of which is interspersed with hills, the ground slope may be and frequently is two or three times the channel slope. If the channel slope is less than one third the ground slope, the error resulting from the assumption that average length of overland flow is equal to the reciprocal of twice the drainage density may in general be neglected.

STREAM FREQUENCY

This is the number of streams, F_s, per unit of area, or

$$F_s = \frac{N}{A} \qquad (4)$$

where N = total number of streams in a drainage basin of A areal units.

Values of drainage density and stream frequency for small and large drainage basins are not directly comparable because they usually vary with the size of the drainage area. A large basin may contain as many small or fingertip tributaries per unit of area as a small drainage basin, and in addition it usually contains a larger stream or streams. This effect may be masked by the increase of drainage density and stream frequency on the steeper slopes generally appurtenant to smaller drainage basins.

COMPOSITION OF DRAINAGE NET

The term "drainage pattern" is used in rather a restricted sense in many books on physiography, implying little more than the manner of distribution of a given set of tributary streams within the drainage basin. Thus, for example, with identically the same lengths and numbers of streams, the drainage pattern may be dendritic, rectangular, or radial. Neither the drainage pattern nor the drainage density, nor both, provide an adequate characterization of the stream system or drainage net in a given basin. There may be various combinations of stream numbers, lengths, and orders which will give the same drainage density, or there may be similar forms of drainage pattern with widely different drainage and stream densities. Something more is needed as a basis for quantitative morphology of drainage basins. The author has therefore coined the expression "composition of a drainage net," as distinguished from "drainage pattern." Composition implies the numbers and lengths of streams and tributaries of different sizes or orders, regardless of their pattern. Composition has a high degree of hydrologic significance, whereas pattern alone has but little hydrologic significance, although it is highly significant in relation to geologic control of drainage systems.

Cotton (1935) and others have used the term "texture" to express composition of a drainage net as related both to drainage density and stream frequency. For quantitative purposes two terms are needed, since two drainage nets with the same drainage densities may have quite different numbers and lengths of streams. Numerical values of drainage density independent of other units are needed for various purposes.

LAWS OF DRAINAGE COMPOSITION

The numbers and lengths of tributaries of different orders were determined for the streams listed in Table 1. The numbers and the lengths of streams varied with the stream order in a manner which suggested a geometrical progression. Plotting the data on semilogarithmic paper it was found (Fig. 3) that the stream numbers fall close to straight lines, and (Fig. 4) the same is true of the stream lengths. From the manner of plotting, these lines are necessarily graphs of geometrical series, inverse for stream numbers of different orders and direct for stream lengths.[1]

From the properties of geometric series, the equation of the lines giving the number N_o of streams of a given order in a drainage basin can be written

$$N_o = r_b^{(s-o)}. \tag{5}$$

From the laws governing geometric series it is easily shown that the number N of streams of all orders is

$$N = \frac{r_b^s - 1}{r_b - 1}. \tag{6}$$

By definition, o is the order of a given class of tributaries, s is the order of the main stream, and r_b is the bifurcation ratio.

The equation correlating the lengths of streams of different orders is, similarly,

$$l_o = l_1 r_l^{o-1}. \tag{7}$$

---

[1] In the figure the lines of best fit were drawn by inspection. Somewhat more accurate lines could of course be obtained by the method of least squares or the correlation method. However, the agreement of the observed points with the lines located by inspection is so close in most cases that little would be gained in accuracy by the use of these methods.

TABLE 1.—Characteristics of the drainage nets of certain stream basins

Stream	Location	Type	Order of Main Stream o	Area Sq. Mi. A	No. of Streams ΣN_o	No. of 1st order streams N_1	Stream Frequency F_s	Drainage Density D_d	Aver. lgth 1st order streams L_1	Bifur-cation ratio r_b	Length ratio r_l	ΣL
(1)	(2)	(3)	(4)	(5)	(6)	(7)	(8)	(9)	(10)	(11)	(12)	(13)
Esopus Creek	Olive Bridge, N.Y.*	Mountains	5	234.	126	90	0.527	0.849	0.99	3.12	2.31	203.2
" "	Lower Area *	Rolling and Plains	7	426.3	361	256	.847	.818	.81	2.27	1.84	348.6
Rondout "	Honk Falls, N.Y.	Mountains	4	105.	58	44	.552	1.07	1.08	3.30	2.64	112.6
Putnam Brook	Weedsport, N.Y.	Glacial, Drumlin	4	27	26	18	.963	1.95	.77	2.46	2.74	52.7
Cold Spring Brook	"	"	4	15.8	25	15	1.58	2.025	.58	2.62	2.66	32.
Crane Creek	"	"	5	45.7	48	31	1.05	2.03	.81	2.22	2.30	92.6
Ganargua Creek	Lyons, N.Y.	"	6	299.	269	166	.899	1.628	.87	2.89	2.30	487.1
Kauka Lake	Foot of Lake, N.Y.	Hilly, Dissected	5	161.†	170	124	1.055	1.665	1.16	3.25	1.96	268.
Seneca "	" " "	"	6	479.†	472	334	.984	1.59	.95	3.15	2.20	762.5
Owasco "	Weedsport, "	"	5	200.	265	191	1.325	1.79	.83	3.91	2.22	358.
Thunder Bay River	Alpena, Mich.**	Glacial – Flat	4	—	44	33	—	—	—	3.00	—	—

*Ashokan Dam to Saugerties, N.Y.   **Data furnished by Prof. C.O. Wisler.   † Land area, excluding lake.

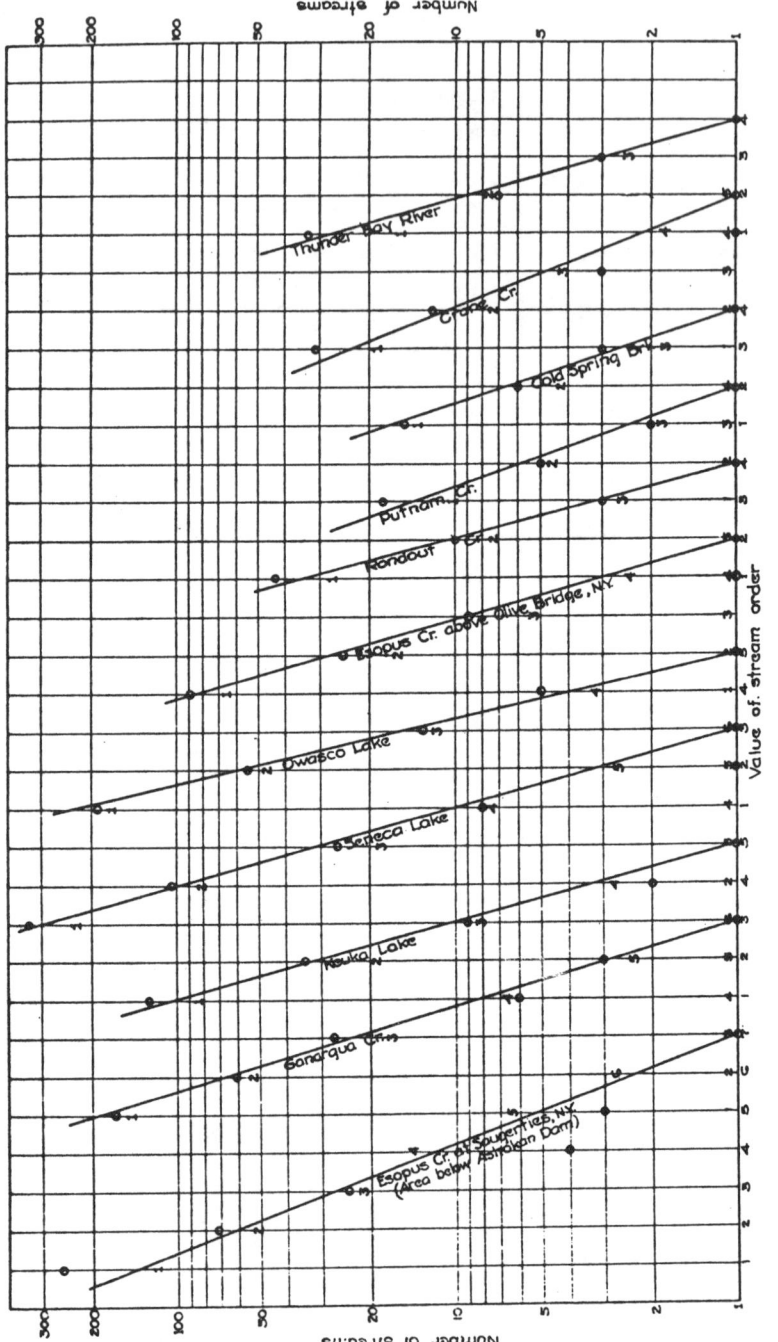

FIGURE 3.—Bifurcation or relation of stream order to number of streams in different drainage basins

These equations may appear formidable, but they are merely the statement in symbolic form of the simple algebraic laws of geometric series. Equations (5) and (7) are the most important and are readily solved by means of logarithms.

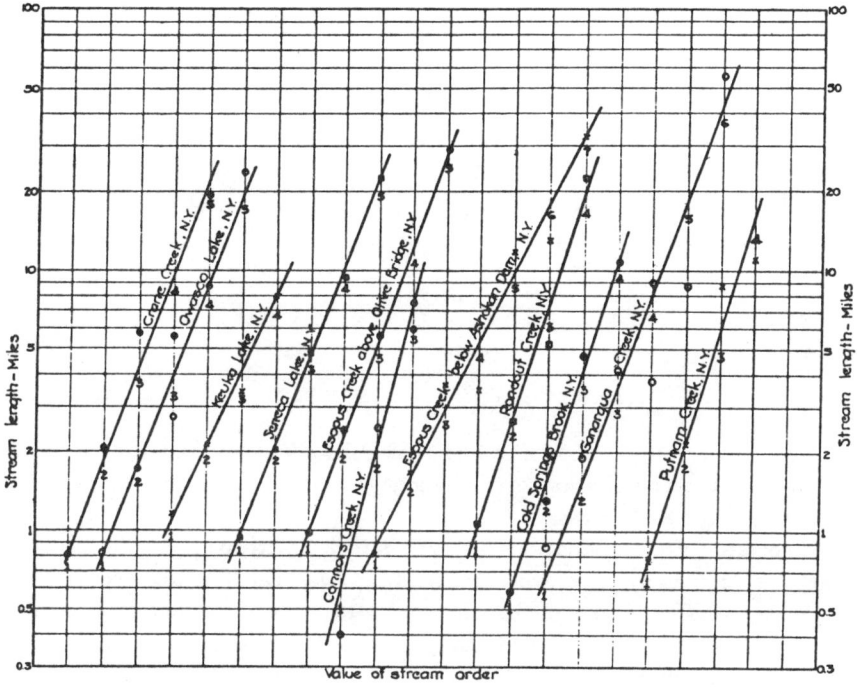

FIGURE 4.—*Relation of stream lengths to stream order in different drainage basins*

As an example, Table 2 shows the observed and computed numbers and lengths of streams of different orders, based on the following values of the variables:

$$r_b = 3.12$$
$$s = 5$$
$$r_l = 2.31$$
$$l_1 = 0.994$$

Actual stream numbers must of course be integers, while the computed numbers may be fractions. Some variation between the computed and observed stream numbers and lengths must be expected, for various reasons. Several drainage basins listed in Table 1 contain large lakes, and the drainage density is less than it would be if the lake did not exist, since there would then necessarily be a stream of the highest order traversing the lake bed. Lower Rondout Creek represents an incomplete drainage basin to which the geometric-series laws do not necessarily apply because it contains a stream or portions of streams of higher order than those originating within this particular area. For some other areas the data were derived from early editions of topographic maps which do not show all the low-order tributaries. In order that the equations shall give correct results the drainage basin must be reasonably homogeneous. This is true of the drainage basin of Esopus Creek above Olive Bridge, which is wholly mountainous. The drainage basin of lower Esopus Creek is, however, rolling and permeable, with great differences in soil, vegetal cover, rainfall, and climate, as compared with the upper basin. Upper Esopus Creek drainage basin

TABLE 2.—*Observed and computed stream lengths and stream numbers*
Drainage basin of Esopus Creek above Olive Bridge, New York.

Stream Order	Number of streams		Average stream length (miles)	
	From topographic maps	By eq. (5)	From topographic maps	By eq. (7)
1	90	94.75	0.994	0.994
2	25	30.37	2.45	2.30
3	9	9.73	5.64	5.31
4	1	3.12	6.00	12.2
5	1	1.00	29.00	28.3

has a much higher bifurcation ratio and stream-length ratio than the lower basin (Table 1), and the composition of the drainage nets is quite different, as shown graphically on Figures 3 and 4.

The importance of these equations lies both in their practical application and in the fact that they represent laws which evolve from physical processes which Nature follows rather closely in the development of stream systems under such diverse conditions as those of upper and lower Esopus Creek. The size of the drainage basin does not enter the equations directly. It is indirectly involved, since the order of the main stream would in general be higher in the larger drainage basin, for two homogeneous drainage basins of different sizes. The order of the main stream is a factor in the equations, and the drainage basin in which the main stream is of the higher order will have, in general, more tributaries of a given order.

The data given in Table 1 cover a wide range of conditions, from precipitous mountain areas, like upper Esopus Creek, and highly dissected areas, like those of Seneca and Owasco lakes, to moderately rolling and flat areas. They cover also drainage basins ranging in size from a few square miles up to several hundred square miles.

The bifurcation ratio (Table 1, column 11) ranges from about 2 for flat or rolling drainage basins up to 3 or 4 for mountainous or highly dissected drainage basins. As would be expected, the bifurcation ratio is generally higher for hilly, well-dissected drainage basins than for rolling basins.

The values of the length ratios (column 12) range from about 2 to about 3; the average is 2.32.

In the examples given in Table 1, the stream lengths were measured to stream tips as shown on U. S. Geological Survey topographic maps. If stream lengths had been measured as extended to watershed lines, the resulting stream-length ratios would have been materially reduced. If c is the average length from the stream tip to the watershed line, and l_1 and l_2 are actual average stream lengths of two successive orders, l_2 being the higher, then, as computed in Table 1,

$$r_l = \frac{l_2}{l_1}. \tag{8}$$

If measured as extended to the watershed lines,

$$r_l' = \frac{l_2 + c}{l_1 + c}. \tag{9}$$

The quantity r_l' will always be less than r_l. The average value of r_l for the streams listed in Table 1 is 2.32. If the stream lengths were extended to the watershed lines, this value would lie between 2.00 and 2.32. The theoretical value of r_l' for streams flowing into larger streams at right angles is 2.00, but r_l' will be greater for streams entering at acute angles, as do most streams on steeper slopes. The distance along the course of a stream from its mouth extended to the water shed line is called "mesh length." The use of this quantity instead of actual stream length is preferable in physiographic studies, and its use leads to closer agreement with the theoretical values.

In Figures 3 and 4, the agreement between the mean lines for the different streams and the observed data is so close that the two following general laws may be stated regarding the composition of stream-drainage nets:

(1) *Law of Stream Numbers:* The numbers of streams of different orders in a given drainage basin tend closely to approximate an inverse geometric series in which the first term is unity and the ratio is the bifurcation ratio.

(2) *Law of Stream Lengths:* The average lengths of streams of each of the different orders in a drainage basin tend closely to approximate a direct geometric series in which the first term is the average length of streams of the 1st order.

Playfair called attention to the "nice adjustment" between the different streams and valleys of a drainage basin but chiefly with reference to their declivities. These two laws supplement Playfair's law and make it more definite and more quantitative. They also show that the nice adjustment goes far beyond the matter of declivities.

TOTAL LENGTH OF STREAMS OF A GIVEN ORDER

Since the total length of streams of a given order is the product of the average length and number of streams, equations (5) and (7) can be combined into an equation for total stream length of a given order.

The total length L_o of tributaries of order o is:

$$L_o = l_1 r_b{}^{s-o} r_l{}^{o-1} \tag{10}$$

The total lengths of all streams of a given order is the product of the number of streams and length per stream. The number of streams is dependent on the bifurcation ratio r_b and increases with stream order, while the length per stream is dependent on the stream length r_l and decreases with increasing stream order. Thus the total lengths of streams of a given order should have either a maximum or a minimum value for some particular stream order. A maximum or minimum may not occur because the stream order required to give the maximum or minimum stream length may exceed the order of the main stream, in which case the total lengths of streams of a given order will either increase or decrease progressively with increasing stream order. An exception occurs where r_b and r_l are equal. Then the total lengths of streams of all orders are the same and equal to $l_1 r_b^s$. The ratio of r_l to r_b is designated ρ and is an important factor in relation both to drainage composition and physiographic development of drainage basins. As will be shown later, the value of the ratio $\rho = \dfrac{r_l}{r_b}$ is determined by precisely those factors—hydrologic, physiographic, cultural, and geologic—which determine the ultimate degree of drainage development in a given drainage basin.

By summation of the total stream lengths for different orders, as given by equation (10), the total stream length within a drainage basin can be expressed in terms of four fundamental quantities: l_1, o_s, r_b, and r_l.

CHANNEL-STORAGE CAPACITY

Natural channel storage is a principal factor in modulating flood-crest intensities as a flood proceeds down a system of stream channels. A knowledge of relative amounts of channel storage at different locations is required for various problems of flood routing and flood control. It can readily be shown that the normal channel storage at a given discharge rate in a given stream channel varies as a simple-power function of the stream length, usually less than the square of the stream length, and this relation can readily be determined from the stage-discharge relation at a gaging station. Equation (10) provides a means of determining total and average stream lengths for each stream order. From this the channel storage provided by each order of streams in the drainage basin can be determined, and by summation the total channel storage in the stream system becomes known. This illustrates the practical application of quantitative physiography to a variety of engineering problems.

Different stream systems may have substantially the same drainage density and yet differ markedly in channel-storage capacity. The higher-order stream channels have larger cross sections and contain more channel storage per unit length than lower-order streams. If r_l/r_b is high, the greater length of larger stream channels may afford greatly increased channel storage per unit of drainage area as compared with a drainage basin with the same drainage density and a lower value of r_l/r_b.

GENERAL EQUATION OF COMPOSITION OF STREAM SYSTEMS

From equation (10) the total length of streams of a given order is:

$$L_0 = l_1 r_b^{s-o} r_l^{o-1}$$

The total length of all streams in a drainage basin with the main stream of a given order s is the sum of the total lengths of streams of different orders, or:

$$\Sigma L = l_1 [r_b^{s-1} + r_b^{s-2} r_l + r_b^{s-3} r_l^2 + \cdots r_b^0 r_l^{s-1}] \tag{11}$$

This equation is cumbersome and can easily be simplified.

Let:

$$\rho = \frac{r_l}{r_b}; \quad r_l = \rho r_b \tag{12}$$

$$L_o = l_1 \rho^{o-1} r_b^{s-1}. \tag{13}$$

Applying subscripts 1, 2, etc., to designate the total lengths of streams of different orders:

$$\Sigma L = L_1 + L_2 + L_3 + \cdots L_s \tag{14}$$

and from (13):

$$\Sigma L = l_1 r_b^{s-1}(\rho^{1-1} + \rho^{2-1} + \rho^{3-1} + \cdots \rho^{s-1})$$
$$= l_1 r_b^{s-1}(1 + \rho + \rho^2 + \rho^3 + \cdots \rho^{s-1}) \tag{15}$$

The term in parentheses is the sum of a geometric series with its first term unity and a ratio ρ and is equal to:

$$\frac{\rho^s - 1}{\rho - 1}.$$

Substituting this value in equation (15):

$$\Sigma L = l_1 r_b^{s-1} \cdot \frac{\rho^s - 1}{\rho - 1}. \tag{16}$$

The drainage density is, from (1):

$$D_d = \frac{\Sigma L}{A}.$$

Substituting the value of ΣL from (16):

$$D_d = \frac{l_1 r_b^{s-1}}{A} \cdot \frac{\rho^s - 1}{\rho - 1}. \tag{17}$$

This equation combines all the physiographic factors which determine the composition of the drainage net of a stream system in one expression. Aside from its scientific interest in this respect, it can also be used to determine drainage density. Values of the factor $(\rho^s - 1)/(\rho - 1)$ can be obtained from Figure 5.

RELATION OF SIZE OF DRAINAGE AREA TO STREAM ORDER

Since equation (17) incorporates all the principal characteristics of the stream system of a drainage basin, it may be considered a quantitative generalization of Playfair's law. It can be written in such a form as to give any one of the quantities l_1, D_d, A, r_b, r_l, and s when the other five quantities are known. If the ratio $\rho < 1$, then, for larger values of s, ρ^s is small, and $\rho^s - 1$ may be taken as -1.0. The equation (17) may then be written:

$$s = 1 + \frac{\log[(1 - \rho)D_d A/l_1]}{\log r_b}. \tag{18}$$

If $\rho > 1$, then, for large values of s, ρ^s is sensibly the same as $\rho^s - 1$, and $\rho^s - 1$ is positive. This leads similarly to the equation:

$$(s - 1) \log r_b + s \log \rho = \log \frac{(\rho - 1) D_d A}{l_1}$$

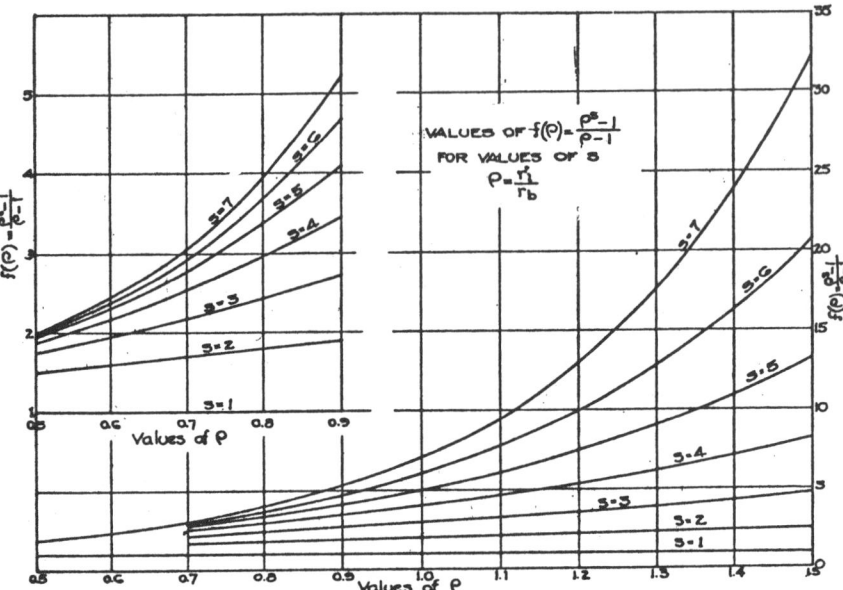

FIGURE 5.—*Diagram of factor $\rho^s - 1/\rho - 1$*

and

$$s = \frac{\log [(\rho - 1) D_d A / l_1] + \log r_b}{\log r_b + \log \rho}. \tag{19}$$

The quantity $\log r_b$ is small relative to $\log (\rho - 1) D_d A$ and may be neglected, so that in either case the order of the main stream developed in a drainage basin of a given area A increases for larger values of s in proportion to the logarithm of the area A. If, for example, with given values of ρ, D_d, r_b, r_l, and l_1, an area of 10,000 square miles is required to develop a stream order s, then, under the same conditions, in a drainage basin of 100,000 square miles, the main stream would be of order one unit higher, and in a drainage basin of 1,000,000 square miles the main stream would be two units higher in order than in an area of 10,000 square miles. This shows at once why stream systems with extremely high orders do not occur—there is not room to accommodate the requisite drainage basins on the solid surface of the earth. The orders of the Mississippi, Amazon, and other large rivers have not been determined accurately, but the Mississippi River quite certainly does not exceed the 20th order.

From equation (17) drainage density should vary inversely as the drainage area A, other things equal. Actually other things are not equal in drainage areas of different sizes, and, although the bifurcation ratio, stream-length ratio, and average length of

1st order streams may be the same in two drainage basins physiographically similar and of different sizes, the order s of the main stream will in general be larger for the larger drainage basin. As a result the drainage density may increase, decrease, or remain substantially unchanged in two similar drainage basins of different sizes.

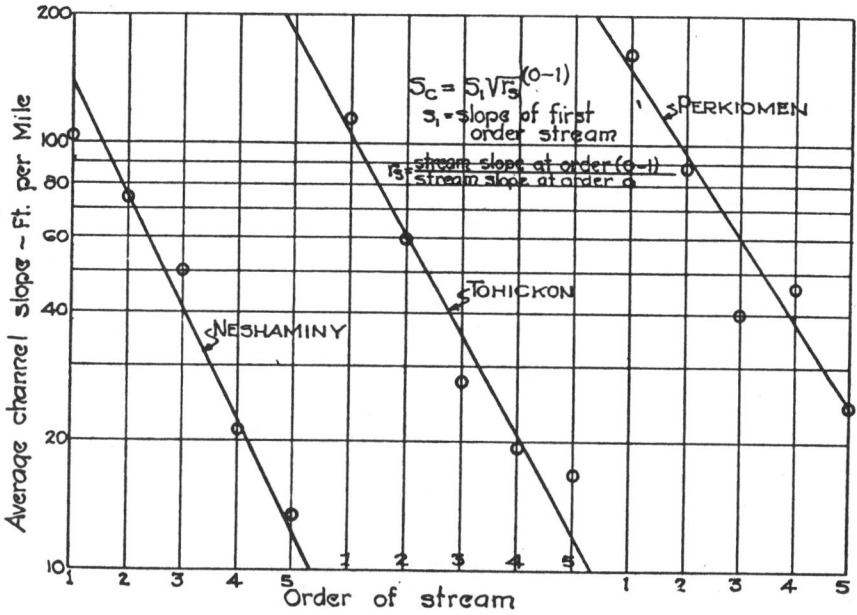

FIGURE 6.—*Law of stream slopes*
Neshaminy, Tohickon, and Perkiomen drainage basins.

LAW OF STREAM SLOPES

In addition to the various quantitative relationships between the different factors involved in drainage composition, expressed by equation (17), there are certain other quantitative relationships. The relation of stream slope to stream order in a given drainage basin is hinted in Playfair's law. As an illustration, the slopes of streams of different orders in the Neshaminy, Tohickon, and Perkiomen drainage basins have been plotted in terms of stream order (Fig. 6), and there is a fairly definite relationship between slope of the streams and stream order, which can be expressed by an inverse geometric-series law.

DETERMINATION OF PHYSIOGRAPHIC FACTORS FOR DRAINAGE BASINS

To determine completely the composition of a stream system it is necessary to know: (1) the drainage area, A, (2) the order s of the main stream, (3) the bifurcation ratio r_b, (4) the stream length ratio r_l, and (5) the length l_s of the main stream or preferably the average length l_1 of 1st order streams. If these data are given, then the drainage density, stream frequency, and other characteristics of the stream system can be determined by calculation, using the equations which have been given.

From equation (5):

$$N_o = r_b^{(s-o)}.$$

If $o = s - 1$:

$$N_{s-1} = r_b.$$

This shows that the bifurcation ratio r_b is equal to the number of streams of the next to the highest order for the given drainage basin.

If the stream numbers for different stream orders are plotted on semilog paper (Fig. 3), the bifurcation ratio r_b can be determined by simply reading from the average line the number of streams of the second highest order.

From equation (7):

$$l_o = l_1 r_l^{o-1}.$$

If $o = 2$,

$$\frac{l_2}{l_1} = r_l.$$

The stream length ratio r_l can therefore be obtained by dividing the average stream length of any order by the average stream length of the next lower order, the values of stream lengths being read from the diagram of stream lengths plotted in terms of stream order. It is preferable to use these data rather than actual measured values, as the number of streams of a given order—particularly the higher order streams—may not be exactly the normal number for the given drainage composition. Stream numbers can, of course, be only integers, and there may be either two, three, or four streams of the second highest order in a given drainage basin where there should be three.

In Table 1 and Figures 3 and 4, the stream lengths and numbers of all orders were determined directly from topographic maps. Where this is done the order s becomes known directly.

In analyzing the drainage net of a stream system it is desirable to trace, with different colors for each order, the stream system from the base map. When the higher-order streams are determined some of the lower-order streams may prove to be the head-water portions of higher-order streams. Figure 7 shows the drainage basin of Hiwassee River above Hiwassee, Georgia, with 1st order streams shown by dotted lines and stream orders indicated by figures.

The determination of stream lengths and orders by direct measurement from maps which are on a sufficiently large scale to show all 1st order streams is so laborious as to be practically prohibitive except for smaller drainage basins.

Fortunately, all the required quantities—l_s, l_1, r_b, r_l, and D_d—can be determined from smaller-scale maps from which the lower-order tributaries are omitted. The maps must show correctly the streams for several of the higher orders. The order of the main stream is of course unknown since it is not in general known which of the lower orders of streams are omitted from the map. The streams shown are assigned orders assuming that the main stream has an unknown order s, the next lower order of stream shown is designated 2, and so on. The number of streams of each assumed

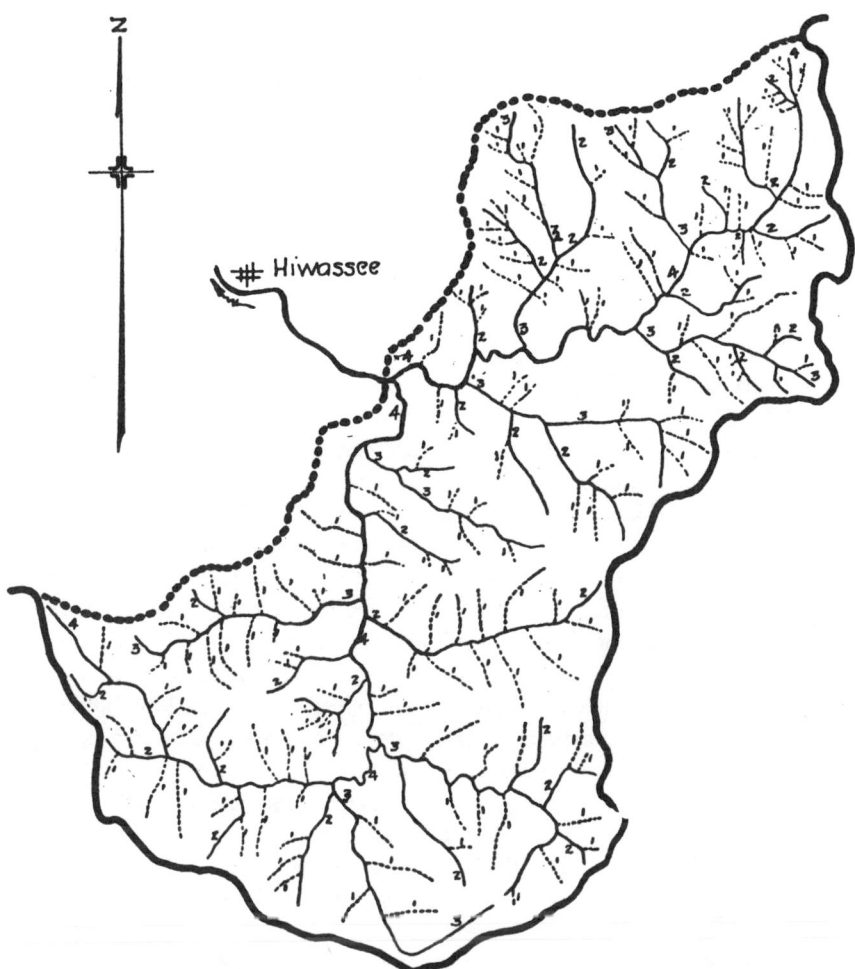

Order	No. of streams	Length miles	Aver. length miles
1	146	72	0.49
2	32	41	1.28
3	9	32.8	3.65
4	2	24.6	12.30
		170.4	

Drainage area = 82.8 sq. mi

D_d = 2.06

FIGURE 7.—*Drainage net, upper Hiwassee River*

order is counted, their stream lengths measured from the map, the results tabulated as follows and plotted as shown by Figure 8A.

Data for Perkiomen Creek

Order	Assumed inverse order	Number of streams	Average stream length (miles)
s	1	1	20.5
$s-1$	2	2	13.75
$s-2$	3	10	3.61
$s-3$	4	32	1.39

If it is assumed that the main stream is of the
4th order, then $l_1 = 1.38$ miles;
5th order, then $l_1 = 0.50$ mile;
6th order, then $l_1 = 0.20$ mile.

Since l_1 is not far from half a mile, the main stream is of the 5th order. From line B (Fig. 8) the number of 2d order streams is 3.15. This is the bifurcation ratio. From line A the lengths of 2d and 1st order streams are, respectively, 1.38 and 0.52 miles. This gives the stream length ratio:

$$r_l = \frac{1.38}{0.52} = 2.70.$$

Data for at least four stream orders are required to determine the order of the main stream from incomplete data by this method. Care must also be used in determining the lines A and B accurately to secure correct results.

The values of the stream lengths as far as known are then plotted on semilog paper (Fig. 8A), in terms of inverse stream orders, a line of best fit drawn to represent the plotted points and this line extended downward to stream length unity or less.

To determine the order of the main stream it is necessary to know the order of magnitude but not the exact value of the average length of streams of the 1st order. The length l_1 of streams of the 1st order is rarely less than a third of a mile, a value which is approached as a minimum limit in mountain regions with heavy rainfall, as in the southern Appalachians. Also it is rarely greater than 2 or 3 miles, values which are approached as maximum limits under some conditions in arid and semiarid regions. Data from which the order of magnitude of l_1 can be determined are always available from some source. In general all that is required is to know whether l_1 is of the order of half a mile, 1 mile, or 2 miles or more. The point at which the stream length shown by the line ab (Fig. 8A) extended downward has a value about the same as the known value of l_1 for the given order indicates the order of the main stream.

This method for determining the order of the main stream is of limited value in some drainage basins, particularly large drainage basins, such as that of the Mississippi River, which are not homogeneous, and where there may be large variations in the length of 1st order streams in different portions of the drainage basins, so that the order of magnitude of l_1 may be difficult to determine. A small portion of a drainage basin, with suitable conditions of high rainfall, steep slopes, etc., may add several units to the value of s for the main stream, although it has little effect on the weighted average value of l_1 for the drainage basin as a whole. For basins which are reasonably homogeneous the method is accurate. Proof of its validity is readily obtained by applying this method to a drainage basin where the values of l_1 and the drainage density D_d have been determined from measurements on a map showing streams of all orders, but using in the determination only the data for streams of higher orders. This was done in preparing Figure 8, which is of the 5th order, although only data for the first four stream orders were used in the computation, it being assumed that l_1 was of an order of magnitude between 1 and 1.5.

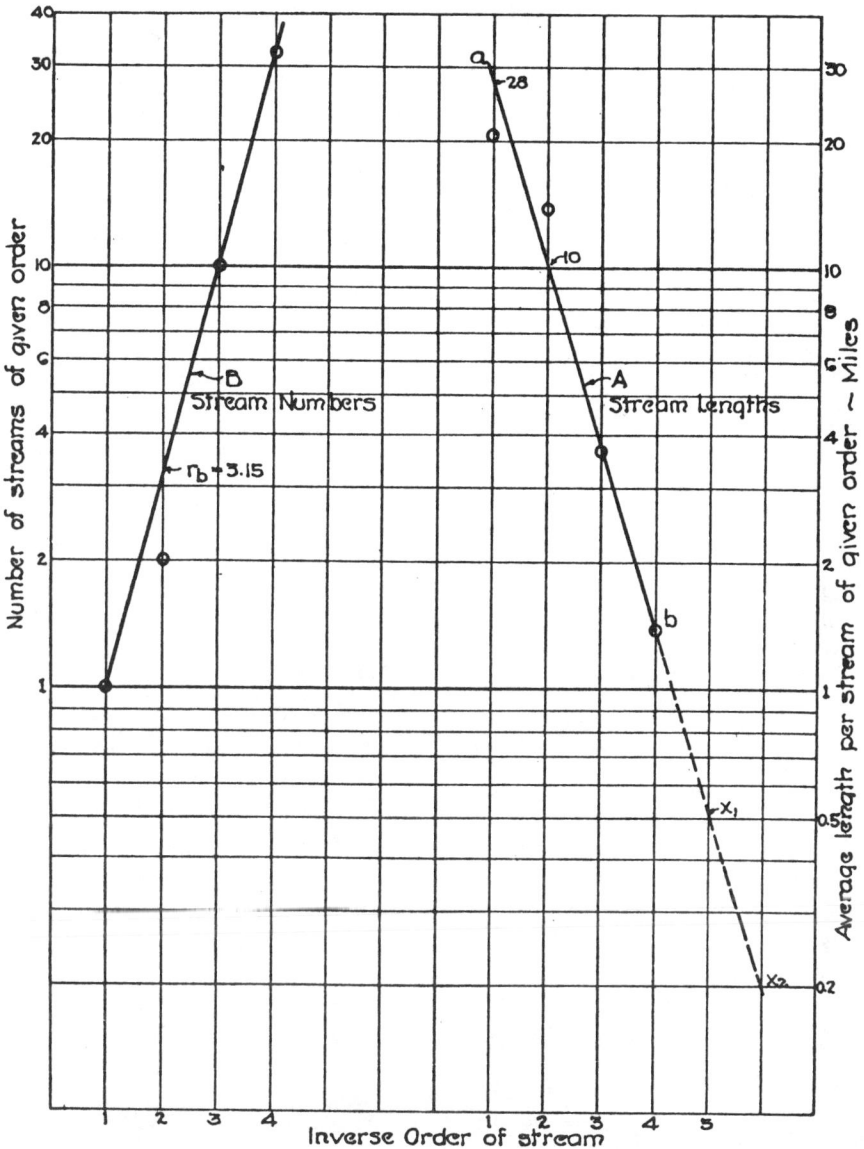

FIGURE 8.—*Graphical determination of stream characteristics*

This determination of s gives also the average length l_1 of 1st order streams. The bifurcation ratio r_b and the stream-length ratio r_l are determined by the slopes of the lines A and B on Figure 8. It is not necessary to know the order of the main stream to determine these quantities. When r_b, r_l, A, s, and l_1 are known, the drainage density can be determined by means of equation (17).

This method of determining s has the advantage that it is at least as accurate when applied to large as when applied to smaller drainage basins. In general, data for more stream orders will be available from a map for a large drainage basin than for a small basin.

Table 3 shows the drainage composition of Neshaminy, Tohickon, and Perkiomen Creek stream systems, derived in the manner described, together with the drainage

TABLE 3.—*Observed and computed drainage densities, Neshaminy, Tohickon, and Perkiomen creeks*

Item		Neshaminy	Tohickon	Perkiomen
1	Stream	Neshaminy	Tohickon	Perkiomen
2	Location	Below Forks	Point Pleasant	Near Frederick
3	Drainage area, square miles	139.3	102.2	152.0
	Computed values:			
4	Stream order from map	5	5	5
5	l_s	35.0	33.0	27.0
6	r_b	3.45	3.00	3.15
7	r_l	2.92	2.85	2.70
8	$\rho = r_l/r_b$	0.85	0.95	0.86
9	$f(\rho)$	3.75	4.50	3.80
10	r_b^{s-1}	141.6	81.0	98.4
11	r_b^{s-1}/A	1.02	0.79	0.65
12	(11) × (9)	3.82	3.56	2.47
13	l_1	0.50	0.53	0.52
14	(12) × (13) = D_d	1.91	1.89	1.28
15	Drainage density from map	1.60	1.91	1.24

densities as computed by equation (17) and as derived from direct measurement from topographic maps.

Drainage densities computed by equation (17) will usually be somewhat higher than those derived directly from maps if stream lengths are measured directly and only to the fingertips of the stream channels, because the stream lengths and mesh lengths are sensibly identical for higher-order streams, whereas there may be 10 to 25 per cent or even 50 per cent difference between stream length and mesh length for low-order streams. In computing drainage density from values of l_1, r_b, and r_l obtained graphically, the computed value corresponds more nearly to drainage density expressed in terms of mesh length than in terms of actual stream length for lower-order streams.

RELATION OF GEOLOGIC STRUCTURES TO DRAINAGE COMPOSITION

The examples of drainage composition shown in Table 1 and in Figures 3 and 4 in nearly all cases represent special or abnormal conditions, such as the presence of large lakes in several of the drainage basins. While this table agrees well with the laws of stream numbers and stream lengths even with such pronounced geologic control of topography as that afforded by the drumlin areas in the Ganargua Creek drainage basin, there are other conditions where geologic controls apparently exert a definite influence on drainage composition. Figure 9 shows two drainage basins the boundaries of which are definitely fixed by geologic structures.

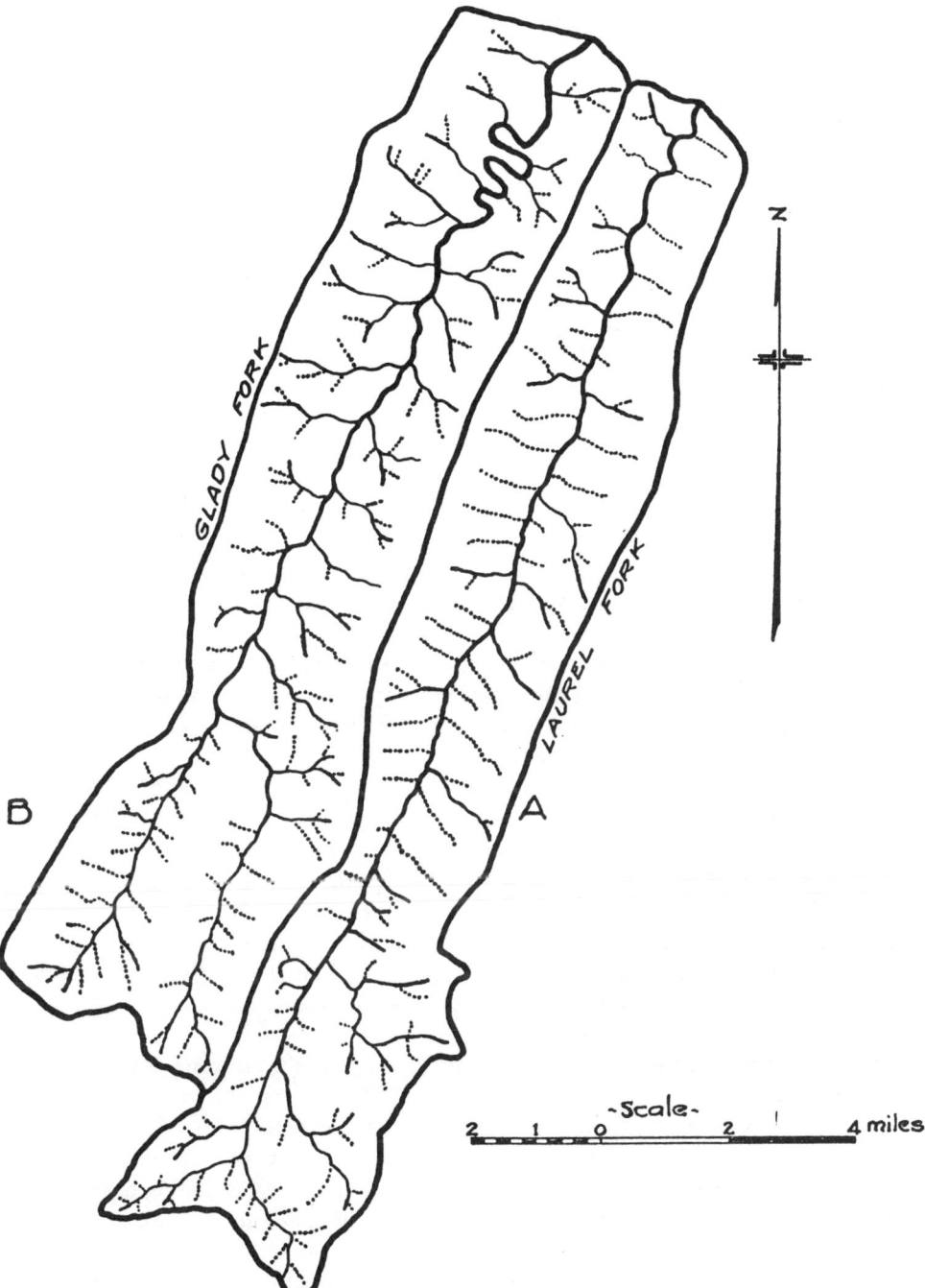

FIGURE 9.—*Drainage patterns of Laurel Fork and Glady Fork, Cheat River drainage basin*
(From Beverly, W. Va., quad., U. S. G. S.)

Data of stream lengths and stream numbers in these basins are as follows:

	Order	Length (miles)	Number of streams	Average length (miles)
Basin "A"—Laurel Fork 50.8 sq. mi.	1	47.80	94	0.51
	2	23.36	26	0.90
	3	4.10	4	1.02
	4	19.62	1	19.62
		94.88		

$$D_d = \frac{94.88}{50.8} = 1.87$$

	Order	Length (miles)	Number of streams	Average length (miles)
Basin "B"—Glady Fork 55.44 sq. mi.	1	38.42	95	0.41
	2	34.76	37	0.94
	3	6.50	6	1.08
	4	20.80	1	20.80
		100.48		

$$D_d = \frac{100.48}{55.44} = 1.82$$

In both streams the law of stream numbers is closely obeyed. The law of stream lengths is approximately obeyed for lower-order streams, but in both basins it is necessary, in order that the area should be drained, that the main stream should have a length sensibly equal to that of the drainage basin; this requires that the length of the main stream should be much greater than it ordinarily would be for a drainage basin of the same order, of normal form.

One may naturally ask whether stream systems in similar terrain and which are genetically similar should not have identical or nearly identical stream composition. Data for the drainage basins of Neshaminy, Tohickon, and Perkiomen creeks in the Delaware River drainage basin near Philadelphia, Pennsylvania, are given in Table 3 and on Figure 10. The drainage patterns of these streams are shown on Figure 11. The physiographic characteristics of these three drainage basins are closely similar.

Tables 4 and 5 show, respectively, drainage composition of streams in the upper Delaware River drainage basin and drainage composition of several small tributaries of Genesee River in western New York.

The streams in the upper Delaware River drainage basin are generally similar, with the exception of Neversink River, in topography, geology, and climate. The various morphologic factors for these basins are of the same order of magnitude although not numerically identical. The tributaries of Genesee River listed in Table 5 represent areas at various locations around the margins of this basin between Lake Ontario and the New York-Pennsylvania State line and comprise a wider range of geologic and topographic conditions than occurs in the Delaware River drainage basins, and there are correspondingly greater variations in the morphologic factors, particularly bifurcation ratio, length of 1st order streams, and drainage density.

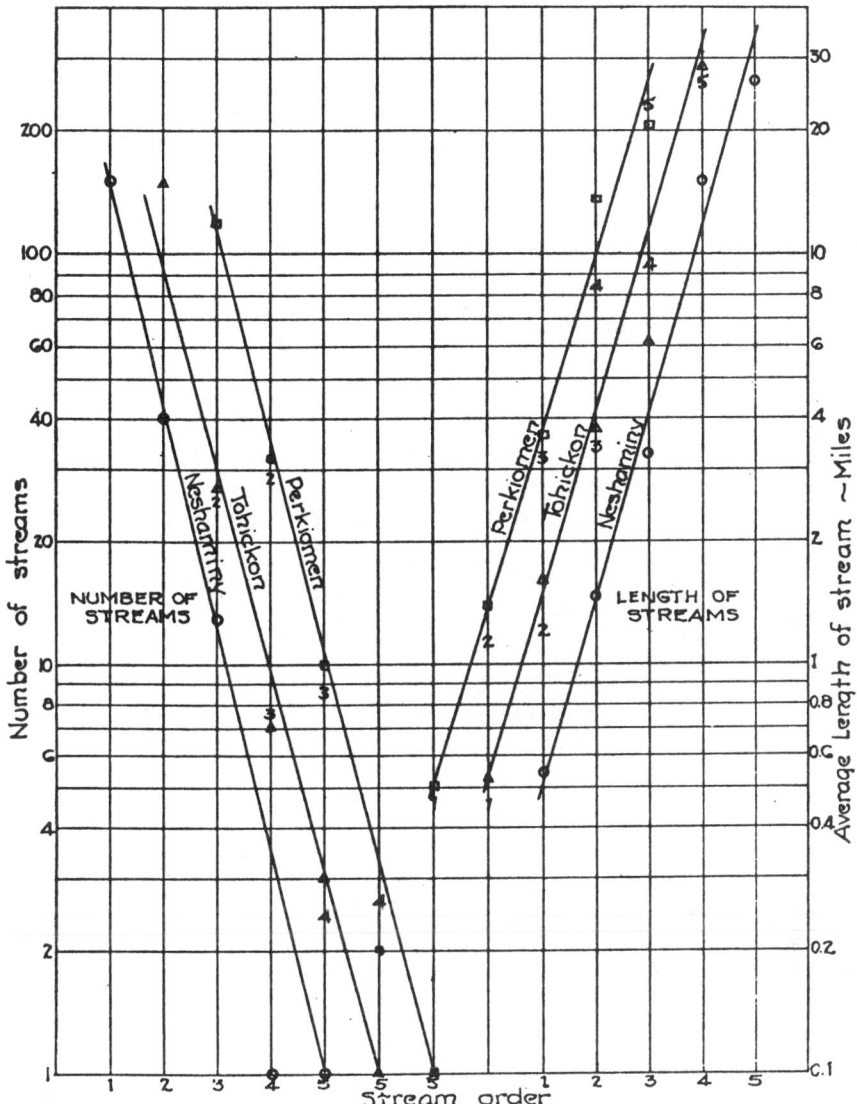

FIGURE 10.—*Stream numbers and stream lengths*
Neshaminy, Tohickon, and Perkiomen drainage basins.

It is found from plotting stream lengths and stream orders, subject to the limiting conditions already described, that both these laws are quite closely followed. Departures from the two laws will, however, be observed, and if other conditions are normal these departures may in general be ascribed to effects of geologic controls. As a rule the law of stream numbers is more closely followed than the law of stream lengths; Nature develops successive orders of streams by bifurcation quite generally

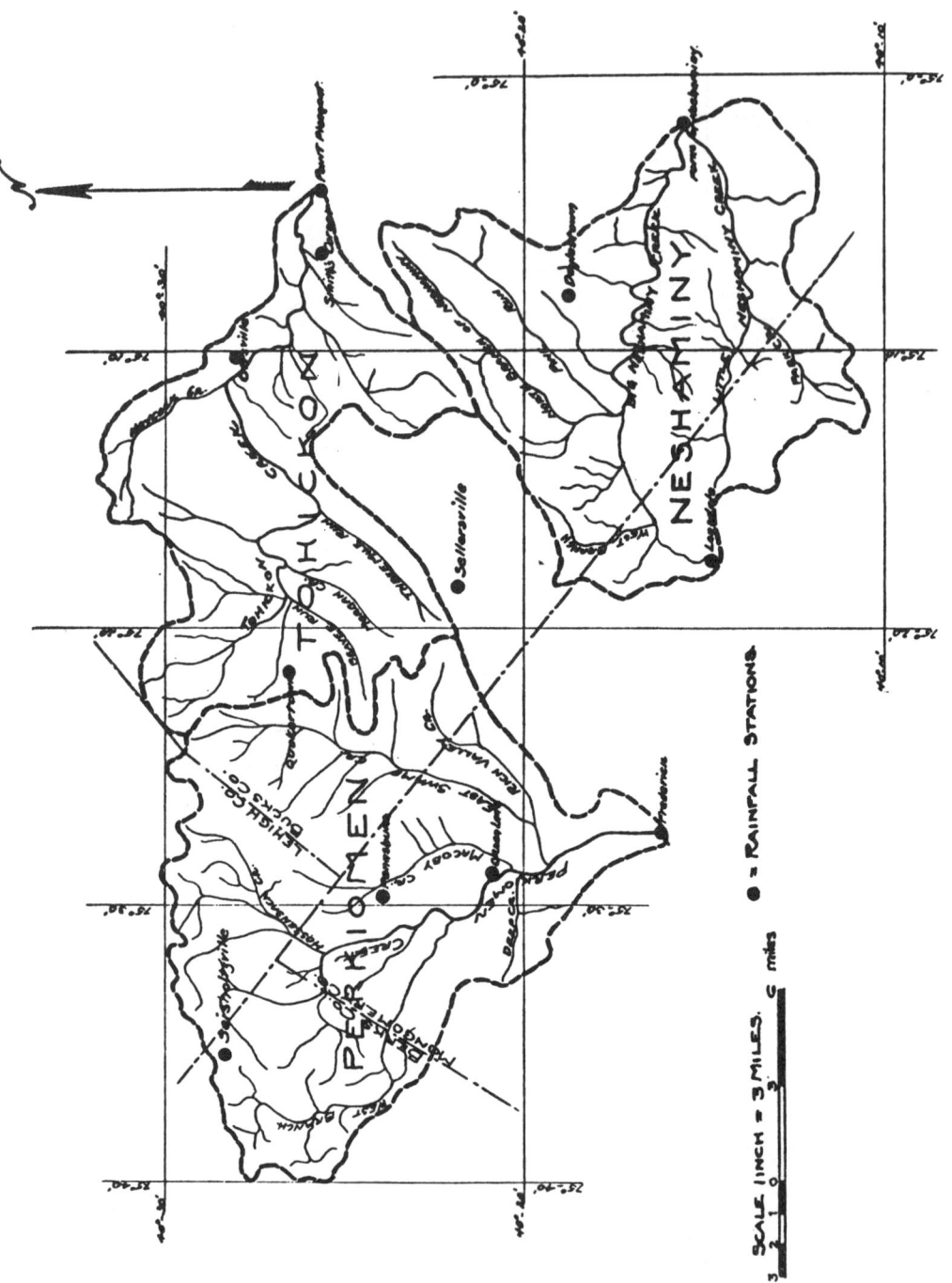

FIGURE 11.—*Drainage patterns*
Neshaminy, Tohickon, and Perkiomen drainage basins.

TABLE 4.—*Physiographic factors for drainage basins tributary to Delaware River*

Item No.	Item	Formula or Method	East Branch Delaware	Beaver Kill	Little Delaware	East Branch Delaware	Willowemoc	Beaver Kill	Neversink
(1)	(2)	(3)	(4)	(5)	(6)	(7)	(8)	(9)	(10)
1	Stream		E. Br. Del. R.	Beaver Kill	Little Del. R.	E. Br. Del. R.	Willowemoc	Beaver Kill	Neversink
2	Location		Fish Eddy	Cooks Falls	Delhi	Downsville	Livingston Manr	Beaverkill	Claryville
3	Drainage Area — Sq. Mi.	A	785	241	50.3	373	63.8	61.0	68.6
4	Length Main Stream Miles	L_s	56.5	28.5	15.0	38.0	14.5	13.0	13.0
5	Width ratio	A/L_s^2	0.246	0.297	0.224	0.258	0.303	0.361	0.406
6	Channel Slope ~ Ft. per mile	S_c $\frac{S_c - S_b}{S_c - S_b}$	338	210	296	406	114	307	336
7	Ground	S_g	1000	779	959	1074	725	1033	1043
8	Slope ratio	S_c/S_g	0.338	0.270	0.308	0.377	0.309	0.297	0.323
9	Order of main stream (a)	n	5	4	4	5	4	3	3
10	Average length 1st order streams (a)	ℓ	0.93	0.72	0.82	0.88	0.60	1.05	0.87
11	Bifurcation ratio (a)	r_b	4.2	5.0	5.0	3.6	2.7	4.5	4.6
12	Stream length ratio (a)	r_ℓ	2.77	3.12	2.59	2.60	2.54	3.51	3.71
13	Ratio ρ (a)		0.66	0.62	0.86	0.72	0.94	0.78	0.80
14	Average length overland flow miles	$L_o = \frac{1}{2D_d}\sqrt{1-(\frac{S_c}{S_g})^2}$	0.346	0.405	0.324	0.317	0.360	0.401	0.435
15	Drainage density	D_d	1.36	1.19	1.47	1.46	1.32	1.19	1.08
16	Latitude		42-00	42-00	42-15	42-10	42-00	42-05	41-55
17	Longitude		74-50	74-40	74-50	74-40	74-40	74-40	74-30

(a) From Bien's Atlas of N.Y. State. Other data are from U.S.G.S. maps

in a uniform manner, regardless of geologic controls. Stream lengths, on the other hand, may be definitely limited by geologic controls, such as fixed boundaries of the outline of the drainage basin.

TABLE 5.—*Physiographic factors—tributaries of Genesee River, western New York*

Item No.	Item		Slader Creek	Gates Creek	Rush Creek	Red Creek	Spring Creek	Stony Creek
(1)	(2)		(3)	(4)	(5)	(6)	(7)	(8)
1	Stream		Slader Crk.	Gates Crk.	Rush Crk.	Red Crk.	Spring Crk.	Stony Crk.
2	Location		Canaseraga	at mouth	Fillmore	at mouth	at mouth	at mouth
3	Drainage Area	Sq. Mi.	15.8	20.2	43.3	25.3	20.3	22.2
4	Width ratio		1.12	1.26	0.39	0.65	1.44	0.89
5	Order main stream	s	3	4	3	3	3	4
6	Stream numbers							
	First order		24	32	38	14	15	25
	Second "		4	10	10	4	4	8
	Third "		1	3	1	1	1	3
	Fourth "			1				1
7	Average stream length —miles							
	First order		0.61	0.49	0.48	1.29	1.22	0.81
	Second "		2.12	0.95	1.15	1.50	2.00	1.28
	Third "		3.75	2.75	10.50	6.25	3.75	3.25
	Fourth "			4.00				5.00
8	Average stream slope – Ft./mi							
	First order		271	89	171	26	37	287
	Second "		121	41	80	18	39	196
	Third "		89	39	49	7	11	169
	Fourth "			25				120
9	Bifurcation ratio	r_b	4.40	3.10	8.10	3.90	3.90	3.10
10	Stream length ratio	r_l	2.21	2.00	4.59	2.57	1.78	1.85
11	Total stream length	mi.	27.00	37.50	60.25	30.25	30.00	38.75
12	Drainage density	D_d	1.71	1.86	1.41	1.20	1.47	1.73
13	Ratio P	r_l/r_b	0.50	0.64	0.57	0.66	0.46	0.60
14	Latitude – north		42-25	42-30	42-25	43-05	43-05	42-30
15	Longitude – west		77-45	77-30	78-05	77-40	78-05	77-40

INFILTRATION THEORY OF SURFACE RUNOFF

GENERAL STATEMENT

The factors and formulas given serve as measuring tools for the quantitative comparison of upland features of drainage basins. Quantitative science develops by the correlation of observed relationships through scientific laws and principles, which may therefore be described as operating tools. Two principal kinds of operating tools are needed in connection with upland erosion: (1) the laws governing the sheet flow of surface runoff, and (2) the laws governing (a) soil resistivity to erosion, (b) erosive force, (c) erosive power of sheet flow, and (d) transporting power of sheet flow. The first of these tools is supplied by the infiltration theory of surface runoff, developed by the author (Horton, 1935; 1937; 1938). Only a few salient features of this theory are pertinent to the present discussion.

The infiltration theory of surface runoff is based on two fundamental concepts:

(1) There is a maximum limiting rate at which the soil when in a given condition can absorb rain as it falls. This is the infiltration-capacity (Horton, 1933).

(2) When runoff takes place from any soil surface, large or small, there is a definite functional relation between the depth of surface detention δ_a or the quantity of water which accumulates on the soil surface, and the rate of surface runoff or channel inflow q_s.

These two concepts, in connection with the equation of continuity or storage equation, form the basis of the infiltration theory. It has hitherto been assumed that surface runoff was some definite fraction of rain. If that were true, then all rains, however low their intensity, should produce runoff. This is not an observed fact.

INFILTRATION-CAPACITY

Infiltration-capacity, f, is governed by physical laws and processes which involve the simultaneous downward flow of water and the upward flow of displaced air through the same system of soil pores (Horton, 1940; Duley and Kelly, 1941) and is used in the sense of a limiting rate of flow, like the capacity of a water pipe.

The infiltration capacity of a given terrain, including soil and cover, is controlled chiefly by (1) soil texture, (2) soil structure, (3) vegetal cover, (4) biologic structures in the soil, especially at and near the surface, including plant roots and root perforations, earthworm, insect, and rodent perforations, humus, and vegetal debris, (5) moisture content of the soil, and (6) condition of the soil surface, whether newly cultivated, baked, or sun-cracked. Temperature is probably also a factor, although its effect is often masked by biologic factors, which also vary with temperature and season.

The infiltration-capacity of a given area is not usually constant during rain but, starting with an initial value f_0, it decreases rapidly at first, then after about half an hour to 2 or 3 hours attains a constant value f_c. The relation of the infiltration-capacity to duration of rain can be expressed accurately by the following equation, with f, f_0, and f_c in inches per hour:

$$f = f_c + (f_0 - f_c)e^{-K_f t} \qquad (20$$

where e is the base of Naperian logarithms, t is time from beginning of rain, in hours, and K_f is a proportionality factor (Horton, 1939; 1940). This equation can easily be derived on the assumption that infiltration-capacity is governed chiefly by the condition of the soil surface and is reduced at the beginning of rain by effects which result from the energy of falling rain and which operate after the manner of exhaustion phenomena. These effects include packing of the soil surface, breaking down of the crumb structure of the soil, swelling of colloids, and the washing of fine material into the larger pores in the soil surface.

As an example typical of many experimental determinations of the change of infiltration-capacity during rain, the values of f have been computed at different times, t, from the beginning of rain for a soil with initial infiltration-capacity $f_0 = 2.14$ in. per hour and which drops to a constant value $f_c = 0.26$ in. per hour in 2 hours. The quantity K_f determines the rate of change of infiltration-capacity during rain for a given rain intensity and in this case has the value 3.70.

t:	0.0	0.2	0.4	0.6	0.8	1.0	1.5	2.0
f, in. per hr.:	2.14	1.16	0.69	0.46	0.36	0.31	0.28	0.26

These values of f show that the infiltration-capacity drops off rapidly at first, then more slowly as it approaches f_c. Between rains drying out of the soil and restoration of crumb structure leads to restoration of f toward or to its initial value. The geomorphic significance of this decrease of infiltration-capacity during rain is illustrated by the fact that if infiltration-capacity remained constant at the high value it usually has at the beginning of rain, then there would be little surface runoff or soil erosion.

It is the minimum value f_c of infiltration-capacity which predominates during most of long or heavy rains which are chiefly effective in producing floods and sheet erosion. Infiltration-capacity is highest and least variable for pure coarse sands. For terrain with such soils, the infiltration-capacity may always exceed the rain intensity, so that even in winter, when the ground is frozen, no surface runoff occurs. The infiltration-capacity concept accounts for the absence of sheet erosion and the meager development of drainage in many sandy regions, even with abundant rainfall.

"Transmission-capacity" is the volume of flow per unit of time through a column of soil of unit cross section, with a hydraulic gradient unity or with a hydraulic head equal to the length of the soil column. Infiltration-capacity and transmission-capacity are related, but they are not identical; the infiltration-capacity is usually less than the transmission-capacity. Under conditions where transmission-capacity prevails, the soil column is fully saturated, and the entire cross section of the void space participates in hydraulic flow. When infiltration-capacity prevails, the soil column is not usually fully saturated, and air must escape upward through the soil as fast as water flows downward into the soil. This occupies a fraction, though usually only a small fraction, of the pore space. Also, because of the surface reduction of infiltration-capacity, already described, the soil surface cannot usually absorb water as fast as water can flow downward through the interior of the soil mass. As a result the soil is not saturated appreciably above its capillary-capacity, even during the heaviest rains. This fact can readily be verified during a rain by picking up a handful of garden soil at a depth of a few inches. As a rule, no water can be squeezed out of the moist soil although, if saturated, water could readily be squeezed out. A common fallacy is the statement that during a storm a severe flood was produced by the soil becoming saturated. This happens only in the case of heavy clay soils. Even if the water table rises to the soil surface, as it sometimes does in swampy areas, the soil can still absorb water, which appears later, somewhere, as ground-water or wet-weather seepage. As an example, in the central New York flood of July 1935, analyses showed that, in the Cayuga and Seneca drainage basins, the terrain maintained an infiltration-capacity of 0.2 to 0.3 inch per hour over the areas where the most intense rainfall and runoff occurred.

Infiltration-capacity of a given terrain can be determined in several ways, either with fully controlled conditions or for a drainage basin as a whole, under natural conditions. Thousands of determinations of infiltration-capacity have been made. These and other similar data, when more fully analyzed and classified, will, it is believed, form one of the most important tools for a quantitative study of drainage-basin morphology.

Rain falling at an intensity i which is less than f will be absorbed by the soil surface as fast as it falls and will produce no surface runoff. The rate of infiltration is then less than the infiltration-capacity and should not be designated "infiltration-capacity." If the rain intensity i is greater than the infiltration-capacity f, rain will be absorbed at the capacity rate f; the remaining rain is called "rainfall excess." This accumulates on the ground surface and for the most part produces runoff, and the difference between rain intensity and infiltration-capacity in such a case is denoted by σ and designated the supply rate ($\sigma = i - f$). For a constant rain intensity i, in inches per hour, the runoff intensity q_s, in inches per hour, approaches the supply rate σ asympototically as a maximum or limiting value as the rain duration increases (Horton, 1939; Beutner, Gaebe, and Horton, 1940). The total surface runoff is approximately equal to the total supply σt_e, where t_e is the duration of rainfall excess.

OVERLAND OR SHEET FLOW

In the minds of most persons the term sheet flow probably implies a greater depth of flow than usually occurs. Sheet erosion is used in contradistinction to channel erosion, and the use of sheet flow to describe overland flow not concentrated in channels larger than rills is appropriate, but it may not imply flow to depths measured in feet or even in inches but rather in fractions of an inch.

Since 1 inch per hour equals approximately 1 second-foot per acre or 640 c.s.m.[2], and an acre is 208 feet square, the surface-runoff intensity q_1 in cubic feet per second from a unit strip 1 foot wide and a slope length l_o will be:

$$q_1 = 0.000023 \, l_o q_s \tag{21}$$

where q_s is the runoff intensity in inches per hour. Discharge = Depth × Velocity, or if δ is the depth of sheet flow, in inches, and v the velocity in feet per second, $q_1 = \dfrac{v\delta}{12}$. It follows that the depth of sheet flow at any point on a slope where the slope length is l_o will be:

$$\delta = \frac{0.000277 l_o q_s}{v}. \tag{22}$$

On a gently sloping lawn, with a length of overland flow of 100 feet and a velocity of a quarter of a foot per second, a depth of surface detention of 0.11 inch will produce 1 inch runoff per hour. Walking over such a lawn while this runoff intensity is occurring, one may not notice that surface runoff is taking place; yet it is this same unobtrusive and almost imperceptible overland flow which, with greater depths and larger volumes and on longer slopes, is largely responsible for carving the landscape of drainage basins into observed forms.

LAW OF OVERLAND FLOW

The velocity of turbulent hydraulic flow is expressed in terms of the Manning formula:

$$v = \frac{1.486}{n} R^{2/3} \sqrt{S} \tag{23}$$

where v is the mean velocity in feet per second, n is the roughness factor, having the same general meaning for sheet flow as for channel flow, R is the hydraulic radius or

---

[2] c.s.m. = cubic feet per second per square mile.

ratio of area of cross section to wetted perimeter. For sheet or overland flow, R becomes identical with the depth δ. S is the slope.[3] Since discharge or volume of flow per time unit per unit width equals the product: Velocity × Depth × Width, the runoff intensity in inches per hour from a strip of unit width, for turbulent flow, can be expressed by:

$$q_s = K_s \delta^{5/3} \qquad (24)$$

where K_s is a constant for a given strip of unit width, having a given slope, roughness, and slope length.

A similar equation:

$$q_s = K_l \delta^2 S \qquad (25)$$

can be derived from Poiseuille's law for nonturbulent or laminar flow.

Overland flow may be either wholly turbulent, wholly laminar, or partly turbulent and partly laminar—patches of laminar flow being interspersed with turbulent flow or vice versa. Since the equations for turbulent and laminar flow are of the same form, it follows that for either laminar or turbulent flow, or for mixed flow, the relation between depth of surface detention and runoff intensity, in inches per hour, should be a simple power function of the depth of surface detention or:

$$q_s = K_s \delta^M \qquad (26)$$

where q_s is the runoff intensity in inches per hour, δ is the depth of surface detention at the lower end of the slope, in inches, K_s is a coefficient involving slope, length of overland flow, surface roughness, and character of flow, and the exponent M has a value of 5/3 for fully turbulent flow.

Except for very slight depths of surface detention, this simple law of surface runoff is remarkably well verified by plot experiments (Fig. 12). The circles (Fig. 12) indicate points derived directly from the hydrograph, and the solid lines the resulting relation curves plotted logarithmically. The points fall almost precisely on the relation lines, indicating an accurate functional relationship between δ_a and q_s.

Except on steep slopes there are always depressions, often small but numerous, on a natural soil surface. If the derived points were plotted for smaller depths than those shown on Figure 12, the corresponding relation lines would curve off to the left, indicating that the power-function relation of q_s to detention depth changes for very slight depths of surface detention. This represents the effect of depression storage. When runoff is taking place, flow through the depressions also occurs, although usually slowly, and hence the full cross sections of the depressions participate in determining the law of overland flow. The runoff becomes zero, however, when the depth of detention is reduced to the depth of depression storage V_d, although water still remains in the depressions. Consequently for slight depths the relation lines curve to the left.

---

[3] Expressed hydraulically as the ratio: fall/horizontal length. For steeper slopes the sine of the slope angle should be used in place of S.

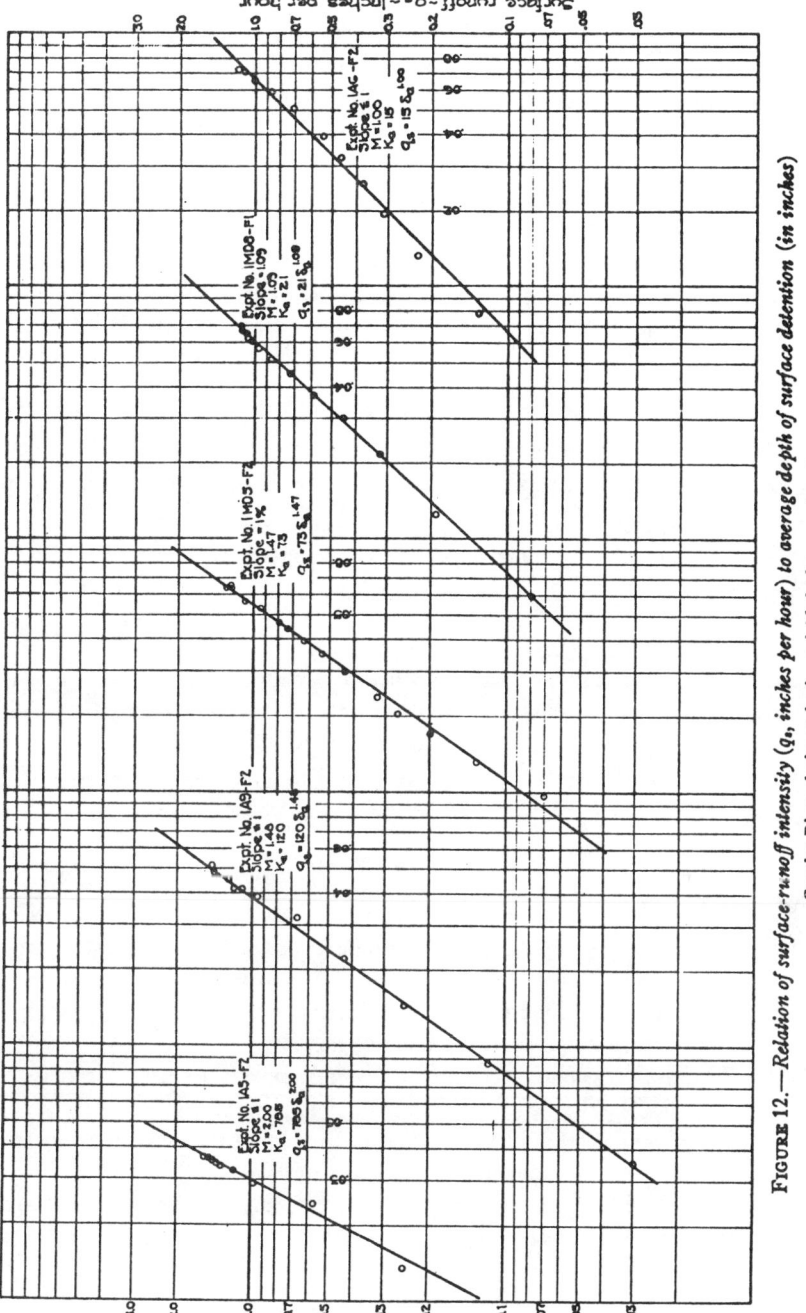

FIGURE 12.—Relation of surface-runoff intensity (q_s, inches per hour) to average depth of surface detention (in inches) Concho River drainage basin sprinkled plot experiments.

INDEX OF TURBULENCE

The exponent in the surface-runoff equation is not the same for all the lines (Fig. 12). This exponent, as indicated by the slopes of the lines, came out 2.0 for the line at the left, 1.48 and 1.47 for the next two lines, and 1.09 and 1.00 for the lines at the right. This suggests that in these cases the flow was not wholly ordinary turbulent flow.

For fully turbulent flow the exponent M should be 5/3, while for fully laminar flow the exponent M should be 3.0.

Where the exponent $M > 5/3$, the larger exponent can readily be explained if part of the overland flow is laminar flow, and this is quite certain to occur where the flow is alternately in thin films on ridges in the form of laminar flow, and through depressions wholly or partly as turbulent flow. An index of turbulence applicable to such a case is expressed by the equation:

$$I = \tfrac{3}{4}(3 - M). \tag{27}$$

If the flow is fully turbulent and $M = 5/3$, this gives $I = 1.0$. If the flow is fully laminar and $M = 3.0$, this gives $I = 0$.

By transposition, the runoff exponent M can be expressed in terms of the index of turbulence:

$$M = 3 - \tfrac{4}{3} I. \tag{28}$$

Turbulence in overland flow increases downslope from the watershed line. In laboratory and field-plot experiments with plot lengths l_o usually less than 25 feet, the flow over surfaces without vegetal cover is usually partially turbulent. On long natural slopes, with l_o much greater—frequently 1000 feet or more—the flow is fully turbulent except for extremely slight depths or close to the head of the slope.

TYPES OF OVERLAND FLOW

The study of overland flow in accordance with the infiltration theory has revealed various phenomena of microhydraulics not commonly present in ordinary channel flow.

Partially turbulent flow described may be considered "mixed flow." In general it consists of turbulent flow interspersed with laminar flow.

If the area on which flow occurs is covered with grass or other close-spaced vegetation, the flow may be "subdivided." Part of the energy available for overcoming resistance is expended on the grass blades and stems, reducing the amount of energy available for expenditure on the soil surface. For the limiting condition of complete subdivision of the flow, all the resistance to flow would be due to the vegetation, and the law of overland flow would be:

$$q_s = K_s \delta_s. \tag{29}$$

The velocity of overland flow would be sensibly constant regardless of the depth of surface detention. Some experiments show substantially this condition. Because of the increased resistance, the depth of surface detention required to carry a given rate of runoff is very greatly increased, and the velocity of overland flow is correspondingly decreased where there is dense cover of grass, grain, or similar vegetation. This is an important fact in relation both to surface runoff and soil erosion.

It is often found in runoff-plot experiments that the hydrograph does not have a smooth surface but is broken into irregular waves or surges. This type of flow may be designated "surge" flow and may be due to several causes:

(1) Under certain hydraulic conditions steady flow cannot occur even on a smooth, unchanging surface (Jeffreys, 1925).

(2) Plant debris, especially of the sand-burr type, may be loosened and carried along with the flow, forming debris dams, behind which the water piles up, and these hold back the water temporarily and then release it in relatively large volumes, producing irregular waves (Beutner, Gaebe, and Horton, 1940).

(3) Active surface erosion may produce a succession of irregular waves due either to mud or mud-and-debris dams of the type last described, to the breaking down of divides between natural depressions, or to the lateral incaving of the walls of gullies (Horton, 1939). Erosion may produce traveling mud dams or mud flows similar to those sometimes produced on a larger scale in mountain canyons by cloudburst storms. Wherever a mud or debris dam is formed, water accumulates behind it until presently the dam moves down the slope with the accumulated water behind it. In case of surge flow or traveling back-water due to debris dams, there is often no consistent relation between depth of overland flow and runoff intensity.

RAIN-WAVE TRAINS

On slopes which are not too flat, shallow flow in the form of a uniform sheet may be hydraulically impossible. The flow then takes the form of wave trains or series of uniformly spaced waves in which nearly all the runoff is concentrated. The author has twice observed such rain-wave trains in intense storms. They occur most commonly in rains of high intensity, particularly those of the cloudburst type, characterized by their ability to tear up sod on slopes and carry fences and other large debris into stream channels. Rain-wave trains occur only under suitable hydraulic conditions and have been described in another paper (Horton, 1939). Observation strongly indicates that rain-wave trains may be important in initiating erosion on sloping lands. If, for example, the waves are 6 feet apart, then each wave contains as much water as would be contained in a length of slope of 6 feet with uniform flow. The successive waves, with their concentration of runoff and energy, can strike sledge-hammer blows on obstructions. They may initiate erosion where it would never occur from the same runoff intensity with steady flow. The difference between the two cases is like that of breaking a rock with a few sledge-hammer blows, when a million taps with a pencil tip would expend the same amount of energy but produce no effect.

The following types of sheet or overland flow take place:

Type of flow	M	I
Pure laminar	3	0
Mixed laminar and turbulent	3 to 5/3	0 to 1
Turbulent	5/3	1.0
Subdivided or superturbulent	5/3 to 1	1 to 1½
Surge flow	Indefinite	—
Rain-wave trains	Indefinite	—

PROFILE OF OVERLAND FLOW

It can readily be shown from the infiltration theory that the profile of sheet or overland flow, or the relation of depth δ of surface detention to the distance x downslope from the watershed line, is expressed by a simple parabolic or power function

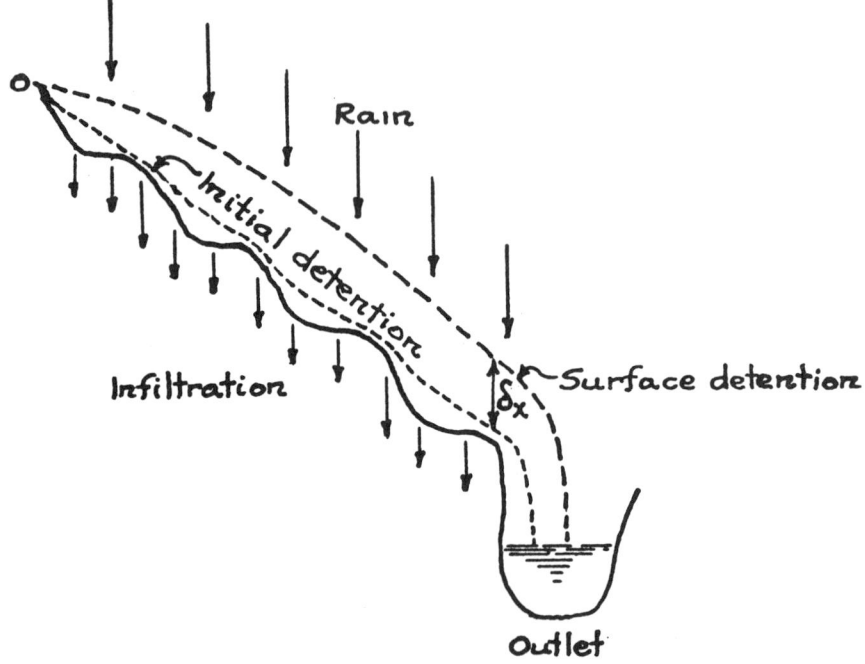

FIGURE 13.—*Half section of a small drainage basin*
Illustrating runoff phenomena. Vertical scale greatly exaggerated.

(Horton, 1938). This relation is illustrated on Figure 13. A similar power function expresses the relation of velocity of overland flow in terms of distance from the watershed line (Horton, 1937). For turbulent flow:

$$\delta_x = \left(\frac{\sigma}{K_s}\frac{x}{l_o}\right)^{3/5} \tag{30}$$

$$v_x = \frac{1.486}{n}\left(\frac{\delta_x}{12}\right)^{2/3}\sqrt{S} \tag{31}$$

and, in general, for δ_x in inches, for any type of flow:

$$\delta_x = \sqrt[M]{\frac{\sigma}{K_s}\frac{x}{l_o}} \tag{32}$$

and for velocity in feet per second:

$$v_x = 0.2836\frac{\sqrt{S}}{n}\left(\frac{\sigma}{K_s}\frac{x}{l_o}\right)^{\frac{2}{3M}} \tag{33}$$

where σ = supply rate, in inches per hour; l_o = total length of slope on which overland flow occurs, in feet; x = distance, in feet, downslope from the watershed line; K_s is a coefficient derived from the Manning formula for turbulent flow, and approximately applicable to other types of flow. Its value is:

$$K_s = \frac{1020\sqrt{S}}{In\ l_o} \tag{34}$$

and the exponent:

$$\frac{2}{3M} = \frac{2}{9.0 - 4I} \tag{35}$$

in which S is the slope, I the index of turbulence, l_o the length of overland flow, and n is a roughness factor, of the same type as the roughness factor in the Manning formula. The equations for turbulent flow are derived directly from the Manning formula and the law of continuity and are rational. The equations for other types of flow are closely approximate. The depth δ_x as given by these equations is the total depth of surface detention, including depression storage. The equations for δ_x fail at points close to the watershed line if, as is often the case, depression storage persists to the watershed line. In the equations both for turbulent and other types of flow, it is assumed that the velocity varies as $\sqrt{S}$, as for turbulent flow. This may not be entirely correct although numerous experiments indicate that, in mixed flow, most of the resistance is that due to turbulence. Theoretically, for laminar flow the velocity should vary directly as the slope, not as $\sqrt{S}$.

These equations apply primarily to steady flow. Experiments show that they are, however, closely approximate during the early stages of runoff, while surface detention is building up to its maximum value. While detailed discussion of the effect of the various factors on surface-runoff phenomena cannot be undertaken here, comparison of the equations shows that the velocity of overland flow increases, while the depth at a given point x decreases, as the slope increases. Increasing roughness of the surface decreases the velocity but increases the depth of surface detention.

The equations for depth and velocity profiles, in conjunction with that for K_s, are of fundamental importance in relation to erosional conditions, since they express the two factors, δ_x and v_x, which control the eroding and transporting power of sheet flow, in terms of the independent variables which govern surface-runoff phenomena. There are six variables: (1) rain intensity, i; (2) infiltration-capacity, f; (3) length of overland flow, l_o; (4) slope, S; (5) surface-roughness factor, n; (6) index of turbulence or type of overland flow, I. To apply these equations to erosion and gradational problems one must also have laws governing the relation of velocity and depth of overland flow to the eroding and transporting power of overland flow.

SURFACE EROSION BY OVERLAND FLOW

SOIL-EROSION PROCESSES

There are always two and sometimes three distinct but closely related processes involved in surface erosion of the soil: (1) tearing loose of soil material; (2) transport or removal of the eroded material by sheet flow; (3) deposition of the material in transport or sedimentation. If (3) does not occur, the eroded material will be carried into a stream.

Every farmer has noticed that the spots most vulnerable to erosion are the steeper portions of the hill or valley slopes, neither at the crest nor at the bottom of the hill

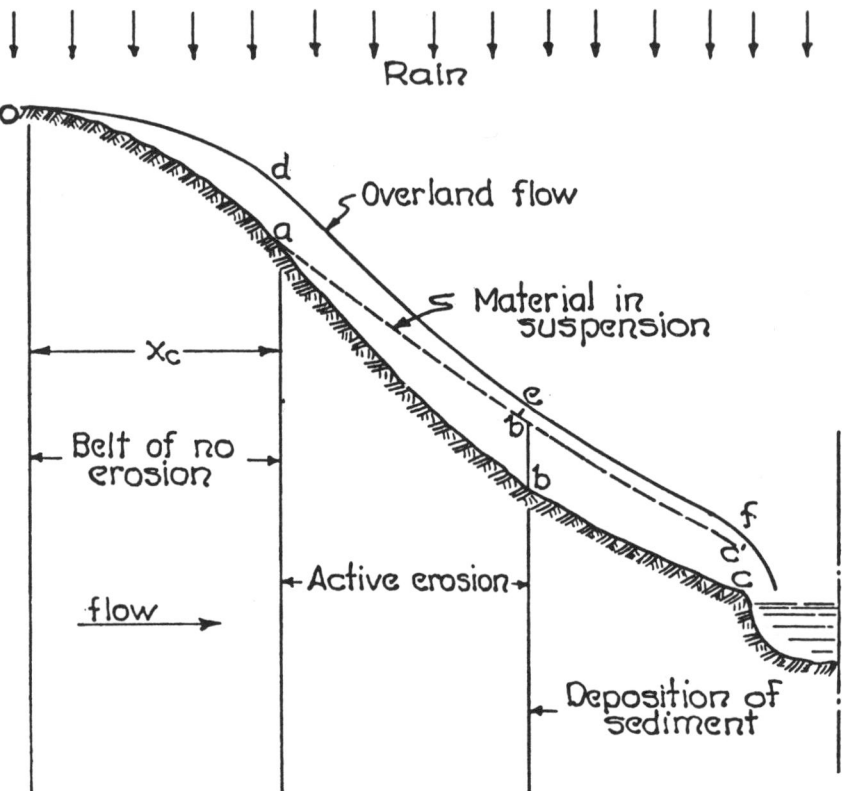

FIGURE 14.—*Half profile of a valley slope*
Illustrating soil-erosion processes.

but intermediate. All soils possess a certain resistivity to erosion, and this resistivity may be increased greatly by a vegetal cover, especially a good grass sod. The underlying soil may have a much smaller resistivity to erosion, and, if the surface conditions are changed by cultivation or otherwise so as to destroy the surface resistance, erosion will begin on land which has not hitherto been subject to erosion.

Figure 14 shows a half profile of a typical stream valley slope, with the vertical scale greatly exaggerated. The line *oabc* represents the soil-surface profile—flat in the region *o*, near the crest, steepest in the region *ab*, about mid-length of the slope, and relatively flat at the foot of the slope, in the region *bc*. The line *odef* represents the surface of sheet or overland flow in an intense rain, the depth of overland flow increasing downslope from *o* toward *f*. In the region *oa* no erosion occurs throughout a distance x_0 from the crest of the slope, and this is called the belt of no erosion. Here the energy of the sheet or overland flow is not sufficient to overcome the initial resistance of the soil surface to erosion, even in the most intense storm. In the belt ab, mid-length of the slope and where the slope is steepest, active erosion occurs.

Beginning at a the amount of material carried in suspension by the overland flow is proportional to the ordinate between the dotted line $ab'c'$ and abc. At a it is zero; at b it is represented by the vertical intercept bb'. Beginning at a a given volume of water, for example, the water flowing over 1 square foot of soil surface, picks up a certain amount of eroded soil matter and carries it in suspension. Passing over the next adjacent square foot of area the same water picks up another increment of soil matter and holds it in suspension, and so on, the amount of material in suspension increasing until at some point b the overland flow is fully charged with material in suspension and can carry no more. Between the point b and the stream channel, no material is carried away because any material picked up must be replaced by an equal quantity of material deposited from that already in suspension. If the slope decreases as shown on the diagram, then the ability of the overland flow to carry away material may decrease, in which case deposition of material or sedimentation on the surface will occur instead of erosion.

RESISTANCE TO EROSION

The physical factors governing soil erosion are: (1) initial resistivity, R_i; rain intensity, i; infiltration-capacity, f; velocity and energy of overland flow or eroding force, F. The breaking down of the soil structure, tearing the soil apart and lifting or rolling soil particles or aggregates, requires the expenditure of energy. Erosion can occur at a given location only where the amount of energy expended as frictional resistance on the soil surface exceeds the amount of energy required to overcome the initial resistance of the soil to erosion. An exception occurs in some cases where the soil is churned up into a semifluid mass by intense rain before surface runoff begins, producing high initial erosion rate. Sustained erosion can occur only where the condition above described is fulfilled.

The term "soil" as related to surface erosion includes not only the soil substance but also the vegetal cover and the structures—physical and biologic—in the surface layers of soil. Soils are of two general classes: (1) indigenous, or those formed by weathering of underlying parent rock, either igneous or sedimentary. Such soils generally prevail outside of glaciated and loess-covered regions. For some types of rock the formation of soil *in situ* is extremely slow. After a shallow surface layer of soil is formed, the formation of additional soil is restricted by the previously formed soil cover, but even in full exposure the rate of soil formation from many types of consolidated and igneous rocks is so slow that when the soil cover has been removed the land becomes worthless. (2) Preformed and transported soils. These consist of rock material comminuted by glacial or aeolian action and transported and deposited. Such soil is often a mixture of transported and indigenous soil material and includes sedimentary soils deposited on lake or ocean floors and afterward exposed. Transported soils, particularly those of glacial origin, are often highly fertile at the time they are laid down, as is evidenced by the growth of thrifty forest vegetation within a few years on soils recently exposed by glacial retreat. Erosion of transported soils, while a serious menace, is not in general so completely destructive as in case of indigenous soils unless the land is gullied and scarified to such an extent as to make cultivation impracticable. With equal runoff intensity the resistance of soil material to erosion generally increases with the fineness of the soil particles or soil texture, the resistance being small for fine uncemented sands but so high for cemented hardpan and tough clay that erosion rarely if ever occurs even on bare soil.

Resistance to erosion is, however, governed more largely by vegetal cover, biologic structures, and physical structure of the soil in the surface layers than by soil structure. A soil which forms a hard crust on drying may be highly resistant to erosion although the same soil when newly cultivated erodes easily. The coherence of soil particles and consequently the resistance to erosion is generally increased by the presence of colloidal matter, particularly that of vegetal origin. Vegetal cover is the most important factor in relation to initial resistance to soil erosion. Its effects on the resistivity of the soil to erosion are complex but include:

(1) Vegetal cover breaks the force of raindrops, thereby reducing the effect of the energy of falling rain in breaking down the crumb structure of the soil and packing the soil surface. For some soils with little coherence, breaking down of the crumb structure by rain impact reduces the soil to a fluid condition, readily susceptible to erosion, while for other soils packing of the soil surface tends to increase the resistance to erosion.

(2) A grass sod operates somewhat like a carpet covering the underlying soil and tends strongly to inhibit erosion.

(3) Fine soil particles adhere to root hairs and plant roots near the soil surface and act strongly as a soil binder. In a forest similar effects are produced largely by the grass cover but are accentuated by differences in soil structure as between natural or undisturbed and cultivated soils and by the presence of an undisturbed humus layer near the soil surface. In addition there is often a dense matting of roots of trees, herbaceous vegetation, and litter within a forest. Some of the runoff may be subsurface runoff and pass through this mat of litter and roots but at so greatly reduced velocity as to inhibit erosion. Factors have been devised which stress the resistivity of soil material to erosion in terms of the chemical and physical composition of the soil. Such factors are, however, inadequate to express the resistivity of a given terrain to erosion because of the predominant effect of vegetation and soil structure and condition, which are not reflected in indexes of the erodibility of the soil material itself.

The resistivity of a given terrain to surface erosion can be expressed quantitatively in terms of the force in pounds per square foot required to institute erosion. This quantity can readily be determined on a given soil surface from measurements of the distance from the watershed line downslope to a point where erosion begins.

Nearly all the factors which control resistance of a soil to erosion also control infiltration-capacity of the soil. At a given point on a given slope and with a given rain intensity the erosion rate is governed by various factors, one of the most important of which is the infiltration-capacity of the soil. In many instances factors which tend to promote a high resistance to erosion also tend to restrict or reduce the infiltration-capacity and vice versa. Consequently open-textured, coarse, sandy soils, such as soils of sand dunes, with little vegetal cover, may never be subject to erosion even in the most intense rains although the soil has little resistance to erosion, because the infiltration-capacity is so high that little or no surface runoff ever occurs.

ERODING FORCE

Erosion by aqueous agencies involves three processes: (1) dislodgment or tearing loose of soil material and setting it in motion. This is called "entrainment." (2) transport of material by fluid motion. (3) sedimentation or deposition of the transported material.

Let x = distance from divide or watershed line, measured on and along the slope (not horizontally); δ_x = depth of overland flow at x, in inches; w_1 = weight per cubic foot of water in runoff, including solids in suspension; α = slope angle; v = velocity of overland flow at x, in feet per second. The energy expended in frictional resistance per foot of slope length on a strip 1 foot wide running down the slope, per unit of time, for steady flow, will be equal to the energy of the volume of water passing over a unit of area per unit of time. This is the product of the weight, fall, and velocity, or:

$$e = w_1 \frac{\delta_x}{12} v \sin \alpha. \tag{36}$$

The energy equals the force times the distance moved. Hence the force exerted parallel with the soil surface per unit of slope length and width is:

$$F_1 = \frac{e}{v} = w_1 \frac{\delta_x}{12} \sin \alpha. \tag{37}$$

Equation (37) is known as the DuBoys formula and is a rational expression which may be used as a basis for determining the eroding force per square foot of soil surface. F_1 is the force available to dislodge or tear loose soil material. Sometimes not all this force is expended in tearing loose soil material; on a grass-covered surface a large portion of it may be expended in frictional resistance on grass blades and stems and have little effect in tearing loose soil material unless or until the grass is broken over by the impact of the overland flow. However, these factors are best included in the resistivity R_i of the soil to erosion.

As shown in connection with surface runoff, for turbulent flow the depth of overland flow at the distance x from the watershed line can be expressed in terms of slope, runoff intensity, and surface roughness:

$$\delta_x = \left(\frac{\sigma nx}{1020}\right)^{3/5} \cdot \frac{1}{S^{0.3}}. \tag{38}$$

The slope S is ordinarily expressed as the tangent of the slope angle α or as $\tan \alpha$. Also, for steady overland flow, $\sigma = q_s$, in inches per hour. Substituting these values in (38):

$$\delta_x = \left(\frac{q_s nx}{1020}\right)^{3/5} \cdot \frac{1}{\tan^{0.3} \alpha}.$$

Substituting this value of δ_x in (37) gives as the total eroding force at x:

$$F_1 = \frac{w_1}{12}\left(\frac{q_s n x}{1020}\right)^{3/5} \cdot \frac{\sin \alpha}{\tan^{0.3} \alpha}. \tag{39}$$

CRITICAL LENGTH x_c—BELT OF NO EROSION

As indicated, erosion will not occur on a slope unless the available eroding force exceeds the resistance R_i of the soil to erosion. The eroding force increases downslope from the watershed line (Equation 39). The distance from the watershed line to the point at which the eroding force becomes equal to the resistance R_i is called the "critical distance" and is designated x_c. Between this point and the watershed line no erosion occurs. This strip adjacent to the watershed line, and immune to erosion, is designated the "belt of no erosion." An expression for the width of the belt of no erosion can readily be obtained from equation (39) by substituting R_i for F_1, making $x = x_c$, and solving the equation for x_c. The runoff is free from sediment where erosion begins, and $w_1 = 62.4$ lbs. per cu. ft.

$$1020\left(\frac{12}{62.4}\right)^{5/3} = 65.0.$$

Substituting this constant in (39) gives:

$$x_c = \frac{65}{q_s n}\left(\frac{R_i}{f(S)}\right)^{5/3} \tag{40}$$

where $f(S)$ is a function expressing the effect of slope on the critical length x_c and is given by the equation:

$$f(S) = \frac{\sin \alpha}{\tan^{0.3} \alpha}. \tag{41}$$

Numerical values of the slope function $f(S)$ are given on Figure 15. For slopes less than 20°, $f(S)$ increases nearly in proportion to the slope. The critical length x_c varies inversely as the runoff intensity q_s in inches per hour, inversely as the roughness factor n, and directly as the 5/3 power of the resistance R_i (equation 40).

Table 6 gives numerical values of x_c for $R_i = 0.01, 0.05, 0.10, 0.20,$ and 0.50 lb. per square foot, for slope angles of 5°, 10°, and 20°, and for four different runoff intensities. These are computed for the roughness factor $n = 0.10$ but can easily be applied to other roughness factors, since the value of x_c is the same if the product $q_s n$ is the same.

Renner (1936) observed the percentages of areas having different slope angles which were subject to erosion in the Boise River drainage basin, Idaho; his results are shown on Figure 16. The extent to which erosion occurred on a given slope increased to a maximum on a 40-degree slope and thereafter decreased to zero approaching a 90-degree slope angle.

A comparison of the results of Renner's observations with the value of the slope function $f(S)$ is also given on Figure 15, and the two curves are in close agreement. Equation (41) for the slope function is rational in that it is based directly on fundamental physical laws and principles with respect to small and moderate slopes. For such slopes both Renner's observations and observations by Fletcher and Beutner (1941) show that for slopes of less than about 20 degrees the amount of erosion generally increases about in proportion to the slope. For steeper slopes the slope

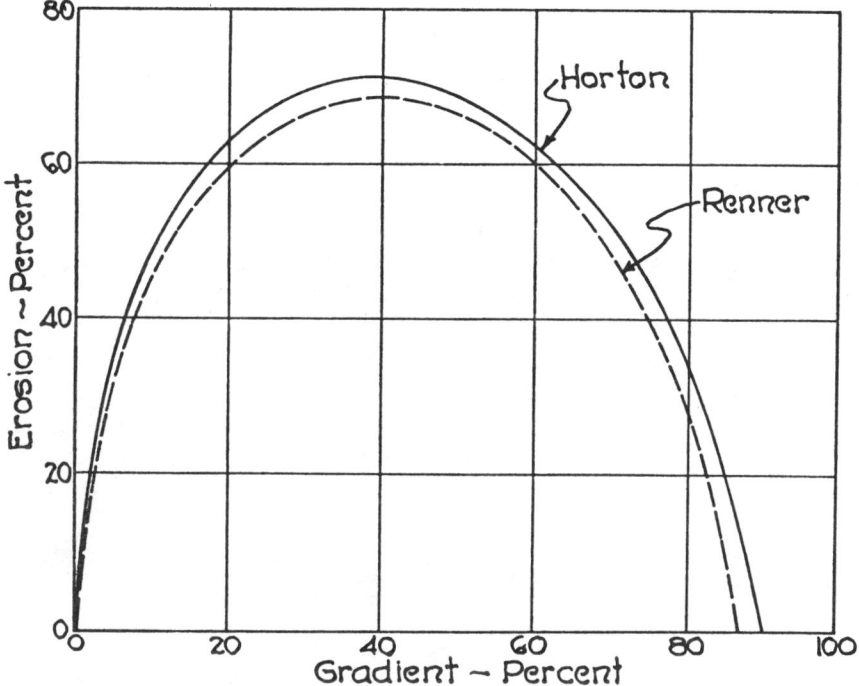

FIGURE 15.—*Horton slope function for surface erosion*

$$f(s) = \frac{\sin\alpha}{\tan^{.3}\alpha}, \text{ where } \alpha = \text{slope angle.}$$

function $f(S)$ must be considered as empirical, but its validity is confirmed by comparison with Renner's observations. This function is predicated on uniform turbulent flow. For very steep slopes such flow cannot occur.

The critical length x_c is the most important factor in relation to the physiographic development of drainage basins by erosion processes and also in relation to erosion control. The value of x_c (Table 6) is highly sensitive to changes in the variables by which it is controlled, in particular the resistance R_i and the runoff intensity q_s. With a newly cultivated bare soil, with R_i small, 0.05 lb. per square foot, for example, with a 10-degree slope and a runoff intensity of 1 in. per hour, the width of the belt of no erosion would be 35.1 feet, whereas on the same terrain, but with a good, well-developed grass sod to protect the soil, and R_i increased to 0.5 lb. per square foot, the belt of no erosion would be 1573 feet wide. The width of the belt of no erosion varies with the rain intensity, and, consequently, regions with frequent storms of high rain intensity are much more subject to erosion, other things equal, than regions with lower rain intensities and less frequent heavy storms. Furthermore, in a given

[*Editor's Note:* Concerning Figure 15, there is no agreement between Horton's slope function which is a maximum at 45 degrees and Renner's findings which show maximum erosion at slopes of 45 percent due primarily to land use.]

TABLE 6.—*Critical length x_c for various values of R_i*

Slope angle α, and runoff intensities q_o, with roughness factor $n = 0.10$: $x_c = \dfrac{65}{q_o n}\left(\dfrac{R_i}{fS}\right)^{5/3}$.

R_i	α (degrees)	q_o, inches per hour			
		0.5	1.0	1.5	2.0
		\multicolumn{4}{c}{$1/q_o n$}			
		20	10	6.67	5.00
.01	5	10.66	5.33	3.55	2.67
.01	10	4.80	2.40	1.60	1.20
.01	20	2.28	1.14	0.76	0.57
.05	5	153.4	76.7	51.16	38.3
.05	10	70.2	35.1	23.41	17.6
.05	20	32.6	16.3	10.87	8.2
.10	5	487.6	243.8	162.61	121.9
.10	10	218.4	109.2	72.84	54.6
.10	20	102.8	51.4	34.28	25.7
.20	5	1535.2	767.6	512.0	383.8
.20	10	689.0	344.5	229.8	172.2
.20	20	325.0	162.5	108.4	81.3
.50	5	7046.0	3523.0	2349.8	1762.0
.50	10	3146.0	1573.0	1049.2	787.0
.50	20	1478.0	739.0	4929.0	369.0

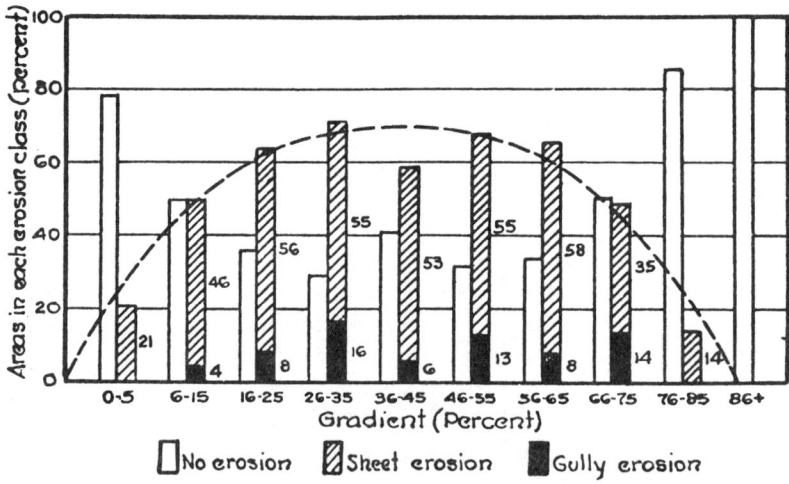

FIGURE 16.—*Gradient and degree of erosion*
(F. G. Renner)

locality, long intervals may elapse during which no erosion occurs on a given slope, than a rain of sufficiently high intensity may cause erosion on a part of the slope.

For example, with a moderately good grass-covered slope, with $R_i = 0.20$ and a slope angle of 5°, a slope 700 feet in length would not be subject to erosion with runoff intensities less than 1 in. per hour, whereas with 2 in. per hour runoff intensity nearly the entire lower half of the slope would be subject to erosion.

Most slopes do not have a uniform gradient from the watershed line to a stream but are flattest near the summit, steepest in the middle portion, and again flat adjoining the stream. For such a slope the belt of no erosion will usually comprise all the upper, flatter portion. If, for example, the slope length is 2000 feet, $q_s = 1.5$ in. per hour, and the mid-portion of the slope has a gradient of 10° and a resistivity of 0.5 lb. per square foot, erosion will begin 1049 feet from the watershed line. If the lower 250 feet of the slope is flatter (its gradient being 5°), then the length of overland flow required to produce erosion with this slope would be 2350 feet. Consequently, no erosion would occur on the lower or flatter portion of the slope. This example shows why erosion is generally confined to the steeper, middle portion of a given slope (Fig. 14).

As another example, on a slope which has been moderately well protected by grass cover, with $R_i = 0.20$, slope angle 5°, and with the limiting maximum value of runoff intensity for the given terrain 2 in. per hour, the width of the belt of no erosion in the maximum storm would be 384 feet. If the slope length was 400 feet, erosion would occur only at the foot of the slope and at rare intervals. If the resistivity of the soil was reduced, for example, by overgrazing, to 0.1 lb. per square foot or half its former value, then the width of the belt of no erosion in a maximum storm would be reduced to 122 feet, and some erosion would occur on the lower part of the slope, while for a runoff intensity of 1 in. per hour the width of the belt of no erosion would be 243 feet. Under the changed conditions some erosion would occur in all storms with runoff intensities exceeding 0.5 in. per hour. An area having a low resistance to erosion and on which erosion occurs over nearly the entire area in the more intense storms becomes partially immune to erosion in lighter storms. Between storms of maximum intensity, resistivity to erosion may be built up by the growth of grass or trees so that when the next succeeding maximum storm occurs the surface resistivity is increased, and the areal extent of erosion greatly diminished. In this manner Nature tends to correct the deleterious effects of surface erosion. Another result of importance is the fact that, on an area subject to erosion only in maximum storms, the total amount of erosion over a given time interval—a century, for example—may be relatively small, while, if the area is subject to erosion in storms with moderate as well as maximum runoff intensities, then because of the much greater frequency of storms of lower runoff intensities, the total erosion will be enormously increased. Not uncommonly the entire surface of the soil is removed in a century or less.

Another factor of importance in relation to erosion is that the soil surface, if protected by vegetation, has commonly a resistance to erosion many times greater than the underlying, unprotected soil. If the surface protection is removed and a maximum storm occurs, erosion will then take place at a rate governed by the lower resistance of the underlying soil. The soil once exposed to direct erosion may then be rapidly removed. Soil removed in the belt of erosion may either be carried away or deposited farther downslope. The manner in which these combined effects develop and control the forms of valley cross sections is considered later.

EROSION RATE

If a factor k_e be introduced in equation (37) to reduce the erosive force to terms of quantity of soil removed, as, for example, the depth in inches of soil material removed per hour, then the erosion rate at the point x would be, making $F_1 = e_r \cdot \frac{1}{k_e}$:

$$e_r = k_e w_1 \frac{\delta_x}{12} \sin \alpha. \tag{42}$$

This equation is rational in form and in fact if the rate at which soil material is torn loose is proportional to the force available from frictional resistance on the soil surface. It relates, however, only to the rate at which soil material can be torn loose and does not take into account the ability of overland flow to transport material in suspension. Equation (42) is limited in its applicability to cases where the erosion rate is less than the transporting power.

TOTAL EROSION AND EROSION DEPTH

Beyond the critical distance x_c and where the overland flow is not loaded to capacity with solid matter in transport, the erosion rate at a given point x may be assumed to be proportional to the net eroding force. Introducing a proportionality factor k_e in equation (39) to reduce erosion force to equivalent depth of solid soil material removed from the surface per unit of time, and making F_1 equal the erosion rate e_r and subtracting the value of F_1 at x_c gives:

$$e_r = \frac{k_e w_1}{12} \left(\frac{q_s n}{1020}\right)^{3/5} f(S) \, (x^{3/5} - x_c^{3/5}). \tag{43}$$

The total erosion between x_c and x is:

$$E_t = \int_{x_c}^{x} e_r \, dx$$

or, letting:

$$B = \frac{5}{8} \frac{k_e w_1}{12} \left(\frac{q_s n}{1020}\right)^{3/5} f(S) \tag{44}$$

and integrating equation (43):

$$E_t = B(x^{8/5} - x_c^{3/5} x). \tag{45}$$

Substituting the slope length l_o for x,

$$E_t = B(l_o^{8/5} - l_o x_c^{3/5}) \tag{46}$$

$$= B l_o (l_o^{3/5} - x_c^{3/5}). \tag{47}$$

The quantity ordinarily measured or measurable in the field is the total erosion per storm. Equation (47) can be used to work back from measured total erosion to the physical characteristics of the terrain which govern erosion rate.

The average depth of erosion between x_c and x is:

$$\frac{E_t}{(l_o - x_c)}.$$

The average depth of erosion is commonly expressed in terms of depth on the entire area, not merely the depth on the part of the area where erosion occurs. When so expressed the average depth of erosion is:

$$E_a = \frac{E_t}{x} = B(l_o^{3/5} - x_o^{3/5}). \tag{48}$$

The coefficient B in this equation is 5/8 of the coefficient of the term containing x in equation (43). Consequently, the average erosion depth over a given area is, for turbulent flow, 5/8 of the erosion depth for the same time interval at the point x.

If the value of x_c is determined from field observation, together with the average erosion depth, the slope length and runoff intensity being known, it becomes possible to determine for a given field or area the erosion force F_1 and the constant k_s. The latter is:

$$k_s = \frac{\text{Erosion depth}}{\text{Eroding force}}.$$

These equations form a practical working basis for determining the erosion constants R_i and k_s and for comparing the erosion conditions on a quantitative basis for different areas.

RELATION OF EROSION TO SLOPE LENGTH

The average depth of erosion on a given slope increases as the 3/5 power of the slope length minus the 3/5 power of the quantity x_c, which is constant for a given slope and storm (equation 48). For example, the relative erosion rates for different slope lengths, with $x_c = 100$ ft. and B taken at unity, are as follows:

l_o =	100	200	500	1000	2000	5000
$l_o^{3/5}$ =	15.85	24.00	40.3	63.10	95.0	166.2
E_a =	0.00	8.15	24.45	47.25	79.15	140.35

For a soil with R_i and x_c each zero, the relative erosion rates would be as shown in the second line of the table—i.e., proportional to $l_o^{3/5}$. The actual relation of erosion to slope length is, however, not quite so simple. In equation (48) and in computing the figures given above, surface-runoff intensity q_s has been assumed constant for all slope lengths. Other things equal, runoff intensity in a given storm decreases somewhat as the slope length l_o increases. Also, if erosion rate is determined as an average for a year or for several storms, the width of the belt of no erosion will vary in different storms. There will in general be more storms producing erosion on the middle and lower than on the upper portions of the slope.

Comparable determinations of soil loss by erosion over a period of 4 to 7 years have been made by the U. S. Soil Conservation Service at several stations, using slope lengths of 145.2, 72.6, and 36.3 feet. Some of the reported results are shown on Figure 17 (Bennett, 1939, p. 152) Differences of soil type, slope and resistivity, rainfall, and infiltration-capacity account for differences in soil loss for a given slope length at the different stations. Variations of these conditions in different portions of the same slope length account for small differences in the relation of slope length to erosion at a given station. In all these experiments the soils were under cultivation, producing corn or cotton, and the values of x_c were small, particularly in summer. The results of these experiments cannot be directly compared with equation (48) because x_c is unknown, and x_c was probably much greater in winter than in summer.

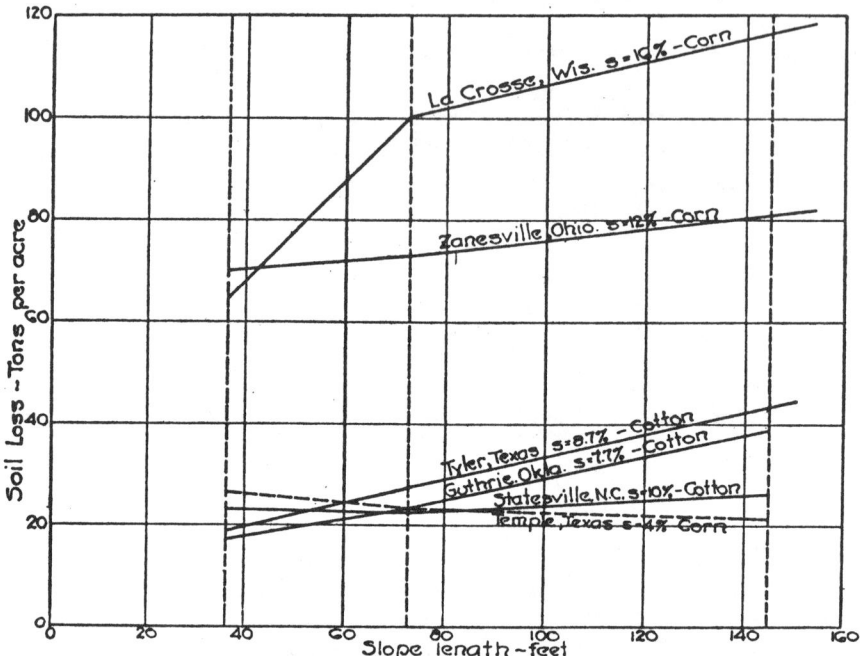

FIGURE 17.—*Relation of erosion to slope length*
From field experiments of U. S. Soil Conservation Service.

Allowing for differences, the results are entirely consistent with equation (48). As shown by equation (48) and also by these experiments, the greater the rain intensity and erosion, the greater, in general, will be the variation of erosion rate with slope length. Rains of low intensity may produce erosion on scattered patches of soil as a result of local variations of infiltration-capacity and resistivity and with little relation to slope length.

RAIN INTENSITY AND EROSION

Maximum rain intensities in a given locality generally occur in storms of the summer thunderstorm type. For such storms the highest rain intensity usually occurs before the middle of the storm and frequently within a few minutes after the beginning of rain. Some soils are easily pulverized when excessively dry but possess coherence through the operation of capillary force when partially dried after a gradual wetting. If an abrupt intense rain occurs on such a noncoherent soil the soil may be beaten into a pasty semifluid mass by rain of high intensity before runoff begins and before the soil surface becomes protected by surface detention. Such a semifluid mass of soil may accumulate on the surface until it becomes sufficiently fluid, or the runoff intensity becomes sufficiently great, to overcome its plastic resistance to flow. It is then carried into the stream by surface runoff *en masse*. The combination of sudden high rain intensity on previously dried soil of low coherence is not uncommon

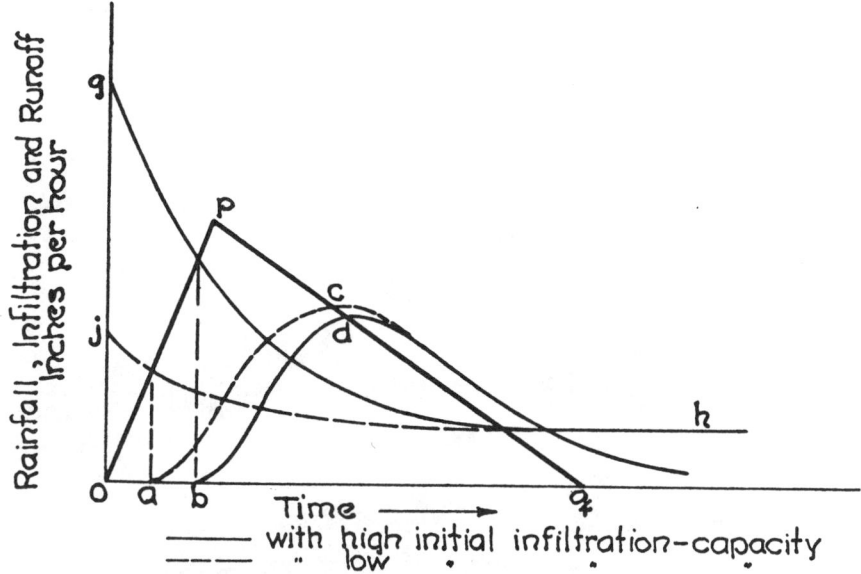

FIGURE 18.—*Relation of initial infiltration-capacity to erosion*

in summer storms in semiarid regions. In addition the initial infiltration-capacity of the dry soil is likely to be abnormally high, and this intensifies the effect of rain impact by increasing the time during which the soil is directly exposed without a protective cover of surface detention.

The combination of the conditions described frequently produces what is referred to as a "cloudburst flood." The term "cloudburst flood" is used because of the characteristics of the flood rather than those of the rain which produces it. Meteorological conditions are, however, involved. Measurements show that, while such floods may carry large volumes of solids, they often carry surprisingly little water as runoff.

In Figure 18 with a rain pattern *opq*, conditions for initially high and initially low infiltration-capacity are shown by solid and dotted lines, respectively. With a high initial infiltration-capacity *og* there is no surface detention or runoff during the interval *ob* during which rain intensity has risen nearly to the maximum. With a previously wet and packed soil and low initial infiltration-capacity *oj*, surface detention and runoff begin earlier at *a*, while the rain intensity is still low. Conditions such as those first described occur both on the upland and in stream channels previously dry. These conditions in both cases usually produce a cloudburst flood, characterized by a wall of turbid water or fluid mixed with debris traveling down the stream channel.

The solid material carried along in cloudburst floods is popularly believed to be derived from stream banks and channels. This may be true, but most of it may be derived from the upland, as is evidenced by the presence of fences, trees, logs, sheep, and other objects derived from the upland enmeshed in the semifluid mass. When

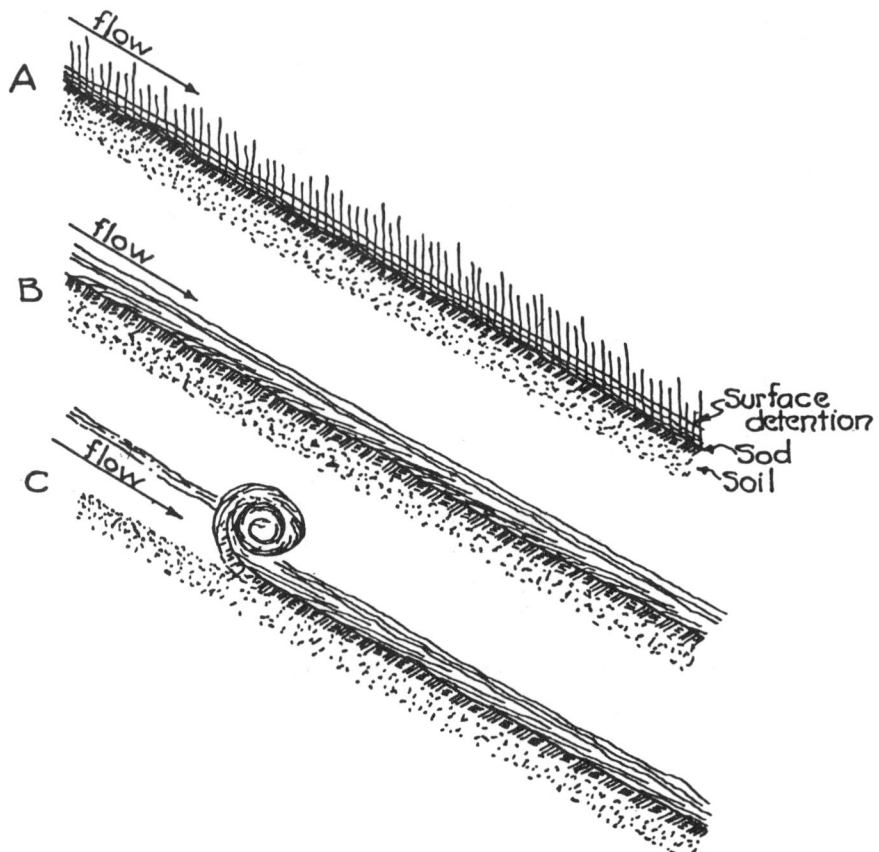

FIGURE 19.—*Erosion of sodded area initiated by the breaking down of grass cover in intense rains*

the flood wave debouches from a mountain canyon, the water in and behind it may escape laterally or by infiltration. The mud flow then slows down, sometimes traveling so slowly that a person could keep pace with it. Finally the mud stops flowing, dries out, and becomes a "fossil cloudburst." Many such "dry floods" may be observed below the mouths of certain canyons in the Wasatch Mountains in Utah. Sometimes one or more fossil floods are superposed pick-a-back fashion. An accumulation of such mud flows may form a debris cone (Horton, 1938; Bailey, 1935; Bailey et al., 1934).

Hydraulic conditions do not permit the occurrence of shallow steady flow on steep slopes (Horton, 1938; Jeffreys, 1925); the runoff water is concentrated in a succession or train of more or less uniformly spaced waves. These waves concentrate the impact of overland flow on irregularities or obstructions on the surface and greatly accentuate surface erosion. While grass or close-growing crops are in general an excellent preventive of erosion, they may contribute to active upland erosion in cloudburst floods. As long as the grass remains standing it decreases the velocity but increases the depth of surface detention, and the resistance to runoff is exerted at right angles to the grass stems and has little tendency to pull the sod loose (Fig. 19). When a certain depth and velocity of overland flow is attained the grass begins to flatten down, like the tipping down of a row of dominoes standing on end by pushing the first domino against its neighbor. Then (Fig. 19B) the pull of frictional resistance is exerted parallel with the surface. A bit of sod projecting more, or which is relatively less firmly held, than its neighbors is torn loose by the impact of the wave train, and erosion begins. A dense grass sod may be torn up and rolled down the slope like a snowball (Fig. 19C). Often several parallel strips of sod are torn off from the same slope, each strip a few feet in width. The less resistant underlying soil is thus exposed to erosion, and a cycle of erosion may begin on a slope which has been immune to erosion for centuries.

Thus the conditions which initiate soil erosion and govern its rate of occurrence are simply and readily expressible in quantitative form in terms of known independent variables under some conditions, while under other conditions the factors are so complicated that neither the necessary and requisite conditions to cause erosion nor the rate of its occurrence can be predicted or expressed accurately in quantitive terms.

TRANSPORTATION AND SEDIMENTATION

The transportation or carrying forward of eroded material by overland flow or in stream channels takes place in various ways: (1) as bed load or material moved chiefly by being rolled or pushed along the ground surface or along the bottom of the stream channel; (2) material called "suspended load," which is held more or less continuously in suspension by upward currents due to turbulence; (3) material held permanently in suspension by molecular agitation, called the "Brownian movement"; (4) solids in chemical solution in the water. The last mode of transport, while it is the most important process in connection with ground-water erosion, is relatively low in order of importance in connection with surface runoff.

Bed load comprises materials of sizes ranging from large boulders down through cobblestones, shingle, pebbles, gravel, and coarse sand. Mud flows are an important process in the transport of material in overland flow. This is, properly speaking, bed load. As the terms are ordinarily applied in connection with the dynamics of streams, there is no sharp line of demarcation between bed load and suspended load, the former term applying to material carried along, on, or near, the solid boundary. Flat fragments, such as shingle, are transported by sliding or as bed load. Round particles, such as fine gravel, may be transported by a combination of rolling, sliding, and jumping. This is also counted as bed load. The process of transport of particles by hopping from point to point in semielliptic arcs has been described (Gilbert, 1914) as "saltation."

In turbulent flow, eddies thrown off at the solid boundary surfaces have an upward component of velocity. At the same time there is a gradual settling of the water between eddies, creating a downward current equal in volume to that of the upward current in the eddies. The upward velocity and the magnitude and frequency of eddies increase with the velocity of flow and with the roughness of the boundary surface. The eddies are slowly dissipated by viscous resistance as they proceed upward. Both experiment and theory indicate that an eddy originates near the boundary of the fluid as a vortex ring system consisting of the vortex ring and its surrounding field. The fluid comprised in the ring retains its identity throughout the life of the ring, as in the case of the familiar smoke ring. The fluid comprising the field is continually changing, like the water in a wave.

Saltation in its simplest form involves the picking up of solid particles by ascending eddies. Those lifted and transported by the field are usually carried only a short distance; those entrapped near the center of the section of a vortex-ring may be carried much farther, until they are thrown out of the ring by centrifugal force. These two processes are more or less distinct although both depend on the laws of vortex motion. For this reason mathematical analyses of bed load and suspension transport without taking vortex motion into account are likely to prove inadequate and unsatisfactory. A given particle may be thrown out of one eddy at a certain level above the channel bottom, picked up by another, and carried forward, and so buffeted about, like a player in a football scrimmage. It may remain a long time in suspension, finally reaching the bottom, only to rest for a moment and then embark on another wild escapade.

Various attempts have been made to derive empirical expressions for rate of erosion from runoff-plot experiments. In general the conditions of the experiments have been such that it cannot be determined certainly whether the results represent the rate at which the given surface runoff could erode soil material or the ability of the sheet flow to transport such eroded material.

Most of the work done on sediment transport has been in connection with stream channels. Turbulent flow consists of laminar flow on which is superposed the effect of the transverse motion of eddies. If the flow is turbulent, then only a minute fraction of the energy consumed would be required to provide an equal mean velocity of laminar flow. The remaining energy becomes, in effect, latent at the boundary by conversion into rotational energy of vortex motion.

Two principal results follow: (1) The mean velocity is reduced from that for laminar to that for turbulent flow; (2) the velocity distribution is changed from that for laminar to that for turbulent flow.

For the usual slight depths of sheet flow the energy actually used in translational motion of the fluid is a much larger fraction of the total energy available than for types of flow commonly occurring in stream channels. The relative roughness is usually much greater for overland flow than for channel flow. Sand particles 0.001 foot in diameter with overland flow 0.01 foot in depth correspond in relative roughness to boulders 1 foot in diameter in a stream channel 10 feet in depth. Because of these and other differences the extent to which experiments and analyses for channels are applicable to sediment transport in sheet or overland flow is an open question.

Channel flow is ordinarily turbulent except in lakes, while overland flow, except on steeper slopes, usually comprises a mixture of areas on which the flow is turbulent and depressions through which the flow may be laminar.

Much more experimental and analytical work is needed on this problem. However, the following facts appear to be well established and suffice for present purposes:

(1) The transporting power of sheet flow increases with the amount of eddy energy due to surface resistance.

(2) Kinetic energy varies as the square of velocity, and transporting power of sheet flow must vary at least as the square—perhaps as some higher power—of the velocity.

(3) There is a maximum or limiting volume of eroded material which can be transported in suspension by a unit volume of overland flow at a given velocity.

ORIGIN AND DEVELOPMENT OF STREAM SYSTEMS AND THEIR VALLEYS BY AQUEOUS EROSION

RILL CHANNELS AND RILLED SURFACE

The first step toward the gradation of newly exposed sloping terrain is the development of shallow parallel gullies wherever the length of overland flow is greater than the limiting critical distance x_c. These are "rill channels," and a surface covered with such channels is a "rilled surface." Rill channels are usually relatively uniform, closely spaced, and nearly parallel channels of small dimensions which are initially developed by sheet erosion on a uniform, sloping, homogeneous surface. They are sometimes described as "shoestring gullies," but the term "gully" as ordinarily understood connotes larger and less regular channels developed by sheet erosion at a later stage. A rilled surface presents a striated appearance in plan and a finely serrated appearance in cross section.

Excellent examples of rilled surfaces may be found in newly made road cuts, on the slopes of highway and reservoir embankments, spoil banks, and mine dumps. In road cuts, water often drains onto the newly made slope from above the edge of the cut, in which case the rill channels extend the full length of the slope. If such drainage does not occur then the rill channels invariably begin at a little distance below the top edge of the cut. Actually the value of the critical length x_c may be very small, a few inches to a few feet, on newly exposed steep slopes, and the fact that the rills do not extend to the top of the slope would not ordinarily be noticed. Where a rilled surface develops on a newly exposed slope the usual result is the development of a deep central master rill or gully, with more or less parallel, shallower, shoestring rills, decreasing in depth and frequency, on both sides of the master gully. These shoestring rills do not generally survive. The deeper ones close to the master gully are absorbed by the master rill by bank caving or are destroyed by the breaking down of the narrow ridges between them. Those more remote are later obliterated when lateral slope has developed sufficiently to permit cross flow.

On some newly exposed lands, with high infiltration-capacity and high resistivity to erosion, the length of slope from the major divide to the downslope edge of the area may never exceed x_c. Under these conditions a rilled surface may not develop. This condition often occurs on sand-dune areas and in some glaciated areas with deep permeable soils, especially where grass or other vegetal cover develops soon after the disappearance of glacial ice. In the latter case x_c on the newly formed surface may exceed the values of l_o pertaining to the drainage of melt-water from the ice sheet, a rilled surface may not develop, and the topography will remain much the same as when the ice disappeared, except that gradation by solution may take place. In desert regions, with suitable relations between the rain intensity, infiltration-capacity, surface resistivity, and the slope, a rilled surface may develop with little or no cross-grading, so that surface gradation may never extend beyond the rill stage.

ORIGIN OF RILL CHANNELS

"Sheet erosion" implies the formation of either a rilled or gullied surface. From the discussion thus far it would appear that overland flow downstream from the critical point x_c on a smooth uniform surface should remove a uniform layer of soil instead of producing a rilled surface.

The question may fairly be asked: Why does a drainage basin contain a stream system? Surface runoff starts at the watershed line as true sheet flow, without channels. Even below the critical distance x_c it should apparently continue as such sheet flow combined with sheet erosion. Why, then, do rill channels develop? The answer is that channels start to develop where there is an accidental concentration of sheet flow. Accidental variations of configuration may provide the requisite initial conditions where a local area has a lateral slope joining a longitudinal slope or where two lateral slopes join and form a trough.

Most cases of active erosion observed at present represent conditions where there is or has been a protective vegetal cover and the initial resistivity of the soil surface is greater than that of the immediately underlying subsoil.

Consider a point upslope from x_c. If, as a result of change in cover conditions, either the resistance R_i or the infiltration-capacity is reduced, the point x_c may move upslope from the given point, which will then be susceptible to erosion. When the remaining protective cover is broken through at a given point, a channel or gully will form which will proceed rapidly upslope, chiefly by headward erosion, because of the lower resistivity of the underlying soil.

This, however, is not the mode of origin of rill channels, which, it must be presumed, often form on new terrain without vegetal cover and with a value of R_i sensibly the same at and to some depth below the soil surface. Slight accidental variations of topography may produce a sag in which the depth of sheet flow is a maximum at the point a (Fig. 20), the line bb' representing the water surface at maximum runoff intensity. As a result of the greater depth at a, erosion will be most rapid at that point, and increased channel capacity will be provided at a, and part of the water which originally flowed in shallower depths on the adjacent area will be diverted into this enlarged channel. This may accelerate the process until the entire flow is concentrated in the rill channel (Fig. 20). This does not involve headward erosion in the ordinary sense. However, when a rill channel has once formed, sheet flow coming down the slope upstream from the head of the rill will be deflected toward and diverted into the rill channel, thus providing a means of rapid headward extension of the rill.

This process of rill formation can often be observed on a cultivated slope during heavy runoff. The size and spacing of the rill channels vary with the slope, runoff intensity, and length of overland flow, ranging from a few inches apart on a cultivated slope to many feet or yards apart on long slopes with low runoff intensity and higher erosive resistance. In some areas in abandoned lake beds or exposed coastal

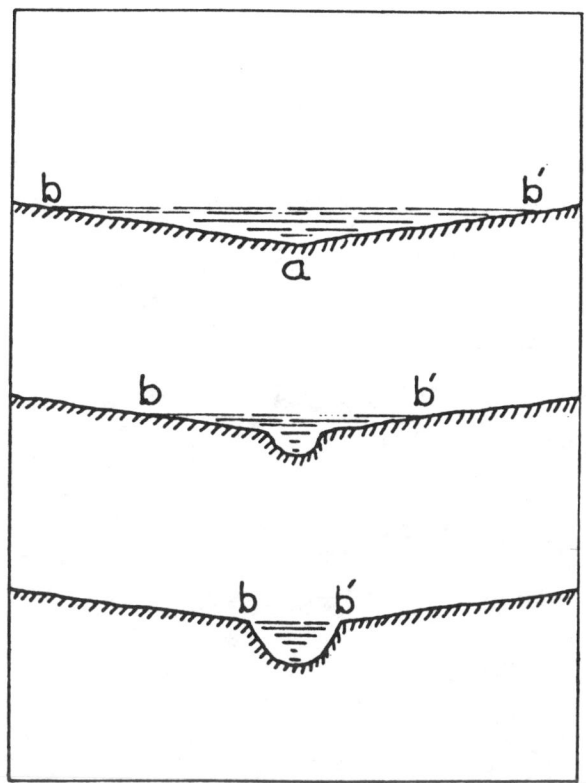

FIGURE 20.—*Successive stages of rill-channel development*

belts in arid regions stream development has never progressed beyond the rill stage. Such an example is given on the Moon Mountain, Arizona-California, quadrangle, U. S. G. S. topographic map (Fig. 21). Later stages of stream-channel development belong to the domain of channel dynamics and involve velocity distribution, silt equilibrium, and other factors which cannot be considered fully here.

The ultimate dimensions of a stream channel are, as indicated by Playfair's law, such that it is adapted to the area which it drains. Stream channels tend to acquire ultimate dimensions such as to carry all or most of the flood waters of the stream. This is largely because most surface erosion and channel erosion occur during floods.

CROSS-GRADING AND MICROPIRACY

A system of parallel shoestring gullies is transformed to a dendritic drainage net as the result of the tendency of the water to flow along the resultant slope lines and

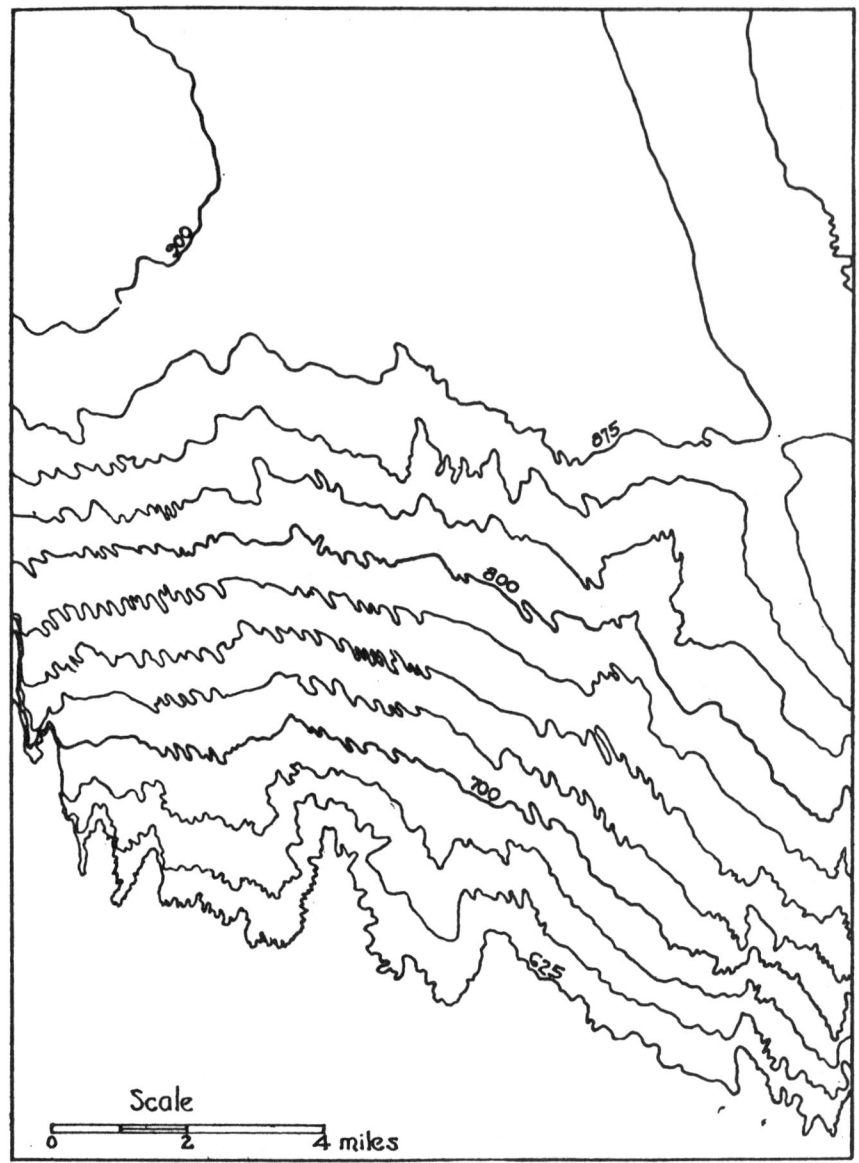

Figure 21.—*Portion of Moon Mountain, Ariz.-Calif., quadrangle, U. S. G. S.*

is a direct consequence of the overtopping and breaking down of intermediate ridges between gullies by overland flow during heavier storms.

The deepest and widest rill develops where the net length $l_o - x_c$ in which erosion can occur is greatest. If x_c varies, this may not occur where the total length l_o of overland flow is greatest. The longest, deepest, and strongest rill channel will be called the "master" rill. Owing to smaller values of $l_o - x_c$, proceeding away from the master rill on each side, the rills will be shallower, or, considering two adjacent rills, the bottom of the one farther from the master rill will be higher.

When a storm occurs exceeding in intensity preceding storms on the newly exposed areas, the divide between two rills may be broken down at its weakest point by (1) caving in of the divide between two rills, diverting the higher into the lower rill; (2) erosion of the divide by the deeper or lower rill, thus diverting the higher rill; (3) overtopping of the divide at the low point by the higher rill, again diverting it into the lower rill. This breaking down of divides between adjacent rill channels and diverting the higher into the lower rills is described as micropiracy. Micropiracy much resembles stream capture by lateral corrasion, but micropiracy results chiefly from water overtopping a low spot in the narrow ridge between two rills. Micropiracy obliterates the original system of rills and their intermediate ridges on a uniform newly exposed surface. The process of erosion, in the course of development of a stream system and its accompanying valleys destroys most of the record of their origin. Ultimately the original slope parallel with the stream is replaced on each side of the stream by a new slope deflected toward the stream. This process is described as "cross-grading."

The initiation of cross-grading is illustrated on Figure 22, which shows a plan of a small area of newly exposed land, $aa'bb'$. The line cc' marks the downslope limit of the belt of no erosion, $aa'cc'$. The critical distance x_c is assumed to vary with slope, infiltration-capacity, and initial resistance to erosion. Rills develop downslope from cc', and their development is followed by cross-grading, as shown on the cross sections taken on the line dd' and numbered 1 to 4, inclusive. The line of resultant slope in each case is in the direction shown by the arrows, and the rills increase in depth and degree of gradation at a given time proceeding away from the lateral boundaries toward the initial or master rill. In section 2 (Fig. 22) the rilled surface has developed, but flat "lands" still persist between rills, and the resultant slope is still parallel with the original slope. In section 3 some rills have combined by cross-grading, creating slight cross slopes, but the overland flow is still carried chiefly by rills parallel with the original slope. In section 4, with increased cross slopes and perhaps a heavier storm, active cross-grading has taken place, especially mid-length of and in the lower portion of the original slope; the direction of overland flow is no longer parallel with the original slope, but overland flow takes place partly in the rills and partly across the intervening ridges, somewhat as shown on a larger scale on Figure 23, and in detail for a single pair of rills on Figure 24. This process can sometimes be observed in heavy storms on lands cultivated nearly but not quite parallel with the contours, the tillage marks corresponding to the original rills above described.

In Figure 22, section 5, the original serrated rilled surface has been obliterated by cross-grading and is replaced by an irregularly roughened surface on which a new rilled surface tends to develop, with flow lines parallel with the resultant slope. This represents the end point of the first stage of valley gradation and stream development. At this stage there exists only one stream in the area, and this follows the course of the initial or master rill. This idealized picture of the cross-grading of a rilled surface is based in part on field observations of eroding slopes and in part on the observed manner of erosion of experimental plots, using artificial rainfall. Under natural conditions the results are rarely so uniform as those shown on Figure 22.

A break across a rill divide may result from numerous causes, such as a rock or

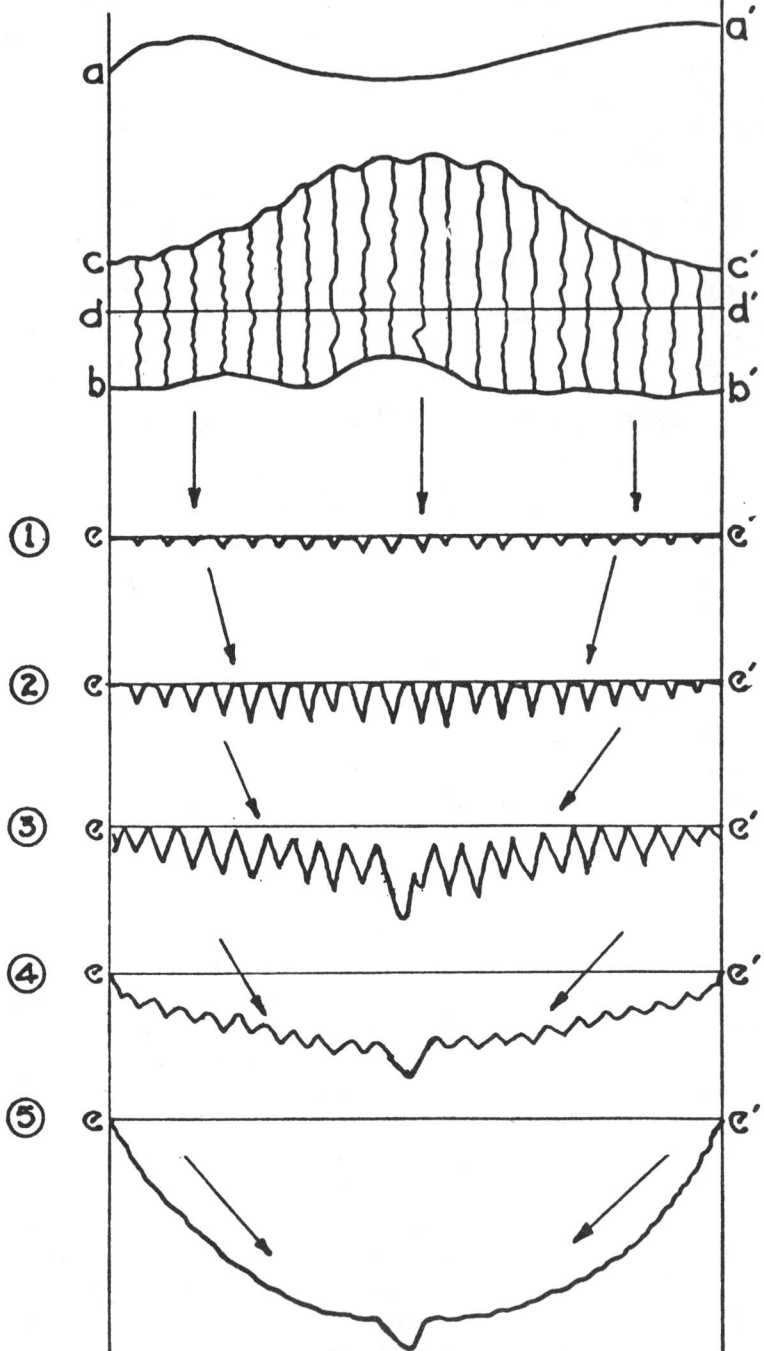

FIGURE 22.—*Development of a valley by cross-grading*

ORIGIN AND DEVELOPMENT OF STREAM SYSTEMS

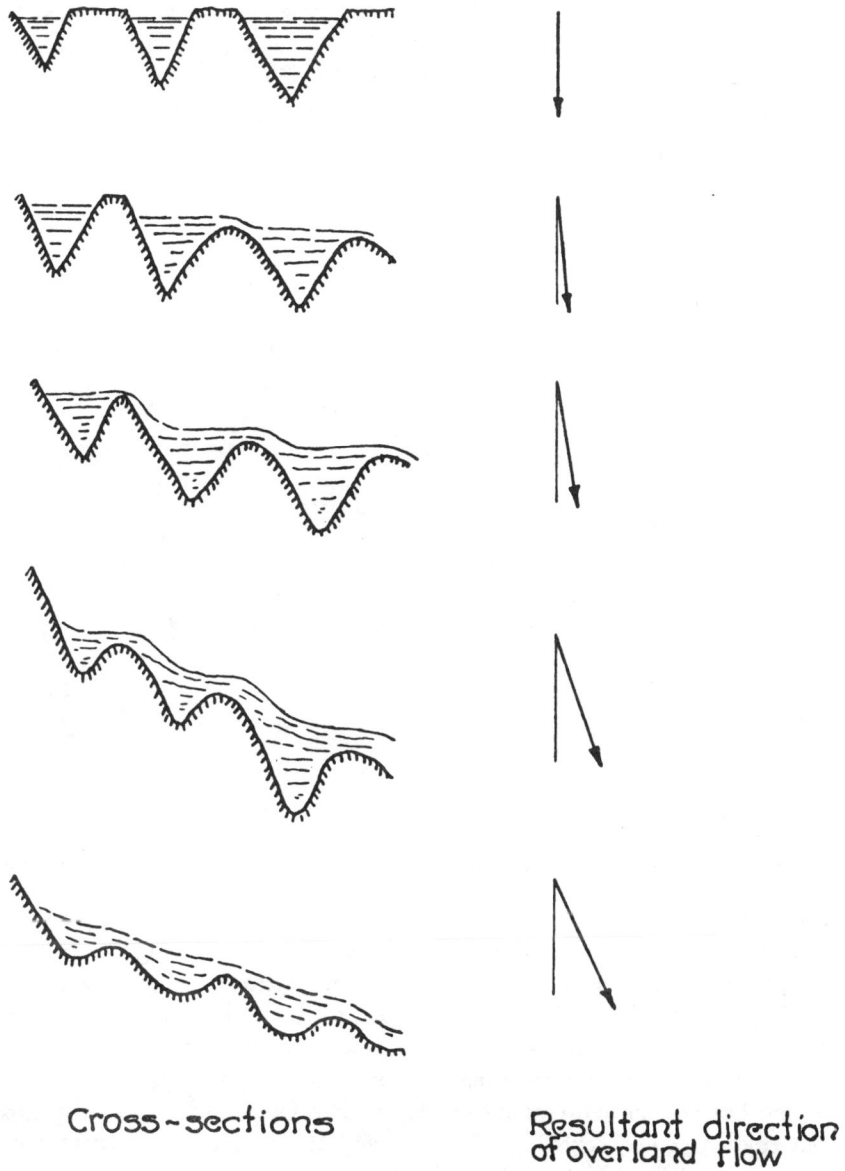

FIGURE 23.—*Successive stages of rill obliteration*

obstruction in the path of the rill, causing back-water upstream therefrom, or the caving in of the bank of the rill, thus obstructing its flow and producing a side outlet. A very common cause is an accidentally low divide between two adjacent rills. In most cases the channel of the diverted rill will be obstructed just downstream from the point of diversion. If the break supplies a free outlet from the diverted rill, then immediately downstream from the break, as at x (Fig. 24, B), there will be little or no flow, and erosion will cease, while erosion just above the break will continue. The channel above x will quickly become deeper than the abandoned rill channel below x. There will be thrust, as indicated by the downslope arrows on Figure 24, at the point of diversion, resulting in a tendency toward the formation of a rounded

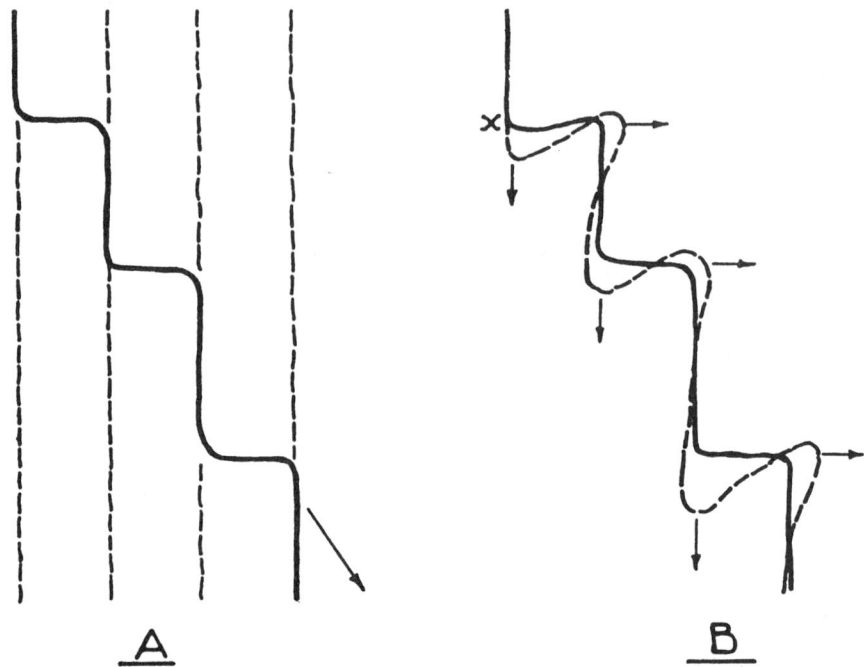

FIGURE 24.—*Development of a stream from a rill system by cross-grading*
(A) Showing angular plan of initial stream; (B) Development of bends by thrust at angles of an initial stream.

bend by impact erosion. Similar thrust across the slope will occur at the second angle of each cross rill diversion, with similar results.

Micropiracy tends to give the resulting stream at first a more or less angular course (Fig. 24) which, as the angular bends are rounded by erosion, will finally develop a more or less tortuous stream course following generally the line of resultant slope. On steeper slopes the resulting stream tends strongly to maintain a straight course and has eroding power sufficient to do so. On flatter slopes, with greatly reduced eroding power, centrifugal force around the initial angular stream bends tends to enlarge their radii. Weak bends merge with stronger ones until ultimately a system of stream meanders is developed. Since the master rill either initially or by cross-grading, on a given slope, ultimately becomes a permanent stream, it appears that conditions favoring the formation of stream bends on flatter slopes are inherent with the origin of the streams.

The elimination of rill systems on gentle slopes is very different from the development of mature meander belts such as those observed in natural streams. Even where stream bends originate in this way the evidence of the intermediate steps is eliminated in the course of the process.

This is not the sole explanation of the origin of stream bends, for individual bends arise from other and accidental causes. The fact that bends occur generally in rather definite systems on flatter stream slopes and in material that is almost perfectly homogeneous seems to require something more than purely fortuitous causes to explain their origin. It is at least significant to find that a stream in its initial stage, as it merges from a rill system, provides the necessary conditions for the development of stream bends on flat slopes.

HYDROPHYSICAL BASIS OF GEOMETRIC-SERIES LAWS OF STREAM NUMBERS AND STREAM LENGTHS

General statement.—A conventionalized illustration is given of the main steps involved in the development of a drainage net, showing the hydrophysical basis of the geometric-series laws of stream numbers and stream lengths. A square area with its diagonal parallel with the direction of slope will be assumed. This roughly approximates in form a typical ovoid drainage basin.

First stage.—On Figure 25, $oabc$ is a uniform surface sloping toward o. The maximum length of overland flow is the diagonal length $l_g = ob$. A storm produces surface runoff of intensity sufficient to reduce the critical distance x_c to some value $bg < l_g$. Sheet erosion can then occur over the area $ofgh$. The runoff intensity q_1, in cubic feet per second per unit width, will increase proceeding from a and c toward the center line nearly in proportion to the length of overland flow. There will be no erosion at f or h. The erosion intensity will increase from these points along the line fh toward the center line. The maximum intensity of surface runoff cannot occur until surface detention is built up to a point where the inflow to and outflow from surface storage are equal. Since the unit runoff intensity q_1, other things equal, increases nearly as the length of overland flow, the critical intensity necessary to induce erosion will be exceeded first along the center line of the area and then progressively later proceeding toward f and h (Fig. 25, a). Sheet erosion will begin first along the diagonal line og, spreading more or less rapidly toward the points f and h. In a typical storm the duration of surface-runoff intensity adequate to produce erosion will also be greatest near the center line of the area, decreasing toward the sides. Consequently, erosion will begin first, be most intense, and last longer near the center line of the area. A series of more or less parallel gullies will develop (Fig. 25, a), decreasing in depth and frequency, proceeding away from og, forming a rilled surface.

In Figure 25a the lines of overland flow are parallel with the central or master rill. Actually as the valley develops, the lines of overland flow will follow the resultant slope lines converging toward the central rill first at acute angles and, as the slope increases, at increasingly larger angles, approaching right angles as a limiting case.

This is the first stage of stream and valley development. It may later produce further upslope erosion in more intense storms. At the end of the first stage (Fig. 25a), there is a single central stream, with a V-shaped valley.

Second stage.—The development of the first stream (Fig. 25a) has divided the area and reduced the maximum length of overland flow on the remaining areas on each side to $\frac{1}{2} l_g$ or to half the initial maximum length of overland flow. Lateral tributaries have not yet developed because (1) Lateral flow cannot occur unless there is a lateral component of slope, and (2) until the lateral slope extends far enough from the main stream so that in a given storm the value of the critical length x_c is reduced

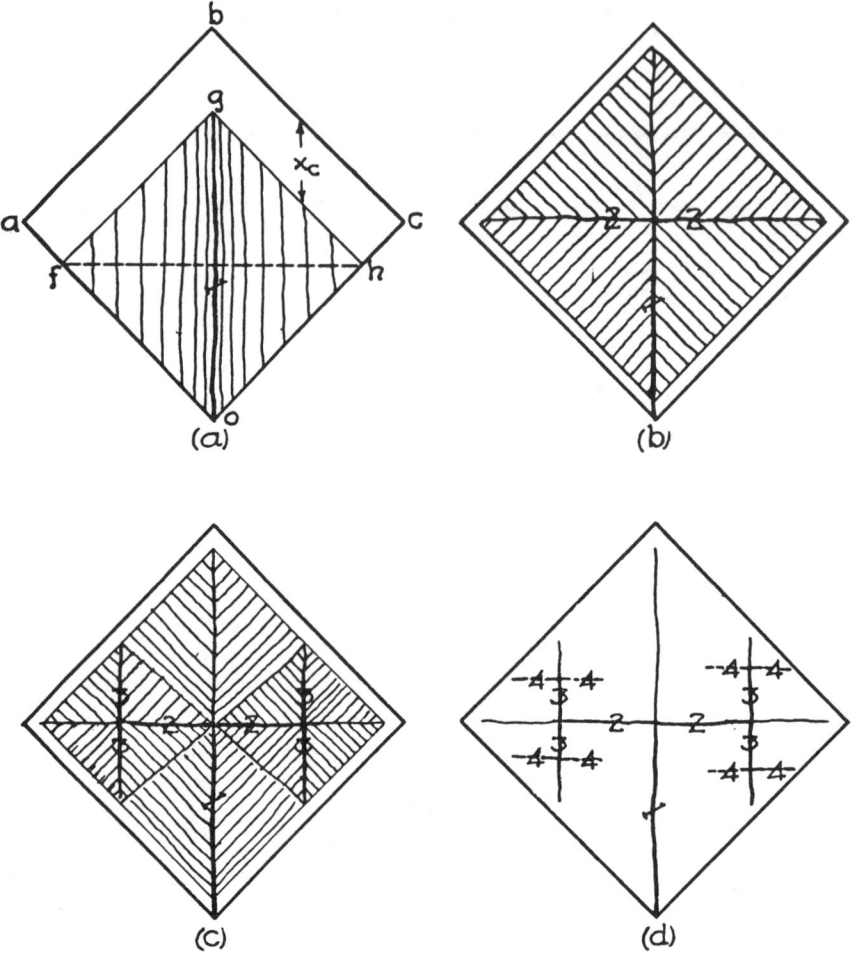

FIGURE 25.—*Development of a drainage net in a stream basin*
(Schematic).

to less than the width of the lateral slope, or in case of complete development of the initial valley, the value x_c must be reduced to less than $\frac{1}{2} l_g$.

The second stage involves the development of lateral tributaries and their valleys. During a storm of sufficient intensity to meet the prerequisites described, a pair of master rills or lateral tributaries with accompanying shoestring gullies will develop, one from each of the equal areas on opposite sides of stream 1 (Fig. 25b). These lateral streams are designated "2," and their development and that of their lateral slopes will follow the same course as in the case of stream 1. At the end of the second stage the area will appear as shown by Figure 25b. In the meantime, further headward development or downward gradation, or both, of the valley of stream 1 may have occurred.

Subsequent stages.—At the end of the second stage the maximum remaining length of overland flow in the area has been again halved or reduced to $\frac{1}{4} l_g$, and a still more intense flood is required to reduce the critical length x_c to a value less than this and permit the development of a third group of lateral tributaries. When such a flood occurs, each of the No. 2 tributaries will develop a pair of lateral tributaries, designated "3" on Figure 25c.

The development of lateral slopes adjacent to a given stream brings in overland flow from additional areas, increasing downstream, and accelerates grading of the main stream and its immediate valley. The stream system at the end of the fourth stage is shown by Figure 25d. The number of the stage of stream development corresponds to the order of the main stream. The main stream is of the 1st order at the end of the 1st stage, and of the 4th order at the end of the 4th stage.

Development of lateral tributaries and the manner in which they develop is the direct consequence of (1) the existence of a critical length x_c of overland flow required to institute erosion; (2) the operation of cross-grading. Drainage patterns, while invariably following the two fundamental geometric-series laws as to stream length and stream numbers, can still develop in an infinite variety of ways.

Two questions naturally arise: (1) What would happen if the newly exposed area was a continuous belt along the coast line, with no lateral boundaries? (2) Why and how are the boundaries of drainage basins developed? This case will be considered later.

ADVENTITIOUS STREAMS

Differences occur between the development of streams under natural conditions and those assumed in the example because:

(1) The drainage area is usually not rectangular but ovoid.

(2) Newly formed tributaries follow in general the resultant slope of the cross-graded areas on which they develop and hence enter the parent stream at more or less acute angles; the steeper the slope, the more acute is the angle of stream entrance. Tributaries thus tend to be longer than if they entered at right angles, and the stream-length ratio is consequently usually greater than 2.0.

(3) There are nearly always variations—sometimes large variations—of infiltration-capacity in different parts of the area.

(4) There are also variations—sometimes extreme—in the initial resistance of the terrain to erosion, as, for example, where part of the area is in consolidated and part in unconsolidated material, or part covered with vegetation and part bare. As a result of these departures from hypothetical conditions, the following results often occur:

In certain parts of the area the length of overland flow to the parent stream or its larger tributaries may be less relative to x_c than on the remaining areas tributary to the last group of streams developed. Then some 1st, and perhaps 2nd, order tributaries develop, entering the main stream or larger streams directly and not through higher-order tributaries. These streams, which result from accidental variations of conditions within the area, may appropriately be designated "adventitious" streams. The development of adventitious streams increases the number of streams of lower orders and tends to make the bifurcation ratio greater than 2.0, as it usually is for natural streams. Lateral slope also increases the length of tributaries and makes the stream-length ratio also greater than 2.0. An increase of r_l is also

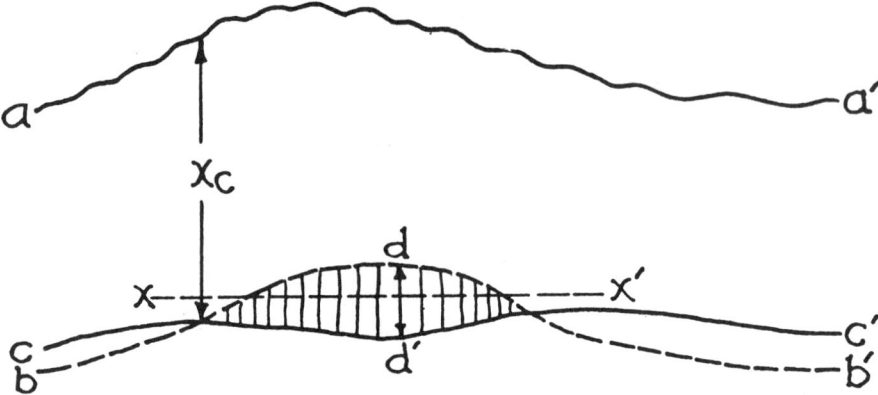

FIGURE 26.—*Beginning of erosion on newly exposed land*

produced because streams do not extend to the watershed line although they may extend by headward erosion to a distance less than the critical minimum value of x_c from the watershed line. Adventitious streams usually increase the bifurcation ratio more than the stream-length ratio is increased by the conditions described, with the result that the ratio $\frac{r_l}{r_b}$ is in general a fraction, and the total length of streams of a given order is not constant but decreases proceeding from the lowest to the highest stream orders. Adventitious streams do not in general develop simultaneously with larger streams in the basin but are developed later as the development of the stream system approaches maturity.

STREAM DEVELOPMENT WITH PROGRESSIVELY INCREASING LAND-EXPOSURE COMPETITION

For illustration the exposure of coast marginal lands will be assumed to be nearly a uniform homogeneous sloping plane, extending from a divide line aa' (Fig. 26) to the new coast line cc'. It is assumed that the soil surface in the newly exposed belt $aa'cc'$ is initially bare and has a certain infiltration-capacity f and a surface resistance to erosion R, such that the critical length x_c required to permit surface erosion to occur in the most intense rain is as shown on the diagram. The dashed line bb' is at a distance x_c from the watershed line aa'. As long as the coast line is within the belt $aa'bb'$, no streams will develop, runoff will be in the form of direct sheet flow to the new coast line, and no erosion takes place. There will be irregularities in the watershed line aa' and in the coast line cc', and, when the length of overland flow exceeds x_c at some point d, erosion will begin at that point. When the coast line has reached the position cc' there will be a small area, as outlined by a dashed line, within which $l_o > x_c$, and within this area sheet flow will produce erosion and a series of rill channels parallel with the direction of the initial slope surface. The first rill channel will be at dd', where the length of overland flow first exceeds the critical length x_c. As the coast line recedes the belt in which erosion can occur will increase laterally and longitudinally, and the system of rill channels will be extended correspondingly

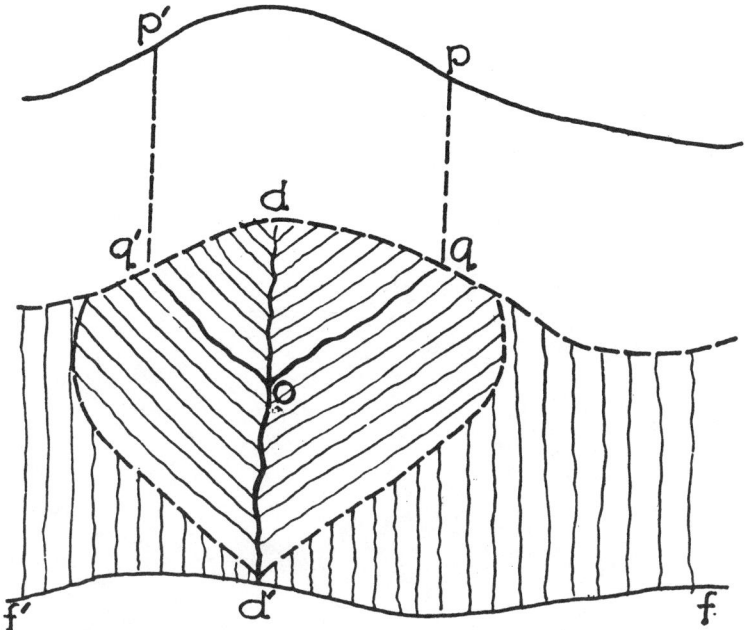

FIGURE 27.—*Development of first pair of tributaries on new stream system*

in both directions. The rill at dd' was first formed and has been longest subject to erosion and will become the master rill. Cross-grading will take place, producing new components of slope toward the rill dd' and obliterating the original rill channels. New rill channels will develop following the new direction of slope, on each side of the original stream dd' (Fig. 27). In general the lengths of these new rill channels will increase proceeding down the slope from the line cc'. At some point o (Fig. 27) a new rill channel will have a greater length oq and greater runoff than rills between o and d. It will have developed earlier than rill channels entering the parent stream dd' between o and d'. It therefore has greater runoff and a longer duration of runoff in which to cut its channel than rills formed farther down the slope. Such rill channels will survive as a tributary stream. Such a rill channel occurs on each side of the parent stream in the vicinity of o, and cross-grading toward these tributary streams will also occur. Cross-grading of the areas adjacent to these two tributaries will produce cross-graded slopes on either side of each tributary (Fig. 28), until there is again a location on each of these areas favorable for the development of tributaries, and new tributaries will develop, usually one on each of the two preceding tributaries, as at m, n, and p (Fig. 28). This process will continue until finally there is no land surface above the mouths of the original tributaries where the length of overland flow exceeds the critical length x_c.

If the coast line recedes farther (Fig. 29), the area upslope from oo' on the right hand side of the stream dd' is tributary to the stream oq. The original rill channels parallel with dd' upslope from oo' have been obliterated, and the runoff from the area

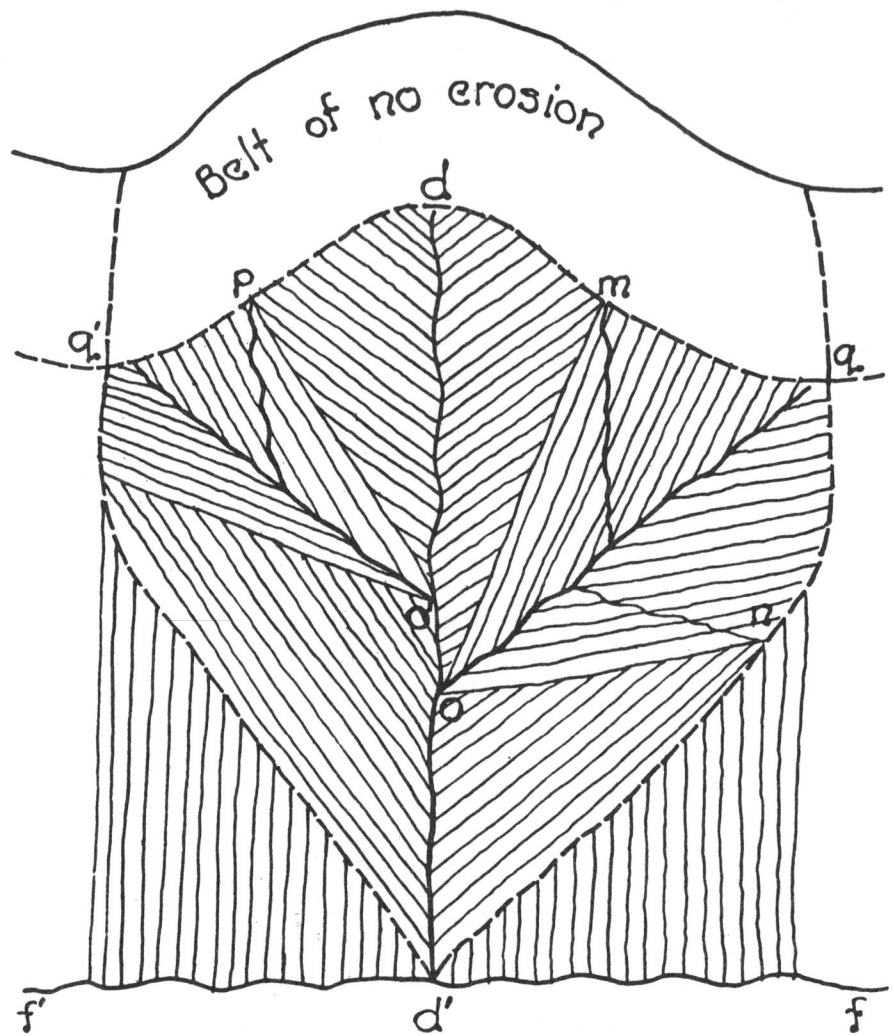

FIGURE 28.—*Lines of flow after cross-grading of first pair of tributary areas*

$oo'o''$ now enters stream oq. As the coast line recedes a new system of rill channels parallel with dd' develops downslope from oo', and the area $oo'd'd''$ will become cross-graded toward dd'. When the length of overland flow within the area $oo'd'd''$ becomes sufficiently great at some point q', a new tributary $q'r$ will develop along the line of the resultant slope, and its basin will in turn be developed by cross-grading. There must be a certain minimum space or intercept between tributaries of the main stream to provide adequate length of overland flow to permit a lower tributary to the main stream to develop. Furthermore, the tributary $q'r$, having developed much later than the tributary oq, will extend its drainage area laterally more slowly than the latter, with the result that the drainage basin will tend to have an ovoid outline (Fig. 29).

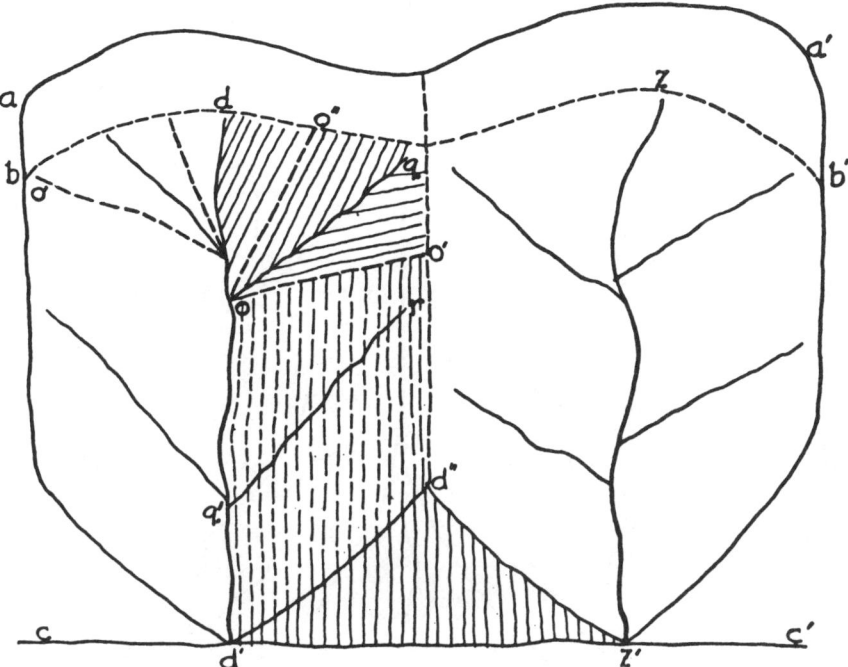

FIGURE 29.—*Development of lower pairs of main tributaries*

Another stream may also develop at zz' in the same manner as the stream at dd'. The development of this stream may have begun either a little earlier or a little later than the stream dd', and the final location of the lateral divide between the two drainage basins will be determined by the conditions of competition. The older stream will absorb the greater part of the area between the two streams. A marginal area of direct drainage $d'd''z'$ is left between the two major drainage basins. If the length of overland flow here becomes sufficient, an intermediate subordinate stream will develop.

The appearance of the final stream systems in the two drainage basins will be somewhat as shown by Figure 30.

Two major factors control the development not only of the drainage basin of a given stream but the systems of drainage basins tributary to a new coast line:

(1) Streams develop successively at points where the length of overland flow becomes greater than the critical length x_c.

(2) Competition results in the survival of those streams which have the earliest start or had the greatest length of overland flow, or both, and which are therefore able to absorb their competitors by cross-grading.

END POINT OF STREAM DEVELOPMENT

Stream development on a newly exposed slope continues until the greatest remaining length of overland flow is less than the critical distance x_c required to institute erosion.

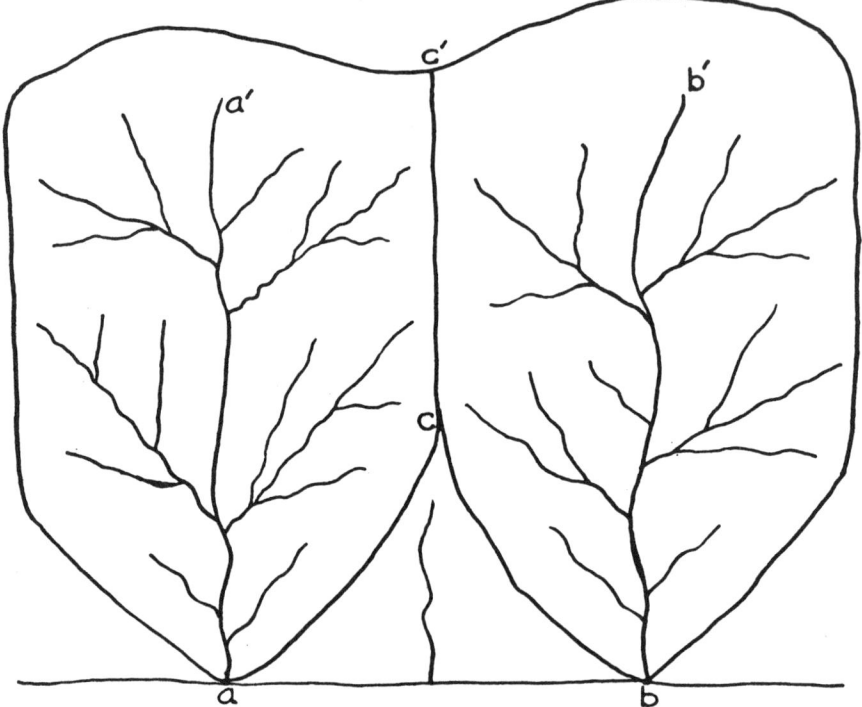

FIGURE 30.—*Final development of two adjacent drainage basins on newly exposed land*

At a certain stage of gradation (Fig. 31) the stream *oa* has developed with a drainage basin *ocd*. Before cross-grading of this area the critical length x_c is, for example, equal to that shown by the line *mm'* on the insert, and this is less than *oa*. After cross-grading of the area *ocd* this critical length is somewhat reduced by increased resultant slope and is now *mn*. The greatest lengths of overland flow on the areas *oca* and *oad* are now along the slope lines *de* and *ce*, but these are both less than *mn*. Hence no additional streams will develop in the area *ocd*.

The upper ends of the streams in a drainage basin will extend at least to the distance x_c from their watershed line, measured in the direction of slope. They may be extended closer to the watershed line by headward erosion, under suitable conditions.

For streams to be perennial at their sources there must be ground-water flow at the head of the stream channel. In regions where there is a permanent ground-water horizon under the drainage basin the most common condition is that the stream is intermittent for a distance downstream from the point where its channel begins. Figure 32 shows the profile at the head of a stream. The watershed line is at *a*, and a definite channel begins at *b*. There is a water table underneath the headwater

ORIGIN AND DEVELOPMENT OF STREAM SYSTEMS

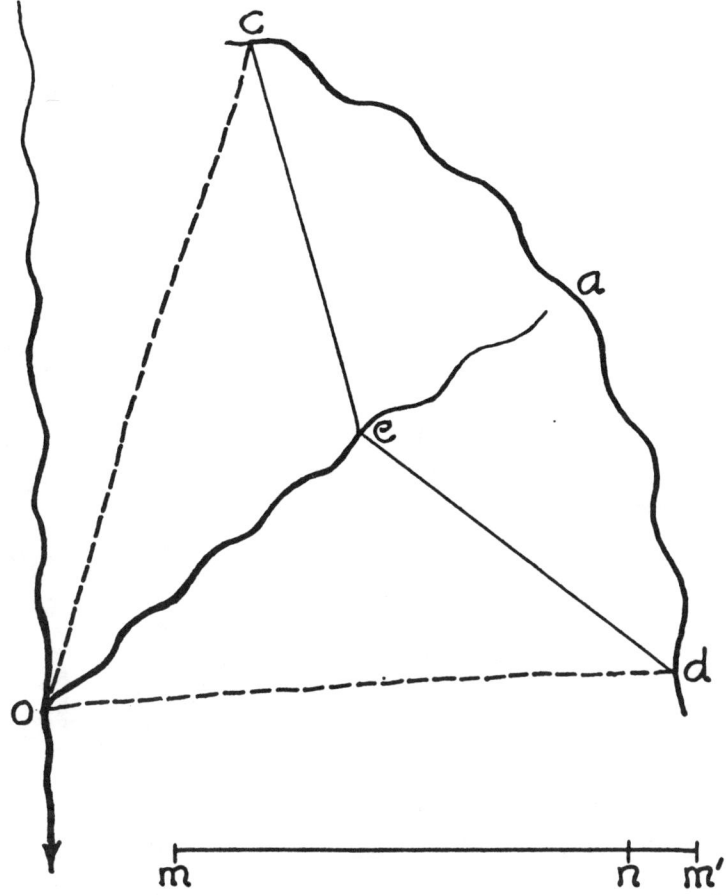

FIGURE 31.—*End point of stream development*

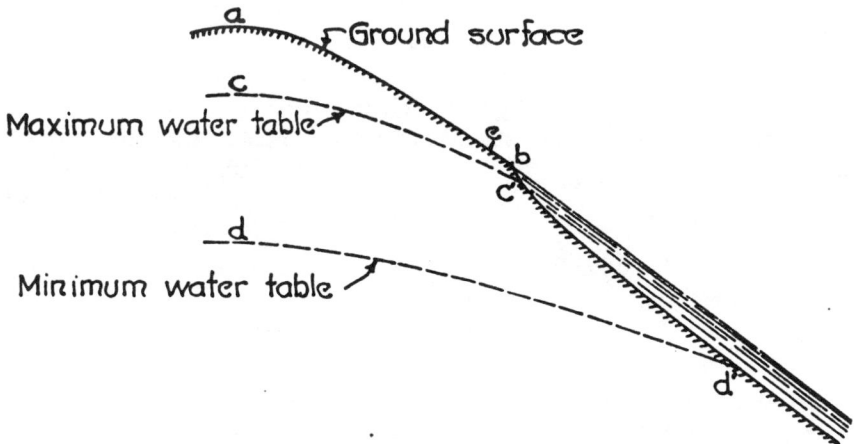

FIGURE 32.—*End point of a definite stream channel*

belt of no erosion ab, the maximum ground-water table is at cc', and the minimum at dd'. Between cc' and dd' the stream is intermittent. At c' part of the infiltration on the upper drainage area enters the stream. At times of maximum surface runoff the ground-water flow may represent a considerable fraction of the total flow. If,

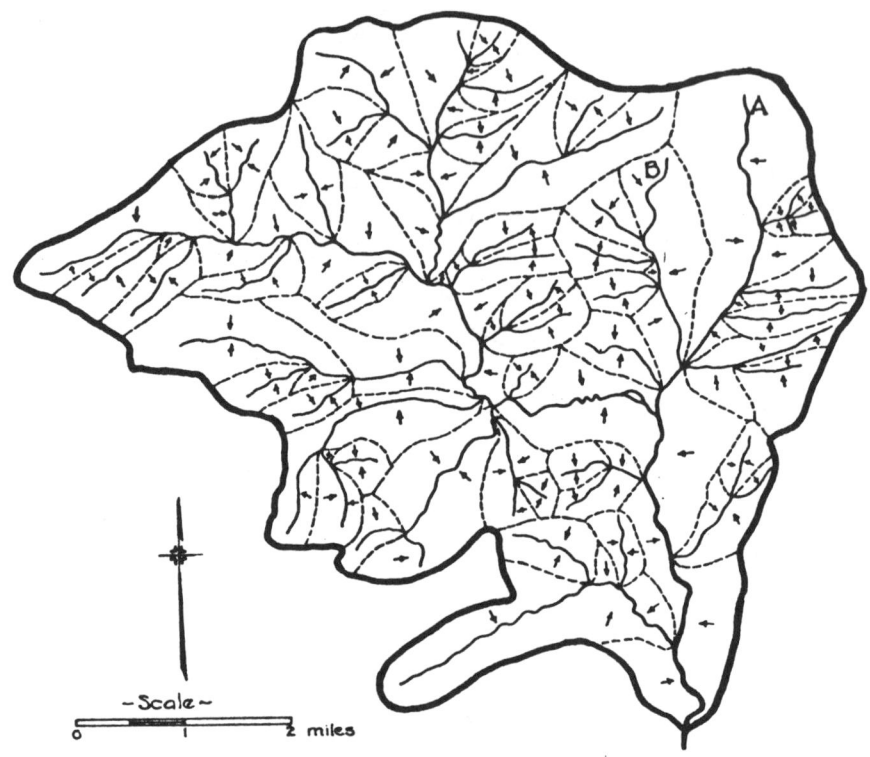

FIGURE 33.—*Drainage basin of Pennypack Creek*
Above Valley Falls, Pa., showing subareas from which surface runoff is derived.

for example, the ground-water flow is one fourth of the total flow at c', then, if the channel extended a little farther upslope to e, the maximum runoff would be reduced one fourth by elimination of ground water. There is therefore an abrupt and sometimes considerable change in the total runoff at about the point where the maximum level of the water table intersects the stream channel. Surface runoff plus ground-water flow can generally extend the channel upstream farther by headward erosion than could surface runoff alone. Hence the channel usually ends near the point where ground-water flow is no longer effective. Ground-water flow at c' is intermittent, but it usually continues much longer than surface runoff and by maintaining the soil at the head of the stream channel moist and soft it promotes extension of the channel by headward erosion and bank caving.

The final results of stream development under natural conditions are illustrated by Figure 33. Some of the streams in the lower part of the basin are clearly adventitious. There are several drainage basins, such as A and B, where tributaries have developed only on one side of the parent stream, leaving, in this case, an isolated plateau in the interfluve area, although the drainage development of the basin is evidently mature.

STREAM-ENTRANCE ANGLES

From geometrical considerations the following equation has been obtained for the entrance angle between a tributary and the higher-order stream which it enters (Horton, 1932)[4]:

$$\cos z_c = \frac{\tan s_c}{\tan s_g}$$

where z_c is the entrance angle between the two streams; s_c is the channel slope of the parent or receiving stream; s_g is the ground slope or resultant slope, which is here assumed to be the same as the slope of the tributary stream.

Values of the entrance angle computed by this equation for different values of the ratio s_c/s_g are as follows:

s_c/s_g =	0.9	0.8	0.7	0.6	0.5	0.4	0.3	0.2	0.1
z_c =	25.5°	36.8°	45.5°	37.0°	60.0°	66.2°	72.3°	78.3°	84.2°

As shown by Table 4, stream slopes are always less than the adjacent ground slope, and tributaries should enter the confluent stream at acute angles when the slopes of the channels of the tributary and confluent streams are nearly the same. The equation takes on the indeterminate form 0/0 if the two slopes s_c and s_g are equal. This means that the two streams will be parallel and will not join. Three cases will be considered for purposes of illustration.

CASE 1—FLAT STREAMS DEVELOPED ON A FLAT AREA: When the parent stream has developed and cross-grading has proceeded to a point where a pair of tributaries develop, the parent stream will in general have cut into the initial surface to some depth, and its stream slope in the vicinity of the debouchure of the tributaries will be materially less steep than the original slope, while the slopes of the tributaries as they approach the parent stream will be materially steeper than the original slope. As a consequence, instead of the ratio s_c/s_g being close to unity, this ratio will seldom have a value greater than 1/2 or 1/3, and the tributaries will not enter the main stream at acute angles, as would be the case if s_c and s_g were nearly equal, but will more generally enter the parent stream at angles of 60° to 80°. On extremely flat surfaces in humid regions a swampy condition often prevails, and stream-entrance angles are but little subject to control by erosion conditions. On semiarid plains where little erosion occurs, acute entrance angles of tributaries to the parent stream may sometimes be observed.

CASE 2—FLAT VALLEY SLOPE WITH MODERATE TO STEEP ADJACENT GROUND SLOPE: Under these conditions the ratio s_c/s_g is nearly always low, and the stream-entrance angles to the main or parent stream are commonly 60° or greater. As the

---

[4] Derivation of this equation is given correctly in the reference cited. Interpretation of the equation as there given is incomplete and not wholly correct.

stream system develops, the slope of the main stream steepens proceeding upstream, and the lateral ground slopes also steepen proceeding upstream. The ratio s_c/s_g may remain sensibly constant, or it may either increase or decrease. Most commonly it decreases to some extent. Quite generally the entrance angles of tributaries to the main or initial stream are quite uniform and range from 60° upward, decreasing somewhat upstream.

CASE 3—TRIBUTARIES ON A STEEP SLOPE: Tributaries developed on the same slope generally run nearly parallel, and if the main valley is relatively flat they will enter the parent stream at an angle of 90°, representing a limiting condition which is approached but not often attained. Tributaries developed on the same lateral slope may of course join and are especially likely to join where drainage development is incipient, as on steep, rocky slopes and in semiarid regions where tributary development has been arrested at the end of the rill stage. Parallel tributaries which join on a steep slope under these conditions commonly have an acute angle of juncture. In this case the ratio s_c/s_g is close to unity.

DRAINAGE PATTERNS

Much has been written regarding the forms of drainage patterns. They are usually classified as dendritic (treelike), rectangular or trellised, radial, and centripetal. The terms radial and centripetal commonly refer to the arrangement of a group of drainage patterns originating at or converging to a common point and do not refer in general to the pattern in an individual drainage basin. All drainage patterns of individual drainage basins are treelike, but different patterns resemble the branchings of different kinds of trees and range from those with branches entering the parent stream nearly at right angles, to those with tributaries nearly parallel and entering their parent streams at small angles. The form of the drainage pattern depends to a large extent on the relation of the slope of the parent stream to the resultant ground slope after cross-grading. If this ratio increases with successive cross-gradings, stream-entrance angles of successive tributaries are somewhat more acute for successively lower-order streams, affording the most usual type of dendritic drainage pattern.

On a relatively flat surface the directions of resultant overland flow after the first cross-grading are nearly at right angles to the initial stream, and the second series of streams developed enter the parent stream nearly at right angles. Cross-grading of the areas tributary to these streams produces but a slight change in the slope ratio s_c/s_g, so that the next order of streams also enters the parent streams more or less nearly at right angles. In this way a rectangular drainage pattern is developed.

If, on a steep, sloping, original surface, the headwater divide forms roughly an arc of a circle, then the first two tributaries developed will enter the parent stream from opposite sides at nearly the same point (Fig. 34). These streams will develop long tributaries nearly parallel with the initial stream, giving rise to a centripetal drainage pattern (Fig. 34).

On flat slopes each successive cross-grading of a given subarea changes the direction of the next stream to develop on the area through an angle approaching 90° as a

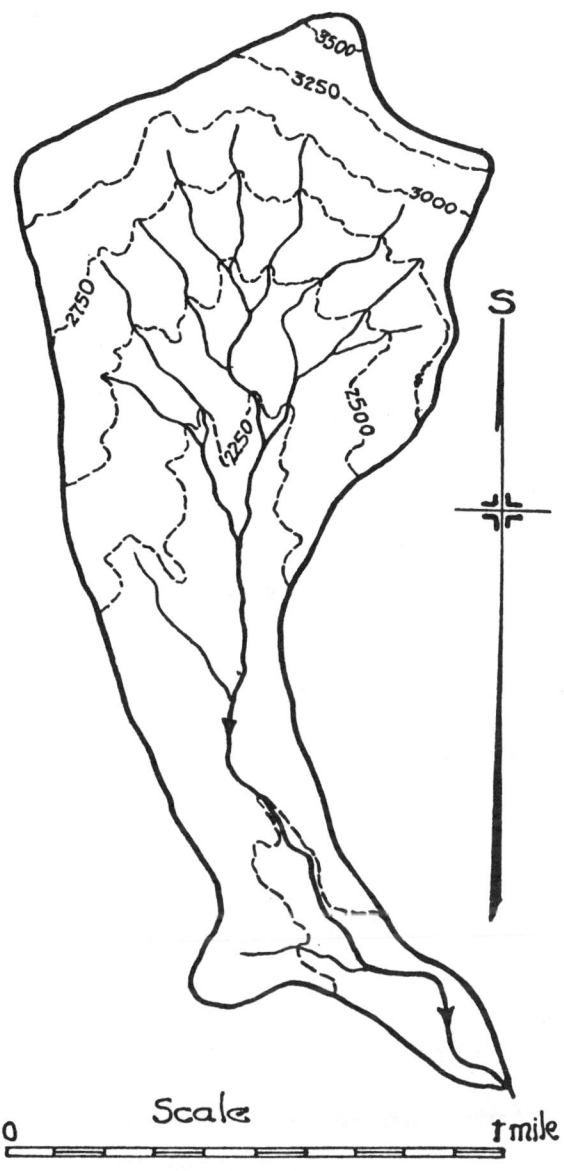

FIGURE 34.—*Centripetal drainage pattern*
Payne Creek, Ga., Mulky Gap quad., U. S. G. S.—T. V. A.

limit and changes the direction of overland flow through a corresponding angle. The direction of resultant cross-graded slope at the end of a given stage becomes the direction of the stream of the next succeeding stage. The directions of streams and of resultant slopes will change through nearly a right angle with each successive stage of stream development and cross-grading, and the directions of streams and of resultant slopes tend generally to be the same in any two stages of stream development which are either both even numbered or both odd numbered.

ASYMMETRICAL DRAINAGE PATTERNS

Because newly developed tributaries enter their parent streams at acute angles, they divide their tributary areas into two parts such that the remaining upslope tributary areas are larger than those on the downslope side, using the terms "upslope" and "downslope" with reference to the two sides of the tributary. Because of inequality of area, width, and slope on the two sides of a tributary, the next lower order of tributaries may develop with two or three tributaries on the upslope side and fewer or none on the downslope side, a common phenomenon, particularly in mountain areas. Since the average elevation of the upslope area is greater than that of the downslope area, this phenomenon is sometimes attributed to increase of rainfall with elevation. It may occur, however, as the result of differences of tributary area and length of overland flow on the upslope and downslope sides of the parent streams, independently of variation of rainfall or runoff on the drainage basin. Burch Creek and Reels Creek drainage basins (Utica, New York, quad., U. S. Geological Survey) afford examples of asymmetrical drainage-basins.

PERCHED OR SIDEHILL STREAMS

In general, streams follow the bottoms of the valleys in which they are located. Small—usually 1st order—streams are occasionally perched precariously on the side slopes of graded valleys of higher-order streams. The course of such a stream is often more nearly parallel with the antecedent slope than with the cross-graded slope. At the foot of the slope the stream often turns abruptly and debouches into the parent stream at nearly a right angle (Fig. 35). Evidently gradation of the valley of the parent stream cd reached the stage shown in the figure before the slope became steep enough to reduce the critical length x_c below the maximum length l_g of overland flow on the right-hand side, and l_g became greater than x_c only when gradation of the valley slope had reached the end point. A weak stream, ab, then developed by micropiracy and cross-grading, but owing to some local cause, such as increased resistivity of the soil to erosion at increased depth below the original surface, this stream was unable to develop a valley of its own by further cross-grading and so remained high above the parent stream on the antecedent rilled surface, until, with increasing volume and slope, it turned nearly a right angle as it entered the parent stream.

REJUVENATED STREAMS; EPICYCLES OF EROSION

In the preceding sections it has been assumed that: (1) Uplift or exposure of new terrain took place continuously though not necessarily at a uniform rate, the region finally becoming stable; (2) the initial resistance R_i of the soil surface to erosion remained constant. The effect of subsequent further elevation or subsequent subsidence of an area on which a stream system has already developed has been extensively discussed in connection with the Davis erosion cycle (Wooldridge and Morgan, 1937) and will not be considered further here. Before leaving the general subject of stream development and valley gradation consideration will be given to the effect of (1) differences between surface and subsurface resistivity to erosion, (2) changes in the surface resistivity to erosion.

The term "rejuvenated stream" is applied to a stream system in which a renewed cycle of erosion begins and which may extend the drainage net after it has reached maturity. Rejuvenation may result from several causes, although in the Davis

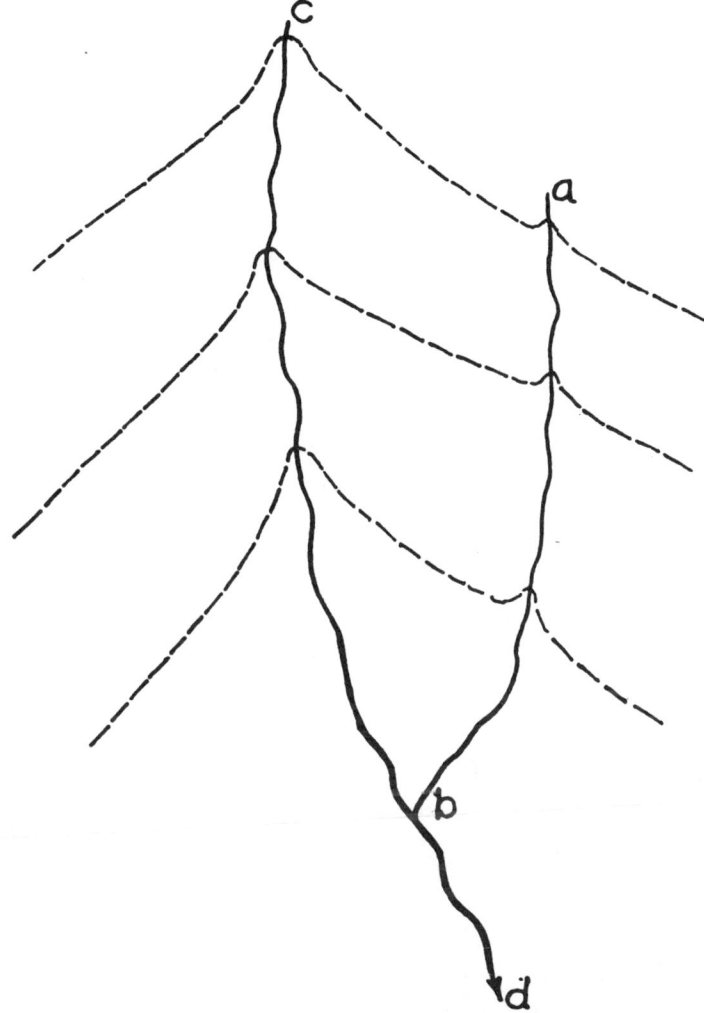

FIGURE 35.—*Perched or hillside stream*

sense the term is applied chiefly where it results from widespread geologic changes such as renewed uplift, folding, and tilting.

Accelerated or decreased erosion may result without any such geologic changes if the original terrain varies in erosional resistivity or infiltration-capacity proceeding downward from the surface. Then, as erosional gradation takes place, changes in the critical length of overland flow x_c will occur, and if these changes are abrupt they may result in important effects, either (1) marked increase in drainage density and extension and number of minor tributaries, if R_i and x_c decrease downward from the surface, or (2) abandonment and fossilization of pre-existing streams and tributaries, if R_i and x_c increase with increased gradation.

A third condition may also bring about changes in erosion rate and stream development which is more common than rejuvenation due to strictly geologic causes. This occurs where, as the result chiefly of climatic or cultural changes, there is a change in the surface-erosional resistivity or infiltration-capacity of the terrain which brings about changes in the critical length x_c and in the consequent development of drainage.

Accelerated erosion due to the removal or replacement of an initially resistant surface by a less resistant surface has been appropriately described by Bailey (1935) as an epicycle of erosion. This term is appropriate since it implies a marked changed in erosional and gradational activity, superposed on the normal erosional conditions. Changes in erosional conditions brought about by dust storms and the formation of loess veneer on soil surfaces, and changes in erosional activity resulting from improper cultivation of the soil, deforestation, fires, or overgrazing of range lands, afford excellent examples of epicycles of erosion.

Where a less permeable and more resistant surface layer of soil or sod overlies weaker or more permeable subsoil, there will be in effect two different values of x_c, one pertaining to the surface layer, the other to the underlying material. This occurs where well-established grass or other vegetal cover overlies a noncohesive sandy soil or where there is a layer of loess or similar fine-textured material, with moderate or high cohesiveness, overlying more permeable and less cohesive material, such as sand.

If the overlying resistive material is broken through, the value of x_c pertaining to the underlying material governs subsequent stream development. In such cases the development of a drainage net is likely to be erratic and sporadic. On much of the area there may be but few streams. This will be true where the larger or surficial value of erosive resistance R_i and critical distance x_c are effective. At other locations where the smaller subsurface values of R_i and x_c have become effective, active and extensive stream development may take place. Extensive plains, for the most part undisturbed by erosion, may be dissected by rapidly growing and irregularly branching systems of gullylike channels. This condition exists in the Pontotoc Ridge region of the Little Tallahatchie, Mississippi, drainage basin, where deep incoherent sand is overlain with a thin veneer of fine uniform loessal silt. In this region x_c for the underlying sand is practically zero, and stream development may extend far above the x_c limit for the surface material as a result of headward erosion. The author has observed gullies in the Pontotoc Ridge region which in some cases have extended not only to but somewhat beyond the topographic boundaries of their drainage basins (Happ et al., 1940). This has resulted from the slumping of masses of earth from the nearly vertical and sometimes undermined scarp formed by the erosion of the deep, incoherent sand.

The destruction of vegetation by smelter fumes early in the present century in the vicinity of Ducktown and Copper Hill, Tennessee, brought about a new erosion cycle. Glenn's early report (1911) and the author's later observations show that forest and hills sometimes protected the sod locally even where the trees were killed, and where the sod was protected no erosion occurred. As described by Glenn (1911, p. 78):

"The erosion starts near the bottom of a slope, and where the soil is porous rapidly cuts a steep-sided gully to a depth of 5 to 12 feet below the surface, where the underlying schist is as a rule still measurably firm. After a gully has reached its limit in depth it widens until its walls coalesce with the walls of adjacent gullies, by which time most of the soil has been removed."

Over much of the denuded area erosion has not been as complete as that above described. Narrow flat lands still persist between the parallel gullies, and uneroded, nearly flat summits of the hills are conspicuous. In some cases the gullies afford excellent examples of cross-grading in progress, with remnants of the antecedent rill surface still visible.

The erosional topography of this region was essentially mature before denudation took place wherever there was a well-established sod cover. The resistivity of the underlying soil to erosion is, however, so small that, lacking protection, the critical distance x_c is reduced nearly but not quite to zero. Consequently the walls between initial parallel ridges on steep slopes have sometimes coalesced, as described by Glenn. Within a few years after destruction of the vegetation the drainage density was increased locally from ten to one hundred fold, and where this occurred the end point of the new erosion cycle was quickly attained.

In the gully formation in the Pontotoc Ridge region in Mississippi and in the vicinity of Ducktown, Tennessee, surface and subsurface resistance R_i differed, the surface resistance being initially greater and the terrain initially stable against erosion. Reduction of surface resistance resulted from improper cultivation in the Pontotoc Ridge region and from partial destruction of vegetal cover by smelter fumes in the Ducktown region, and an active epicycle followed in each case. The formation of arroyos on overgrazed land affords another example of an epicycle of erosion where the value of x_c is less for underlying soil than for undisturbed surface cover.

DRAINAGE-BASIN TOPOGRAPHY

MARGINAL BELT OF NO EROSION; GRADATION OF DIVIDES

In addition to controlling the drainage density and the composition of the drainage pattern and fixing the end point of development of a stream system on a given area, the critical distance x_c and the belt of no erosion which it produces govern the degree of gradation which can occur on a given area and the extent of gradation along and adjacent to both exterior and interior watershed lines or divides.

If the angle between the watershed line aa' (Fig. 36A) and the direction of overland flow is A, then for a given critical length x_c there will be a belt of no erosion on the given side of the watershed line having a width

$$w_s = x_c \sin A.$$

This marginal belt of no erosion $aa'cc'$ is relatively permanent. It is widest, other things equal, where the direction of overland flow is most nearly normal to the watershed line; this is usually around the headwaters of an exterior divide. The width of the marginal belt of no erosion decreases for a given x_c as the direction of overland

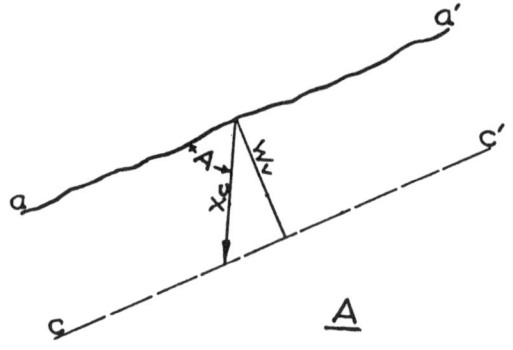

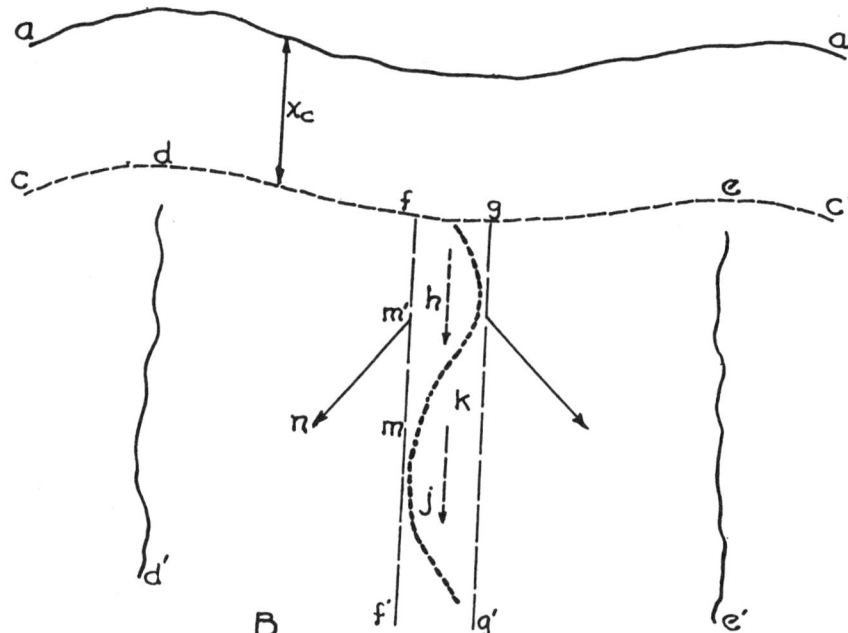

FIGURE 36.—*Belts of no erosion*

(A) Width of belt of no erosion.   (B) Longitudinal belt of no erosion.

flow becomes more nearly parallel with the direction of the divide, a condition which commonly occurs along lateral segments of the main divide surrounding a drainage basin.

If aa' (Fig. 36B) represents the exterior divide at the head of a newly exposed area, then, with sufficient newly exposed surface, streams will develop, starting at d and e. The entire slope from cc' to the outlet is subject to sheet erosion. Cross-grading begins adjacent to the streams and spreads laterally until there remains a narrow belt $ff'gg'$ not yet cross-graded. Dashed arrows indicate directions of overland flow antecedent to, and solid arrows the corresponding directions with, cross-grading. This belt has, however, been previously subject to sheet erosion since it lies downslope from the headward belt of no erosion, and the direction of overland flow is parallel with the slope. The profile of the belt $ff'gg'$ is concave, and it lies, except at its ends, considerably below the original slope. The narrow belt $ff'gg'$ is still subject to cross-grading. Slight variations in surface conditions will divert most of the surface runoff at a given location, as at h, into one stream or the other. The divide between the streams will move away from the stream into which the diversion occurs. The direction of overland flow on the diverted area will swing around until it is more or less parallel with that on the adjacent cross-graded slope, and a belt of no erosion will develop on the side of the divide on which diversion occurs. This belt will have a width $x_c \cos A$, where A is the runoff angle between the diverted surface runoff and the antecedent slope. This angle will vary from zero to $mm'n$, and the width of the belt of no erosion on the given side will vary accordingly.

At some other location, j, the stream ee' will gain the advantage in competition with gg', and the watershed line will be deflected toward ff'. As a result the watershed line will become sinuous, as shown by the dashed line on Figure 36B. Intermediate between h and j the streams will divide the runoff more or less equally. The watershed line will cross the center of the belt $ff'gg'$, but at this location most of the runoff will have been diverted at h, and there will be less erosion than at either h or j. As a consequence of the competitive development of divides the width of the belt of no erosion will vary from point to point, governed locally by the slope, the direction of overland flow, and the amount of previously undiverted surface runoff originating within the belt of no erosion. The watershed line will be sinuous in plan and profile, and the watershed ridge will be broken up into a series of irregularly spaced hills, often with flat crestal plateaus, and adjacent hill crests will be at about the same elevations. The hills will be separated by saddles, and both will be rounded not only as a result of the manner of their development by aqueous erosion but also by secondary processes, such as earth slips and rain-impact erosion.

A favorable location for flat-top, interfluve hills is at the junction of a longitudinal and a cross divide. Such junctions commonly occur where there is an angle or bend in the parent divide. Under these conditions the flat-top hill usually has an arm extending out onto the interior divide. Flat-top hills and plateaus may also occur at intermediate locations where there is a relatively wide belt of no erosion.

On Figure 37, aa' and bb' are adjacent tributary streams which developed more or less simultaneously on the same side of the parent stream and which flow nearly parallel, and crosswise of the original slope. When these streams have developed on an antecedent slope, cross-grading will occur, spreading laterally on both sides of each stream. Dashed arrows show directions of overland flow on the antecedent surface, and solid arrows show the corresponding directions after cross-grading by the streams aa' and bb'. Most of the surface runoff on the area $aa'm$ will be diverted from the parent stream into stream aa' by cross-grading.

Downslope from aa' this stream can divert only the runoff from the area $aa'c$. Overland flow on the area upslope from the watershed line ac will be parallel to this line, while on the area $acde$ the antecedent direction of overland flow will still persist, and a belt of no erosion will develop. The stream bb' will receive the runoff from the area $bb'f$ on the upslope side and from the area $bb'g$ on the downslope side. The areas aca' and bgb' will have been subject to at least two cross-gradings, and as a result the direction of overland flow on these areas will have been turned nearly

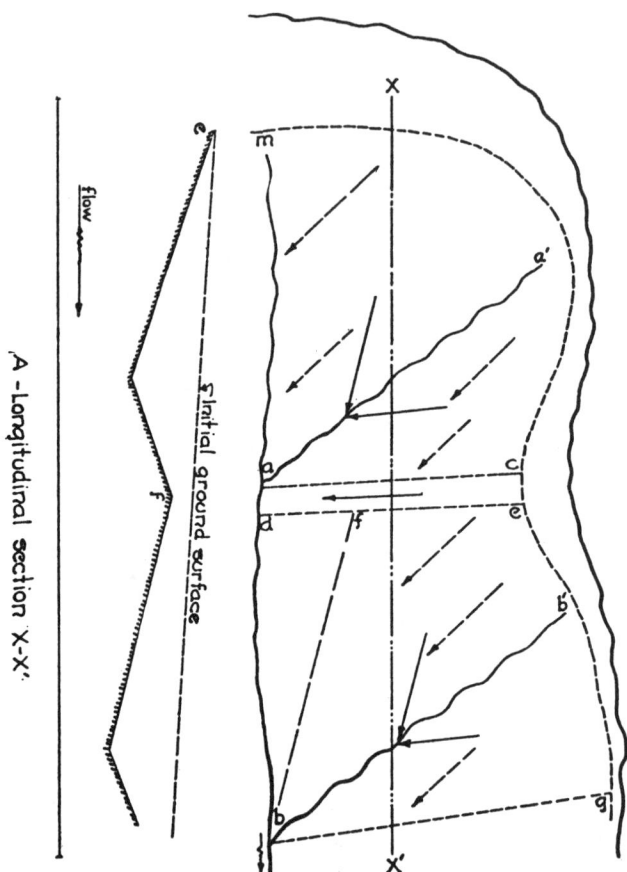

FIGURE 37.—*Belt of no erosion at a cross divide*

through a right angle. The direction of overland flow on such areas downslope from streams running crosswise of the original slope may, of course, have either a downslope component (Fig. 37), or it may have its direction of flow reversed with respect to the original slope. This happens if the direction of overland flow is deflected through more than 90°.

A longitudinal section along the line xx' is shown on Figure 37A. The initial surface is shown by a dashed line. In spite of the fact that e is higher than f, the resultant slope is not materially different on the wide and narrow sides of the valley, a fact often noticed on topographic maps.

It has been shown that the occurrence of a belt of no erosion along an interior divide between streams parallel with the original slope is contingent on the development of components of flow across the divide by micropiracy. A belt of no erosion usually occurs on each side of the divide but is relatively narrow in relation to x_c, and in the vicinity of saddles between crestal hills it may have been subject to crossgrading during its development. If a divide runs crosswise of the drainage basin (*acde*, Fig. 37), the belt of no erosion will temporarily be subject to longitudinal erosion, but presently, as a result of erosional competition, hills and saddles will develop, breaking up the longitudinal components of overland flow into elements each less than x_c, as in case of a longitudinal divide. A crosswise belt of no erosion will usually be wider, and the interfluve hills and plateaus developed thereon will usually be larger and with flatter tops, than in case of a divide running parallel with the original slope.

As a drainage system develops, additional belts of no erosion are introduced along the new interior divides, thereby reducing the portion of the total area over which sheet erosion can occur, other things equal. These later divides have been longer subject to gradation than those developed earlier, and they are generally at lower elevations relative to the original surface. Streams that ultimately become the higher-order streams of the drainage basin usually develop early in the erosion cycle, and their divides are usually higher relative to the original surface than those of lower-order tributaries.

A belt of no erosion once developed persists throughout subsequent stages of gradation although subject to variations in width with subsequent cross-grading of the adjacent terrain.

If the drainage basin of a tributary is narrow and steep on one side, with overland flow at right angles to the divide, the belt of no erosion may extend from the watershed line to the stream on that side, while a flatter slope or overland flow at an acute angle on the opposite side may permit erosion over all or a part of the area on that side.

Discussion thus far has related chiefly to the earlier stages of gradation of a drainage basin where the length of overland flow is generally much greater than the critical length x_c. At later stages of stream development the critical length x_c is decreased by successive subdivisions, with the birth of new tributaries, until finally there remains little or no intermediate length of overland flow between the belts of no erosion and the streams.

A practical illustration of a belt of no erosion is afforded by slope terracing. This introduces a system of artificial watershed lines or cross divides on the terraced slope such that the remaining lengths of overland flow are everywhere less than the critical distance x_c, and the area between a terrace divide and the next one downslope constitutes a belt of no erosion.

Nature accomplishes a similar result in the development of a drainage-basin system by the successive development of tributaries of lower orders, thereby cutting down the length of overland flow until it does not exceed the critical length x_c anywhere within the drainage basin. The drainage net is then complete. Since the development of tributaries of successively lower orders in a stream system does not go on indefinitely, the drainage density approaches a finite limit. Drainage density seldom exceeds 3.0 and commonly is 1.0 to 2.0 in humid regions where soil erosion is active, as shown in column 9 of Table 1.

INTERFLUVE HILLS AND PLATEAUS

Within almost any drainage basin approaching maturity, especially with steeper slopes, relatively flat-topped interfluve hills and plateaus are scattered over the

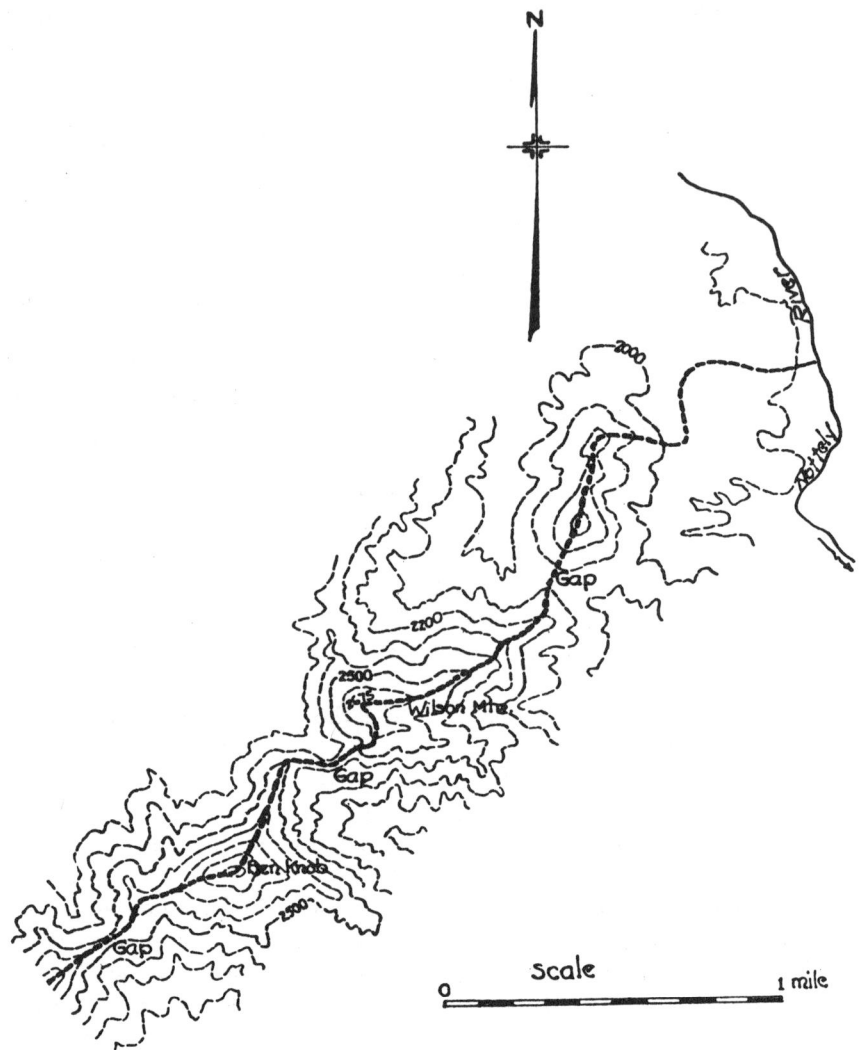

FIGURE 38.—*Topography of an interior cross divide*
(Coosa Bald, Ga., quad., U. S. G. S.—T. V. A.)

area. These hills and plateaus are not, as sometimes described, ungraded areas. They represent remnants of antecedent slopes and are therefore areas on which gradation was arrested when the adjacent streams and valleys developed.

Interior divides running crosswise of the drainage basin have broader belts of no erosion and are more permanent than either interior divides or lateral segments of the main divide running parallel with the original slope. Interior, flat-top, residual hills are most common along the lines of transverse divides and at their junctions with interior longitudinal divides. Figure 38 shows the topography of a transverse divide in the drainage basin of Nottely River between Wold and Crumby creeks (Coosa Bald, Georgia, quad., U. S. G. S.—T. V. A. map), with interfluve hills (knobs and balds) and saddles (gaps).

TABLE 7.—*Flat tops in upper Esopus Creek drainage basin*

Distance above Olive Bridge (miles)	Elevation of flat tops (feet)	Approximate area (Sq. miles)	Average elevation, major divide (feet)	Elevation of stream (feet)
(1)	(2)	(3)	(4)	(5)
3.5	1500	0.09	1600	600
3.7	2000	0.61	1650	610
5.7	2500	0.11	2200	640
6.4	2500	0.03	2350	660
7.35	3500	0.06	3000	700
7.35	2500	0.51	3000	700
9.2	3500	0.05	3150	740
9.2	2500	0.23	3150	740
9.2	2500	0.16	3150	740
9.7	2500	0.24	3400	760
10.6	3000	0.15	3300	800
10.6	2500	1.00	3300	800
11.9	3000	1.85	3100	900
13.4	2500	0.14	3200	950
14.3	3500	0.05	3450	1000
18.3	3000	0.04	3200	1500

The topographic map of Esopus Creek drainage basin above Olive Bridge in southeastern New York shows several residual flat-top hills and plateaus, all located along the lines of interior cross divides (Phoenicia, Kaaterskill, Margaretville, and Slide Mountain quads., U. S. G. S. topographic maps). Column 2 of Table 7 shows the elevations of the highest closed 500-foot contours. They are given in order of occurrence proceeding upslope from Olive Bridge. Column 3 shows the area within the contour. Above these contours the summits are relatively flat. Column 4 gives the average elevation of the main divide at the same cross section. Flat tops on the same cross divide are usually at nearly the same elevations, but those on adjacent divides, even where at nearly the same distance upslope, may be of quite different elevations; the one located on the divide which developed later is usually lower.

At the head of the drainage basin the marginal belt of no erosion along the main divide represents a portion of the original surface. While the elevations of the summits of flat tops usually increase upslope, those on lateral divides near the head of the drainage basin have been graded somewhat and are consequently usually at a lower elevation than the divide at the head of the basin.

Interior interfluve hills and plateaus are, however, not always lower than the adjacent peripheral divide because of conditions of exposure of the original surface. If the original surface was warped upward or is domelike, interfluve hills may rise above the adjacent main divide. The occurrence of flat-top, interfluve hills and plateaus results from the development of a divide between two streams under competitive conditions and requires only aqueous erosion. The locations and sizes of such hills may, however, be governed by secondary causes.

There may be, on a newly exposed region, local areas where high infiltration-

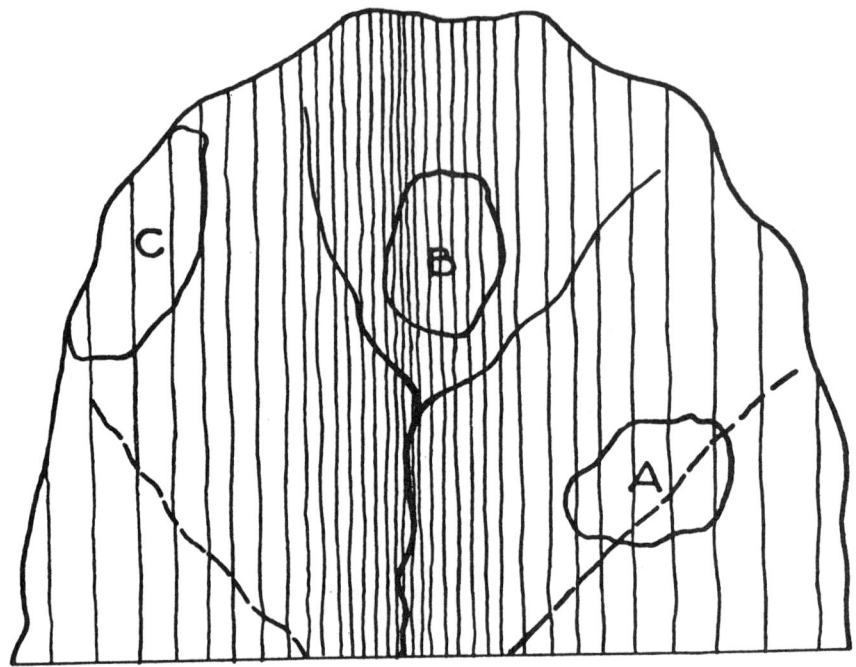

FIGURE 39.—*Origin of ungraded or partially graded interfluve hills and plateaus*

capacity f, large initial resistance R_i, or local flatness of slope, or these combined, make the critical distance x_c abnormally large—for illustration, 100 times as great as for the region in general. On Figure 39, A, B, and C represent such local areas, the lines of overland flow in the first stage of gradation being as indicated. The area A receives such intense runoff at its upslope edge that it may be eroded actively in the first stage of gradation. Area B may be little eroded or eroded only on its downslope portion, while area C will be immune to erosion. In the second and subsequent stages of stream development, none of these areas may be subject to erosional gradation, but they will remain as flat-topped interfluve hills, some of them at the elevations of the original surface, others at somewhat lower elevations.

CONCORDANT STREAM AND VALLEY JUNCTIONS

A new stream develops on a pregraded slope extending away from the parent stream. Thus the new stream enters the parent stream concordantly. A tributary valley has, in general, steeper side slopes and a smaller value of x_c than its parent stream valley. Hence a tributary valley usually grades faster than the coincident gradation of its parent valley, and, although younger, if its stream does not initially enter the parent stream concordantly, it ultimately reaches the grade of the parent stream and debouches into it concordantly. If unrestricted, the tributary stream would cut below the level of its junction with the parent stream, but, since it cannot discharge below the grade of the parent stream at the junction and can easily maintain its grade at the level of the latter, it continues to discharge into the parent stream concordantly, as stated by Playfair's law.

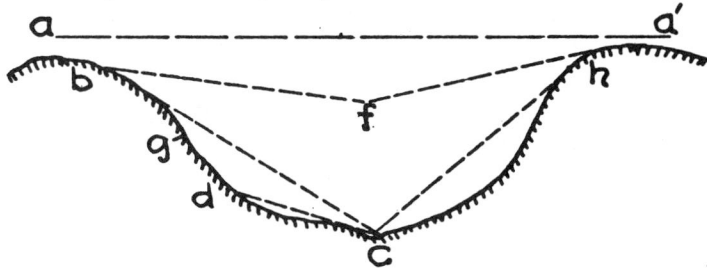

FIGURE 40.—*Gradation of stream valley*

STREAM-VALLEY GRADATION

Stream and valley gradation are closely related. The stream supplies a means of disposal of eroded material from the valley and fixes the minimum level of valley gradation. The valley tributary to the stream supplies the runoff that grades the stream. Stream and valley gradation proceed together, but valley gradation tends to lag behind stream gradation.

In the development of the stream and valley of a first-order tributary, runoff volume adequate to produce erosion close to the stream will be most frequent, and this part of the valley will be cut down rapidly to stream level, progressing backward toward the divide. At any point in the drainage basin downstream from the critical point x_c, the overland flow is increasingly charged with material in transport. This cuts down the transporting power of the overland flow, and near the foot of the slope the eroding and transporting power may be in equilibrium, and further erosion may thereby be inhibited.

Figure 40 represents the cross section typical of a mature tributary stream valley produced by aqueous erosion. As a result of cross-grading and re-cross-grading, the initial surface aa' was cut down to bfh when the stream developed. At each side is a belt of no erosion. For homogeneous material, valley side-slope erosion will not stop at the line bc. If erosion continues until the profile on the left-hand side is bd, the sheet flow, charged with eroded material, arriving at d must be disposed of. Slope is required to carry the water from d to c, and the segment dc will not be graded below this minimum slope. A steeper slope may be maintained from d to c because of sedimentation if the sheet flow from above d is overcharged with sediment with respect to its transporting power at the reduced slope dc.

For turbulent flow the critical length x_c varies inversely as the surface-runoff intensity q_s. Rainstorms range from those with intensities less than infiltration-capacity, and which produce no surface runoff, up to the maximum intensity possible in the given locality. In lighter storms x_c will extend to the stream at c, and no erosion will occur, although there may be runoff. In moderate storms x_c will extend to some point between b and c, and erosion will occur only on the lower portion of the slope. Only in maximum storms will x_c be limited to ab, with consequent erosion or sedimentation throughout the length bc.

As a result of the combination of (1) decreasing frequency of higher rain intensities, (2) the existence of a marginal belt of no erosion, and (3) inhibition of valley-bottom erosion by limited transporting power of overland flow, the valley cross section takes on an ogive or S-shaped form, with a point of contraflexure at some point g and a valley cross section below this point commonly nearly parabolic in form.

A maximum storm produces erosion increasing downslope from x_c and simultaneously provides material in suspension proceeding downslope and thereby increases the width of the belt of sedimentation adjoining the stream. The combined effect is in general to concentrate the steepest-slope angle in the middle portion of the slope extending from the stream to the watershed line.

The slope angle in the belt of steepest slope does not continue to increase indefinitely as erosion continues but approaches a value corresponding to the slope angle for maximum erosion. Obviously if the slope became steeper, the erosion would be decreased instead of increased. This limiting angle of erosion may be either greater than or less than the angle of repose of the soil material when wet. If it is less than the angle of repose, then the slope will ultimately become stable at about the angle of maximum erosion. If the angle of slope for maximum erosion is greater than the angle of repose, the slope will become subject to earth slips as well as to erosion, and each earth slip exposes new material, thus favoring continued erosion. Earth slips do not occur on all mature slopes—in fact, their occurrence is relatively uncommon, and this fact is explained in the manner above described but cannot readily be explained if it is assumed, as has frequently been done heretofore, that erosion continues to increase indefinitely with increase of slope. Land slips, if they occur, generally tend to prolong the process of valley-slope gradation and often decrease the width of the belt of no erosion.

When a first-order tributary develops, the length of overland flow along the course of the rill channels on the antecedent surface is generally less than x_c over most of the area tributary to the stream. Hence only limited subsequent cross-grading can occur. Such cross-grading as does occur is often local and feeble and is usually confined mostly to belts closely adjoining the stream on either side. This often results in a relatively deeply incised V-shaped stream channel adjoined on either side by uneroded upland. In this case the stream and its valley are not truly concordant in the upper reaches of the stream although they become concordant at the mouth of the stream, in accordance with Playfair's law.

Valley gradation, while it may lag behind stream development, is usually close to completion when the latter is completed. As a result the drainage basins of 1st order streams are likely to be born mature or nearly so and yet have the appearance of youth in the physiographic sense, in that their slopes are steep and their immediate stream valleys narrow and V-shaped in cross section. They remain perpetually youthful in appearance not because there is no remaining gradation which could be accomplished but because there are no adequate tools available by which it can be accomplished.

Transverse and longitudinal valley profiles both result from the same hydrophysical processes operating on the same terrain, and they are closely related. The resultant slope is the result of cross-grading at a given point. The transverse slope is the component of resultant slope at a right angle to the stream. Longitudinal slope is the component of the resultant slope parallel with the stream.

Many attempts have been made, beginning with Unwin (1898), to derive an equation for a stream profile. The stream profile at bankful stage is essentially the same as the longitudinal profile of the valley bottom. None of the existing equations is wholly satisfactory. A rational equation for stream profiles must take into account not only the laws of hydraulics and channel dynamics but those also of sheet erosion and storm frequency.

TYPICAL OVOID FORMS OF DRAINAGE-BASINS

Whether a drainage basin develops from the head downstream or from the mouth upstream depends on the manner in which the land becomes exposed. On lands exposed by vertical uplift of oceanic marginal lands, the streams would obviously develop from the headwaters downstream, while on lands exposed by retreat of glacial ice, the drainage basin and stream system may develop from the mouth upstream. In either case the points of first emergence ultimately become either the heads or mouths of principal streams because they provide the greatest lengths of overland flow and hence the points at which l_o first becomes greater than x_c. There is an angle or bend at the point of first emergence which remains as a landward-directed bend at the head of the main divide. Hence the head of a main drainage basin is usually somewhat rounded.

As a result of change of direction of overland flow by cross-grading (Fig. 30), the area acb between two parallel streams never becomes tributary to these streams as long as the coast line remains at ab. Areas tributary to rilled surfaces which initially drained part of the area $aa'bb'$ directly to ab will be captured by competitive cross-grading, and in this way the drainage basin of the streams aa' and bb' will expand laterally until a permanent competitive divide cc' is established.

Similar competitive divides will be established to the right of stream bb' and to the left of stream aa'. An area of direct drainage similar to abc always remains between each two streams, with the result that the final forms of drainage basins are usually pear-shaped or ovoid.

Elsewhere (Horton, 1941) the author has shown that the average of the forms of drainage basins of several of the larger rivers of the world is a slightly asymmetrical, pear-shaped figure with the apex at the outlet end. Also, assuming valley cross sections to be approximately parabolic, it is shown (Horton, 1941) that the trace of the intersection of a parabaloid with its apex upstream and a plane inclined to the axis of the parabaloid is a similar ovoid figure. Hence the hydrophysical laws and processes, as a consequence of which valley sections are generally similar to, though not actually parabolas, lead directly to the development of drainage basins of typical form.

In spite of modifications of individual drainage basins by geologic structures, the semblance to the typical ovoid form is still preserved in most drainage basins.

DEVELOPMENT OF LARGE DRAINAGE BASINS

In this paper, lands exposed by changes of a coast line have been used for purposes of illustration. If an entire continent was exposed progressively the principles outlined would still apply and would lead to the formation of stream systems dependent on the manner of exposure and the position of the continental divide. In North America, for example, upwarping of the northern portion of the continent and a gentle tilt of the remainder to the southeast would provide requisite conditions for the existing stream systems.

Waters of the earth are probably chiefly of volcanic origin. If at an early stage in geologic history a continent was exposed at a time when rainfall was light and oceans shallow or wanting, and there followed a period of gradually increasing rainfall, then a stream system would develop with a tendency to greater concentration in large drainage basins than occurs where exposure of the area is gradual.

If an entire continent was exposed before erosion begins, the maximum length of overland flow for the whole continent would occur at the start. Smaller coastal basins would be more limited in their extension inland than with a gradual exposure of a coastal margin. For such a continental exposure a few major drainage basins would generally develop, draining the greater part of the area. The development of each drainage basin would take place in much the same manner as one of the larger basins on a progressively exposed coastal margin. Smaller coastal streams would be restricted in tributary area by competition with the stronger major streams in much the same way in the two cases.

There would be certain differences in the operation of the processes of erosional gradation for the continental area:

(1) Even an approximation to homogeneity of the area with respect to the controlling factors: rain intensity, infiltration-capacity and surface resistivity to erosion, would not be likely to occur.

(2) A single maximum storm covering the whole of the continent is improbable. A major continental drainage basin would apparently develop in sections, and the sections would combine one after another when erosion attained requisite stages. There would be opportunities for large-scale stream piracy, such as appears to have occurred in the upper part of the area originally tributary to Colorado River.

The development of great drainage basins, such as those of the Mississippi, Amazon, and Nile rivers, must be treated on an individual basis. The recognized persistence of major stream systems, barring such catastrophic events as lava overflows or glacial submergence, suggests that a re-examination of the early erosional history of the earth in the light of hydrophysical processes is well worth while.

DAVIS STREAM-EROSION CYCLE

In discussions of the Davis erosion cycle the cycle is in effect assumed to begin after the development of at least a partial stream system. The initial development of streams is considered to be either fortuitous or governed by antecedent geologic structures and is largely taken for granted (Davis, 1909). The author has considered stream development and drainage-basin topography wholly from the viewpoint of the operation of hydrophysical processes. In the Davis theory the same subject is treated largely with reference to the effects of antecedent geologic conditions and subsequent geologic changes. The two views bear much the same relation as two pictures of the same object taken in different lights—the results are not necessarily in conflict; each supplements the other. The hydrophysical concept appears to be more fundamental because it carries back to the original, newly exposed surface. In comparing the two viewpoints much depends on the meanings given to the words "youth," "youthful," "maturity," and "peneplain."

As commonly applied in connection with the Davis erosion cycle, "young" and "youthful" relate to an area where there has been but little erosional gradation. "Mature" means that all or nearly all the gradation which can result from the operation of existing agencies has been accomplished. On an area where infiltration-capacity and surface resistance are sufficiently high, little erosional gradation may be possible under existing conditions. Such an area is actually mature although from its surface appearance it would be classified as youthful.

Incipient drainage and low drainage density are accepted as *prima facie* evidence of youth. Extensive dissection and high drainage density on a given area are accepted as necessary and sufficient proof of maturity. These are usually sufficient, but they are not necessary as conditions precedent to maturity. An area of low drainage density and with little dissection may have been born mature if the original infiltration-capacity, initial resistance, and rainfall appurtenant to the drainage area were such that the length of overland flow was not sufficient to induce erosion on any part of the area. Such an area represents erosionally mature topography. It is mature not because much has been accomplished but because nothing more can be accomplished without rejuvenation by geologic agencies or as a result of climatic and cultural changes.

Conversely, but less frequently, an area occurs with moderate dissection and a fairly high drainage density which is in a youthful stage because of a small value of x_c. Such areas are most likely to be found where an epicycle of erosion is in progress. That they are not common elsewhere indicates that gradational development of drainage basins is far more generally complete with respect to pre-existing conditions than is usually supposed or assumed. The error has resulted largely from a careless and, as now appears, unjustifiable application of the Davis concepts rather than from error in the concepts themselves if correctly interpreted. For example, areas on the Highwood, Illinois; Rochester, New York; Dunlop, Illinois; Oak Orchard, New York; Parmelee, North Carolina; Anson, Texas; Fargo, North Dakota-Minnesota; and Oberlin, Ohio, topographic sheets of the U. S. Geological Survey have all been referred to as representing "stream erosion—youth" (Salisbury and Atwood, 1908). In spite of the low drainage density and lack of dissection of most of the areas referred to, the stream systems are all or nearly all relatively mature because of high infiltration-capacity, flat slopes, or high surface resistance to erosion or a combination of these factors. Erosion on many of these areas, if it occurs, must be ascribed to conditions which have induced either a local or general epicycle of erosion over part or all of the area.

There is often need to refer to actual age. Tributary streams and their basins are usually younger, in this sense, than higher-order streams or basins. The terms "little dissected" and "well dissected" seem preferable to "youthful" and "mature" in describing erosional status. There may then be areas which are (1) young, little dissected; (2) young, well dissected; (3) old, little dissected; (4) old, fully dissected.

The term "peneplain" seems inappropriate. The ultimate surface of erosion within a main basin boundary is neither "almost a plane," as the prefix "pene" implies, nor is it usually as close to being a plane as was the original surface area from which it has been derived. It seems better to call it a "base surface" in comparison with the original surface below which the stream basin is developed. The base surface at its downslope end is at "base level" in the usual sense. The base surface is, however, generally concave upward except along divides, and its margins intersect the initial surface around the upstream portion of the drainage area. The ultimate base surface is, under ideal conditions, closely similar to a segment of a parabaloid cut by a plane which is not parallel with the axis of the parabaloid. The parabaloidal surface is ribbed with ridges which represent the divides between streams.

Wooldridge and Morgan (1937) use "the invaluable concept of the cycle of erosion as initiated by W. M. Davis." They state (p. 184):

"Some writers have argued that the cycle of erosion can never have run its full course and that the peneplain is an unrealized and unrealizable abstraction."

They invoke geologic factors, as others have done, to complete the erosion cycle—in other words, the Davis erosion cycle is not completed by erosion *per se*. The hydrophysical concept neither denies the effects nor invokes the operation of uplift. It carries the matter of basin development only so far as it can be carried by purely erosional processes, and it shows that there is a definite end point to the development of streams and valleys by aqueous erosion and leaves room for the survival of hills and plateaus between valleys and which are not subject to further erosion or peneplanation. It provides a better foundation than has heretofore existed for the interpretation of the effects of changes of geologic conditions in relation both to the subsequent march of the erosion cycle and in relation to changes of drainage patterns and drainage composition occasioned thereby.

In accordance with the hydrophysical concept most of the observed gradation of divides takes place before the streams which are separated by the given divide are developed—in other words, the terrain where the divide is located is graded in advance at a time when sheet erosion is taking place along or across the line which subsequently becomes the divide. Interfluve hills and plateaus are remnants of this pregraded surface, and when once formed they are permanent features of the topography. The whole concept of ultimate development of a peneplain appears to be founded on the idea that grading of interior divides continues indefinitely and is accomplished by the streams they separate, whereas in the case of the hydrophysical concept the gradation is already, for the most part, accomplished when the adjacent streams originate.

On drainage basins near maturity, erosion occurs only sporadically and in local patches where local conditions provide requisite length of overland flow, runoff volume, slope, and low local resistivity to erosion, or where an epicycle of erosion occurs.

As regards advanced stages of erosional development of a drainage basin the principal differences between the Davis concept and the hydrophysical concept seem to be that the former does not provide any definite end point of erosional development, whereas the hydrophysical concept does provide a definite end point for both stream and valley development.

The physical and mathematical treatment of the subject establishes rational quantitative relationships between the interpretation of observed phenomena accurately and with confidence in the correctness of the results. For example, it has long been known that there is a relation between surface erosion and slope, but without knowing the physical basis of this relationship gross errors could easily be made in interpreting valley-slope erosion. The equations may be applied to the study of individual cases. Such possible applications are of infinite variety and must be left to others. It has been shown that some of the equations are of practical use in engineering problems, as in determining flood-crest modulation by channel storage and the requisite spacing of soil conservation terraces.

In conclusion, what has been given is a framework or outline of drainage-basin development along hydrophysical lines rather than the completed picture. It is hoped that the reader will find that a new, more definite, and more quantitative meaning has been imparted to Playfair's law and the Davis concept of the stream erosion cycle. It is also hoped that the reader will find stimulation to further study and research.

REFERENCES CITED

Bailey, R. W., Forsling, C. L., and Becraft, R. J. (1934) *Floods and accelerated erosion in northern Utah*, U. S. Dept. Agric., Misc. Pub. 196.

Bailey, R. W. (1935) *Epicycles of erosion in the valleys of the Colorado Plateau Province*, Jour. Geol., vol. 43, p. 337–355.

Bennett, H. H. (1939) *Soil conservation*, McGraw-Hill Book Co., New York, 958 pages.

Beutner, E. L., Gaebe, R. R., and Horton, R. E. (1940) *Sprinkled plat runoff and infiltration experiments on Arizona desert soils*, Am. Geophys. Union, Tr., p. 550–558.

Cotton, C. A. (1935) *Geomorphology of New Zealand, Part I*, p. 57.

Davis, W. M. (1909) *Geographical essays*, Ginn & Co., Boston, 777 pages.

Duley, F. L., and Hayes, O. E. (1932) *The effect of the degree of slope on run-off and soil erosion*, Jour. Agric. Res., vol. 45, no. 6, p. 349–360.

―――, and Kelly, L. L. (1941) *Surface condition of soil and time of application as related to intake of water*, U. S. Dept. Agric., Circ. 608, Washington, D. C.

Fletcher, Joel E., and Beutner, E. L. (1941) *Erodibility investigations on some soils of the Upper Gila watershed*, U. S. Dept. Agric., Tech. Bull. 794, Washington, D. C.

Gilbert, G. K. (1880) *Geology of the Henry Mountains [Utah]* (2d ed.), U. S. Geog. Geol. Survey Rocky Mtn. region (Powell), 170 pages.

――― (1914) *The transportation of debris by running water*, U. S. Geol. Survey, Prof. Paper 86.

Glenn, L. C. (1911) *Denudation and erosion in the Southern Appalachian region and the Monongahela basins*, U. S. Geol. Survey, Prof. Paper 72.

Gravelius, H. (1914) *Flusskunde*, Goschen'sche Verlagshandlung, Berlin, 176 pages.

Happ, S. C., Rittenhouse, G., and Dobson, G. C. (1940) *Some principles of accelerated stream and valley sedimentation*, U. S. Dept. Agric., Tech. Bull. 695, Washington, D. C.

Horton, Robert E. (1932) *Drainage basin characteristics*, Am. Geophys. Union, Tr., p. 350–361.

———— (1933) *The role of infiltration in the hydrologic cycle*, Am. Geophys. Union, Tr., p. 446–460.

———— (1935) *Surface runoff phenomena, Part I: Analysis of the hydrograph*, Horton Hydrologic Laboratory, Voorheesville, N. Y., 72 pages.

———— (1937) *Hydrologic interrelations of water and soils*, Soil Sci. Soc. Am., Pr., vol. 1, p. 401–429.

———— (1938) *The interpretation and application of runoff plat experiments with reference to soil erosion problems*, Soil Sci. Soc. Am., Pr., vol. 3, p. 340–349.

———— (1938) *Rain-wave trains*, Am. Geophys. Union, Tr., p. 368–374.

———— (1939) *Analysis of runoff plat experiments with varying infiltration-capacity*, Am. Geophys. Union, Tr., p. 693–711.

———— (1940) *An approach toward a physical interpretation of infiltration-capacity*, Soil Sci. Soc. Am., Pr., vol. 5, p. 399–417.

———— (1941) *Sheet erosion—present and past*, Am. Geophys. Union, Tr., p. 299–305.

Jeffreys, Harold (1925) *The flow of water in an inclined channel of rectangular section*, Philos. Mag., ser. 6, vol. 49, p. 793–803.

Middleton, H. E., and Slater, C. S. (1932) *Physical and chemical characteristics of the soils from the erosion experiment stations*, U. S. Dept. Agric., Tech. Bull. 316, Washington, D. C.

Neal, Jesse H. (1938) *The effect of the degree of slope and rainfall characteristics on runoff and soil erosion*, Mo. Agric. Exper. Sta., Res. Bull. 280.

Renner, F. G. (1936) *Conditions influencing erosion on the Boise River watershed*, U. S. Dept. Agric., Tech. Bull. 528, Washington, D. C.

Salisbury, R. D., and Atwood, W. W. (1908) *The interpretation of topographic maps*, U. S. Geol. Survey, Prof. Paper 60.

Tarr, R. S., and Martin, Lawrence (1914) *College physiography*, Macmillan, New York, 837 pages.

Tipton, R. J. (1937) *Characteristics of floods in the Southern Rocky Mountain region*, Am. Geophys. Union, Tr., vol. 2, p. 592–600.

Unwin, W. C. (1898) *Hydromechanics*, Encyl. Brit., 9th ed.

Wooldridge, S. W., and Morgan, R. S. (1937) *The physical basis of geography*, Longmans, Green & Co., London, 435 pages.

ERRATUM

Page 283, third line from the bottom should read: ". . . The poorly drained basin has a drainage density of 0.73; the well-drained one, 2.74 or four times as great."

5

Copyright © 1957 by the American Geophysical Union

Reprinted from *Am. Geophys. Union. Trans.* 38(6):913-920 (1957)

Quantitative Analysis of Watershed Geomorphology

Arthur N. Strahler

Abstract—Quantitative geomorphic methods developed within the past few years provide means of measuring size and form properties of drainage basins. Two general classes of descriptive numbers are (1) linear scale measurements, whereby geometrically analogous units of topography can be compared as to size; and (2) dimensionless numbers, usually angles or ratios of length measures, whereby the shapes of analogous units can be compared irrespective of scale.

Linear scale measurements include length of stream channels of given order, drainage density, constant of channel maintenance, basin perimeter, and relief. Surface and cross-sectional areas of basins are length products. If two drainage basins are geometrically similar, all corresponding length dimensions will be in a fixed ratio.

Dimensionless properties include stream order numbers, stream length and bifurcation ratios, junction angles, maximum valley-side slopes, mean slopes of watershed surfaces, channel gradients, relief ratios, and hypsometric curve properties and integrals. If geometrical similarity exists in two drainage basins, all corresponding dimensionless numbers will be identical, even though a vast size difference may exist. Dimensionless properties can be correlated with hydrologic and sediment-yield data stated as mass or volume rates of flow per unit area, independent of total area of watershed.

Introduction—Until about ten years ago the geomorphologist operated almost entirely on a descriptive basis and was primarily concerned with the history of evolution of landforms as geological features. With the impetus given by *Horton* [1945], and under the growing realization that the classical descriptive analysis had very limited value in practical engineering and military applications, a few geomorphologists began to attempt quantification of landform description.

This paper reviews progress that has been made in quantitative landform analysis as it applies to normally developed watersheds in which running water and associated mass gravity movements are the chief agents of form development. The treatment cannot be comprehensive; several lines of study must be omitted. Nevertheless, this paper may suggest what can be done by systematic approach to the problem of objective geometrical analysis of a highly complex surface.

Most of the work cited has been carried out at Columbia University over the past five years under a contract with the Office of Naval Research, Geography Branch, Project NR 389-042 for the study of basic principles of erosional topography. References cited below give detailed explanations of techniques and provide numerous examples taken from field and map study.

Dimensional analysis and geometrical similarity—We have attempted to base a system of quantitative geomorphology on dimensional analysis and principles of scale-model similarity [*Strahler*, 1954a, p. 343; 1957]. Figure 1 illustrates the concept of geometrical similarity, with which we are primarily concerned in topographical description. Basins A and B are assumed to be geometrically similar, differing only in size. The larger may be designated as the prototype, the smaller as the model. All measurements of length between corresponding points in the two basins bear a fixed scale ratio, λ. Thus, if oriented with respect to a common center of similitude, the basin mouths Q' and Q are located at distances r' and r, respectively, from C; the ratio of r' to r is λ. In short, all corresponding length measurements, whether they be of basin perimeter, basin length or width, stream length, or relief (h' and h in lower profile), are in a fixed ratio, if similarity exists.

All corresponding angles are equal in prototype and model (Fig. 1). This applies to stream junction angles α' and α, and to ground slope angles β' and β. Angles are dimensionless properties; hence the generalization that in two geometrically similar systems all corresponding dimensionless numbers or products describing the geometry must be equal.

Studies of actual drainage basins in differing environments show that in many comparisons in homogeneous rock masses, geometrical similarity is closely approximated when mean values are considered, whereas in other comparisons, where geologic inhomogeneity exists, similarity is def-

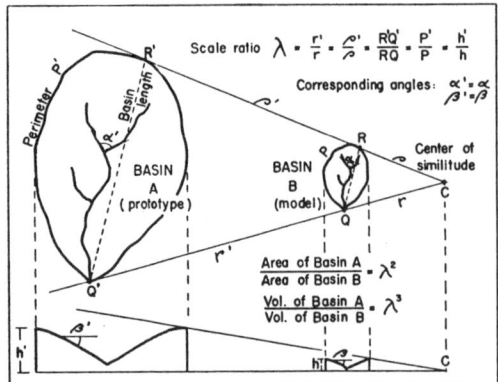

FIG. 1. – Principles of dimensional analysis and geometrical similarity applied to drainage basins

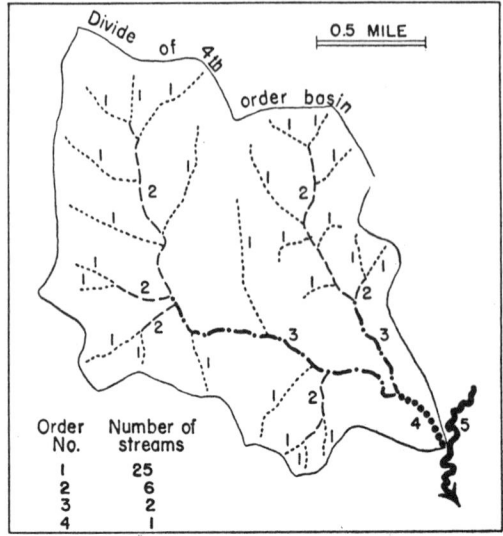

FIG. 2 – Method of designating stream orders (Strahler, 1954a, p. 344)

inately lacking [*Strahler*, 1957]. One advantage of using the principles of similarity as a basis of operations is that it focuses attention upon (a) linear scale differences that are independent of form or shape properties, and (b) form differences existing independently of size differences.

The remainder of this paper describes certain landform properties that are dimensionless; others that have dimensions of length or length products and which serve as scale-of-size indicators. Apart from systematizing the analysis, this information is useful in formulation of rational equations relating geomorphic properties to various related or controlling factors with which a significant regression may be expected.

Order analysis—The first step in drainage basin analysis is order designation, following a system only slightly modified from *Horton* [1945, p. 281–282] (Fig. 2). Assuming that the channel-network map includes all intermittent and permanent flow lines located in clearly defined valleys, the smallest finger-tip tributaries are designated Order 1. Where two first-order channels join, a channel segment of Order 2 is formed; where two of Order 2 join, a segment of Order 3 is formed; and so forth. The trunk stream through which all discharge of water and sediment passes is therefore the stream segment of highest order.

Any usefulness which the stream order system may have depends upon the premise that on the average, if a sufficiently large sample is treated, order number is directly proportional to relative watershed dimensions, channel size, and stream discharge at that place in the system. Also, because order number is dimensionless, two drainage basins differing greatly in linear scale can be equated or compared with respect to corresponding points in their geometry through use of order number. The first step in drainage-network analysis is the counting of stream segments of each order. This is followed by analysis of the way in which numbers of stream segments change with increasing order.

Bifurcation ratio—*Horton's* [1945, p. 291] law of stream numbers states that the numbers of stream segments of each order form an inverse geometric sequence with order number. This is generally verified by accumulated data [*Strahler*, 1952, p. 1137; *Schumm*, 1956, p. 603] and is conveniently treated as shown in Figure 3. A regression of logarithm of number of streams of each order (ordinate) on stream order (abscissa) generally yields a straight-line plot with very little scatter [*Maxwell*, 1955]. Even though the function relating these variables is defined only for integer values of the independent variable, a regression line is fitted; the slope of the line, or regression coefficient b is used. The anti-logarithm of b is equivalent to Horton's bifurcation ratio r_b and in this case has the value of 3.52. This means that on the average there are three and one-half times as many streams of one order as of the next higher order.

One might think that the bifurcation ratio would constitute a useful dimensionless number for expressing the form of a drainage system. Actually the number is highly stable and shows a small range of variation from region to region or environment to environment, except where powerful geologic controls dominate. *Coates* [1956, Table 3]

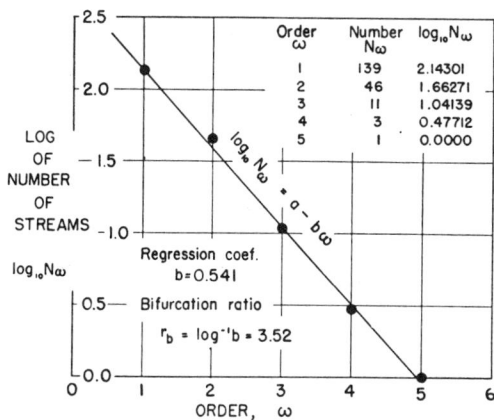

FIG. 3 – Regression of number of stream segments on stream order; data from *Smith* (1953, Plate 8).

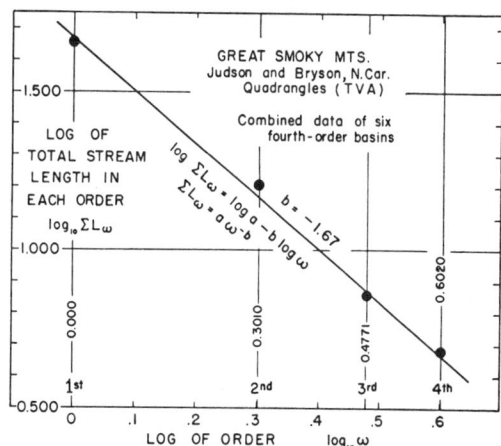

FIG. 4 – Regression of stream length on stream order; stream lengths of six fourth-order basins have been summed for each order to reduce effects of chance variations.

found bifurcation ratios of first-order to second-order streams to range from 4.0 to 5.1; ratios of second-order to third-order streams to range from 2.8 to 4.9. These values differ little from *Strahler's* [1952, p. 1134].

Frequency distribution of stream lengths—Length of stream channel is a dimensional property which can be used to reveal the scale of units comprising the drainage network. One method of length analysis is the measurement of length of each segment of channel of a given stream order. For a given watershed these lengths can be studied by frequency distribution analysis [*Schumm*, 1956, p. 607]. Stream lengths are strongly skewed right, but this may be largely corrected by use of logarithm of length. Arithmetic mean, estimated population variance, and standard deviation serve as standards of description whereby different drainage nets can be compared and their differences tested statistically [*Strahler*, 1954b].

Relation of stream length to stream order—Still another means of evaluating length relationships in a drainage network is to relate stream length to stream order. A regression of logarithm of total stream length for each order on logarithm of order may be plotted (Fig. 4). Again, the function is defined only for integer values of order. Several such plots of length data made to date seem to yield consistently good fits to a straight line, but the general applicability of the function is not yet established, as in the case of the law of stream numbers.

The slope of the regression line b (Fig. 4) is the exponent in a power function relating the two variables. Marked differences observed in the exponent suggest that it may prove a useful measure of the changing length of channel segments as order changes. Because this is a non-linear variation, the assumption is implicit that geometrical similarity is not preserved with increasing order of magnitude of drainage basin.

Drainage basin areas—Area of a given watershed or drainage basin, a property of the square of length, is a prime determinant of total runoff or sediment yield and is normally eliminated as a variable by reduction to unit area, as in annual sediment loss in acre-feet per square mile. In order to compare drainage basin areas in a meaningful way, it is necessary to compare basins of the same order of magnitude. Thus, if we measure the areas of drainage basins of the second order, we are measuring corresponding elements of the systems. If approximate geometrical similarity exists, the area measurements will then be indicators of the size of the landform units, because areas of similar forms are related as the square of the scale ratio.

Basin area increases exponentially with stream order, as stated in a law of areas [*Schumm*, 1956, p. 606], paraphrasing Horton's law of stream lengths.

Schumm [1956, p. 607] has shown histograms of the areas of basins of the first and second orders and of patches of ground surface too small to have channels of their own. Basin area distributions are strongly skewed, but this is largely corrected by use of log of area. Area is measured by planimeter from a topographic map, hence represents projected, rather than true surface area. Estimation of true surface area has been attempted where surface slope is known [*Strahler*, 1956a, p. 579].

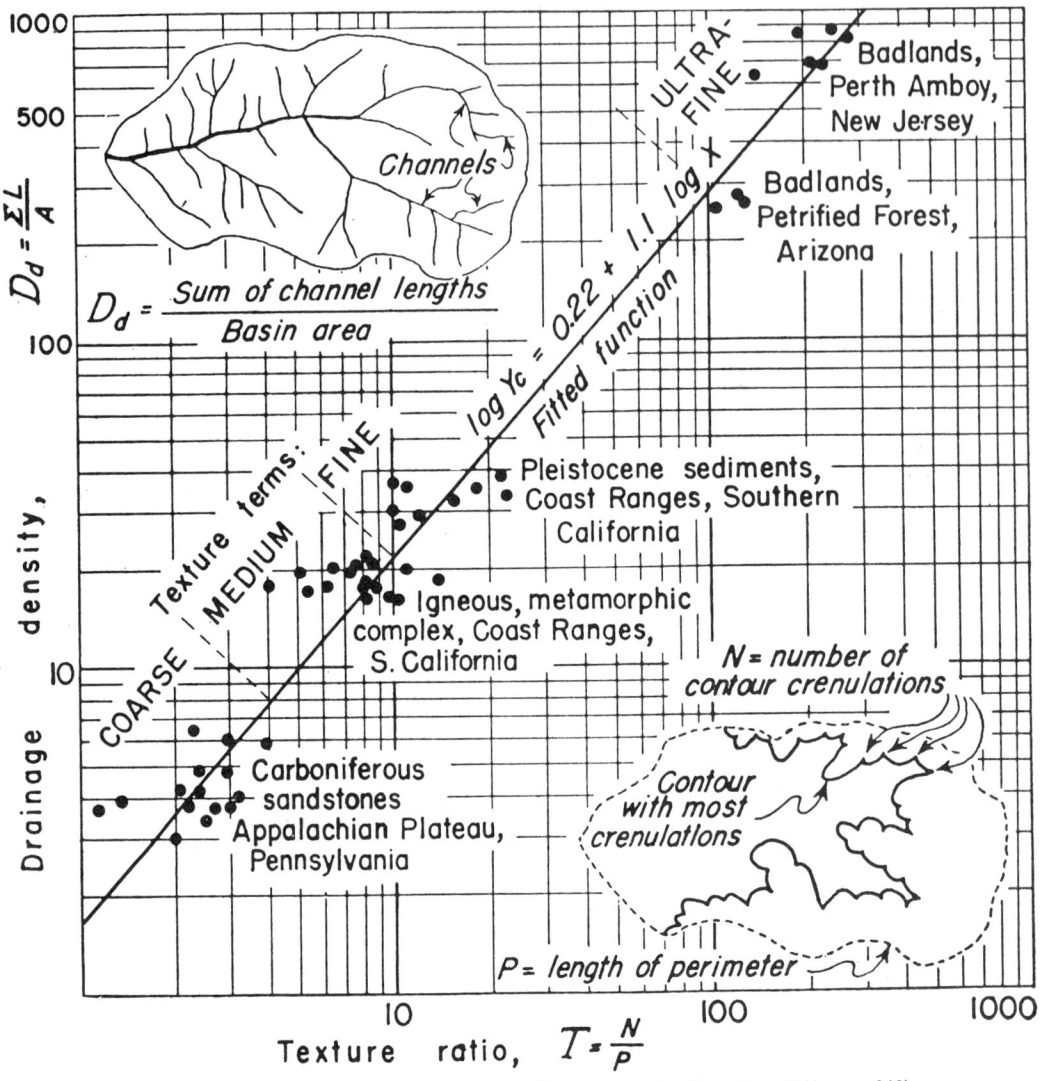

Fig. 5 – Definitions of drainage density and texture ratio (Strahler, 1954a, p. 348)

Drainage density and texture ratio—An important indicator of the linear scale of landform elements in a drainage basin is drainage density, defined by *Horton* [1945, p. 283]. The upper left-hand corner of Figure 5 shows the definition of drainage density as the sum of the channel lengths divided by basin area. Division of length by area thus yields a number with the dimension of inverse of length. In general, then, as the drainage density number increases, the size of individual drainage units, such as the first-order drainage basin, decreases proportionately.

Figure 5 shows the relation between drainage density and a related index, the texture ratio, defined by *Smith* [1950]. Because the contour inflections on a good topographic map indicate the existence of channels too small to be shown by stream symbols, their frequency is a measure of closeness of channel spacing and hence also correlates with drainage density.

Drainage density is scaled logarithmically on the ordinate of Figure 5. The grouped points in the lower left-hand corner of the graph represent basins in resistant, massive sandstones. Here the streams are widely spaced and density is low. The next group of points encountered represents typical densities in deeply weathered igneous and metamorphic rocks of the California coast ranges. In the extreme upper right are points for badlands, where drainage density is from 200 to 900 miles of channels per square mile [*Smith*, 1953; *Schumm*, 1956, p. 612].

Because of its wide ratio of variation, drainage density is a number of primary importance in landform scale analysis. One might expect that sediment yield would show a close positive relationship with drainage density. A rational theory of the relation of drainage density to erosion intensity, predicting the morphological changes to be expected when ground surface resistance is lowered by land use, has been outlined by *Strahler* [1956b].

Constant of channel maintenance—Schumm [1956, p. 607] has used the inverse of drainage density as a property termed constant of channel maintenance. In Figure 6 the logarithm of basin area (ordinate) is treated as a function of logarithm of total stream channel length (abscissa). Stream length is cumulative for a given order and includes all lesser orders; it is thus the total channel length in a watershed of given order. Length in this case is projected to the horizontal plane of the map; true lengths would be obtained by applying a correction for slope.

An individual plotted point on the graph represents a given stream order in the watershed, as numbered 1 through 5. Using data of the three examples given by Schumm, the sets of points fall close to a straight line of 45° slope; thus the relationship is treated as linear even though plotted here on log-log paper. If the logarithm of the intercept is read at log stream length = 0, and the antilog of this intercept is taken, we obtain the constant of channel maintenance C which is actually the slope of a linear regression of area on length.

The value of $C = 8.7$ in the Perth Amboy badlands means that on the average 8.7 sq ft of surface are required to maintain each foot of channel length. In the second example, Chileno Canyon in the California San Gabriel Mountains, 316 sq ft of surface are required to maintain one foot of channel length.

The constant of channel maintenance, with the dimensions of length, is thus a useful means of indicating the relative size of landform units in a drainage basin and has, moreover, a specific genetic connotation.

Maximum valley side slopes—Leaving now the drainage network and what might be classified as planimetric or areal aspects of drainage basins, we turn to slope of the ground surface. This brings into consideration the aspect of relief in drainage basin geometry. One significant indicator of the over-all steepness of slopes in a watershed is the maximum valley-side slope, measured at intervals along the valley walls on the steepest parts of the contour orthogonals running from divides to adjacent stream channels.

Maximum valley-side slope has been sampled by several investigators in a wide variety of geological and climatic environments [*Strahler*, 1950; *Smith*, 1953; *Miller*, 1953; *Schumm*, 1956; *Coates*, 1956; *Melton*, 1957]. Within-area variance is relatively small compared with between-area differences. This slope statistic would therefore seem to be a valuable one which might relate closely to sediment production.

Mean slope curve—Another means of assessing the slope properties of a drainage basin is through the mean slope curve [*Strahler*, 1952, p. 1125–1128]. This requires the use of a good contour topographic map. The problem is to estimate the average, or mean slope of the belt of ground surface lying between successive contours. This may be done by measuring the area of each contour belt with a planimeter and dividing this area by the length of the contour belt to yield a mean width. The mean slope will then be that angle whose tangent is the contour interval divided by the mean belt width. Mean slope of each contour interval is plotted from summit point to basin mouth. Curves of this type will differ from region to region, depending upon geologic structure and the stage of development of the drainage system. If the mean slope for each contour belt is weighted for per cent of total basin surface area, it is possible

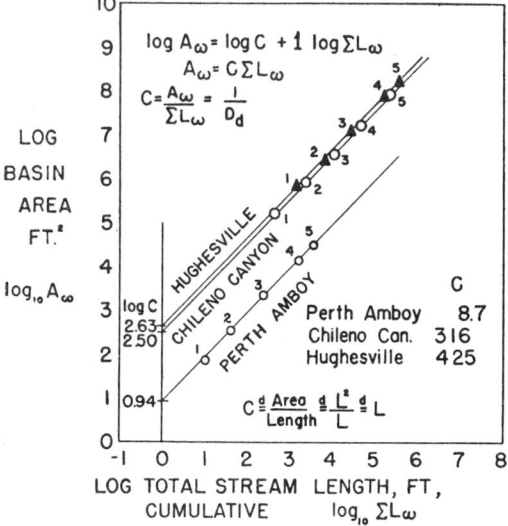

Fig. 6 – Constant of channel maintenance, C. Data replotted on logarithmic scales from *Schumm* (1956, p. 606)

to arrive at a mean slope value for the surface of the watershed as a whole.

Slope maps—Another means of determining slope conditions over an entire ground surface of a watershed is through the slope map [*Strahler*, 1956a]. (1) A good topographic map is taken. (2) On this map the slope of a short segment of line normal to the trend of the contours is determined at a large number of points. These may be recorded as tangents or sines, depending upon the kind of map desired. (3) These readings are contoured with lines of equal slope, here called isotangents. (4) The areas between successive isotangents are measured with a planimeter and the areas summed for each slope class. (5) This yields a slope frequency percentage distribution. Because the entire ground surface has been analyzed, the mean, standard deviation, and variance are treated as population parameters, at least for purposes of comparison with small samples taken at random from the same area.

Lines of equal sine of slope, or isosines, may also be drawn. The interval between isosines on the map becomes the statistical class on the histogram. Sine values are designated as g values because the sine of slope represents that proportion of the acceleration of gravity acting in a downslope direction parallel with the ground surface.

Rapid slope sampling—The construction of slope maps and their areal measurement is extremely time-consuming. Experiments have shown that essentially the same information can be achieved by random point sampling [*Strahler*, 1956a, p. 589–595]. Both random coordinate-sampling and grid sampling have been tried. In the random-coordinate method a sample square is scaled in 100 length units per side. From a table of random numbers the coordinates of sample points are drawn for whatever sample size is desired. The grid method does much the same thing, but is not flexible as to sample size.

Point samples, which are easy to take, were compared with the frequency distribution measured from a slope map. Noteworthy is the extremely close agreement in means and variances, and even in the form of the frequency distributions, including a marked skewness. Tests of sample variance and mean are discussed by *Strahler* [1956a].

Chapman [1952] has developed a method of analyzing both azimuth and angle of slope from contour topographic maps. Although based on petrofabric methods and designed largely for use in geological analysis of terrain, the method might be applied to a watershed as a means of assessing both slope steepness and orientation simultaneously.

Relief ratio—*Schumm* [1956, p. 612] has devised and applied a simple statistic, the relief ratio, defined as the ratio between total basin relief (that is, difference in elevation of basin mouth and summit) and basin length, measured as the longest dimension of the drainage basin. In a general way, the relief ratio indicates overall slope of the watershed surface. It is a dimensionless number, readily correlated with other measures that do not depend on total drainage basin dimensions. Relief ratio is simple to compute and can often be obtained where detailed information on topography is lacking.

Schumm [1954] has plotted mean annual sediment loss in acre feet per square mile as a function of the relief ratio for a variety of small drainage basins in the Colorado Plateau province [Fig. 7]. The significant regression with small scatter suggests that relief ratio may prove useful in estimating sediment yield if the parameters for a given climatic province are once established.

Hypsometric analysis—Hypsometric analysis, or the relation of horizontal cross-sectional drainage basin area to elevation, was developed in its modern dimensionless form by *Langbein* and others [1947]. Whereas he applied it to rather large watersheds, it has since been applied to small drainage basins of low order to determine how the mass is

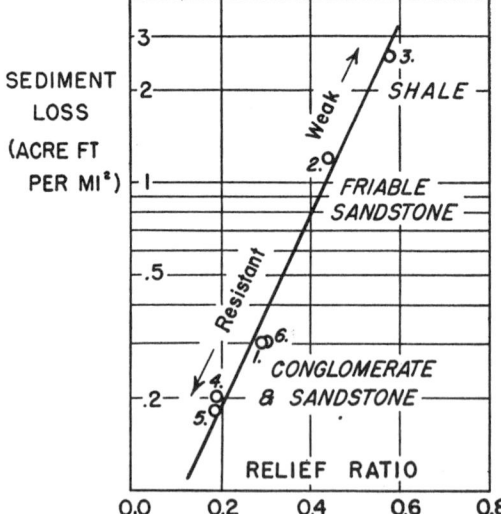

Fig. 7 – Regression of sediment loss on relief ratio, after *Schumm* (1954, p. 218)

distributed within a basin from base to top [*Strahler*, 1952; *Miller*, 1953; *Schumm*, 1956; *Coates*, 1956].

Figure 8 illustrates the definition of the two dimensionless variables involved. Taking the drainage basin to be bounded by vertical sides and a horizontal base plane passing through the mouth, the relative height is the ratio of height of a given contour h to total basin height H. Relative area is the ratio of horizontal cross-sectional area a to entire basin area A. The percentage hypsometric curve is a plot of the continuous function relating relative height y to relative area x.

As the lower right-hand diagram of Figure 8 shows, the shape of the hypsometric curve varies in early geologic stages of development of the drainage basin, but once having attained an equilibrium, or mature stage (middle curve on graph), tends to vary little thereafter. Several dimensionless attributes of the hypsometric curve are measurable and can be used for comparative purposes. These include the integral, or relative

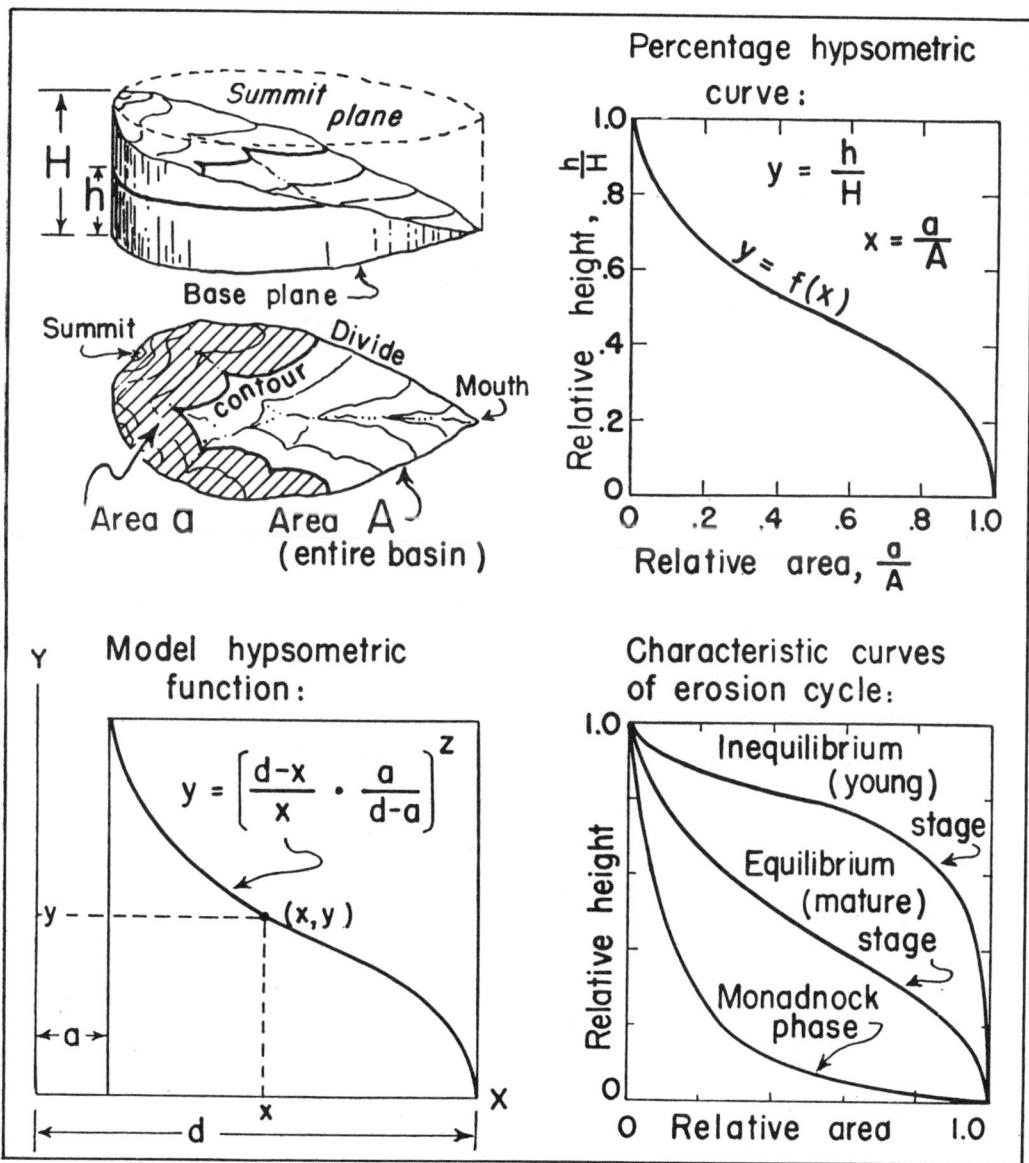

FIG. 8 – Method of hypsometric analysis (Strahler, 1954a. p. 353)

area lying below the curve, the slope of the curve at its inflection point, and the degree of sinuosity of the curve. Many hypsometric curves seem to be closely fitted by the model function shown in the lower left corner of Figure 8, although no rational or mechanical basis is known for the function.

Now that the hypsometric curves have been plotted for hundreds of small basins in a wide variety of regions and conditions, it is possible to observe the extent to which variation occurs. Generally the curve properties tend to be stable in homogeneous rock masses and to adhere generally to the same curve family for a given geologic and climatic combination.

Conclusion—This paper has reviewed briefly a variety of geometrical properties, some of length dimension or its products, others dimensionless, which may be applied to the systematic description of drainage basins developed by normal processes of water erosion. Among the morphological aspects not mentioned are stream profiles and the geometry of stream channels. These, too, are subject to orderly treatment along the lines suggested. The examples of quantitative methods presented above are intended to show that, complex as a landscape may be, it is amenable to quantitative statement if systematically broken down into component form elements. Just which of these measurements or indices will prove most useful in explaining variance in hydrological properties of a watershed and in the rates of erosion and sediment production remains to be seen when they are introduced into multivariate analysis. Already there are definite indications of the usefulness of certain of the measures and it is only a matter of continuing the development of analytical methods until the most important geomorphic variables are isolated.

References

CHAPMAN, C. A., A new quantitative method of topographic analysis, *Amer. J. Sci.*, **250**, 428–452, 1952.

COATES, D. R., *Quantitative geomorphology of small drainage basins in southern Indiana*, Of. Nav. Res. Proj. NR 389-042, Tech. Rep. 10 (Columbia Univ. Ph.D. dissertation), 57 pp., 1956.

HORTON, R. E., Erosional development of streams and their drainage basins; hydrophysical approach to quantitative morphology, *Bul. Geol. Soc. Amer.*, **56**, 275–370, 1945.

LANGBEIN, W. B., AND OTHERS, Topographic characteristics of drainage basins, *U. S. Geol. Surv. Water-Supply Paper 968-C*, 157 pp., 1947.

MAXWELL, J. C., The bifurcation ratio in Horton's law of stream numbers, (abstract), *Trans. Amer. Geophys. Union*, **36**, 520, 1955.

MELTON, M. A., *An analysis of the relations among elements of climate, surface properties, and geomorphology*, Of. Nav. Res. Proj. NR 389-042, Tech. Rep. 11 (Columbia Univ. Ph.D. dissertation), 102 pp., 1957.

MILLER, V. C., *A quantitative geomorphic study of drainage basin characteristics in the Clinch Mountain area, Virginia and Tennessee*, Of. Nav. Res. Proj. NR 389-042, Tech. Rep. 3 (Columbia Univ. Ph.D. dissertation), 30 pp., 1953.

SCHUMM, S. A., The relation of drainage basin relief to sediment loss, *Pub. International Association of Hydrology*, IUGG, Tenth Gen. Assembly, Rome, 1954, **1**, 216–219, 1954.

SCHUMM, S. A., Evolution of drainage systems and slopes in badlands at Perth Amboy, New Jersey, *Bul. Geol. Soc. Amer.*, **67**, 597–646, 1956.

SMITH, K. G., Standards for grading texture of erosional topography, *Amer. J. Sci.*, **248**, 655–668, 1950.

SMITH, K. G., *Erosional processes and landforms in Badlands National Monument, South Dakota*, Of. Nav. Res. Proj. NR 389-042, Tech. Rep. 4 (Columbia Univ. Ph.D. dissertation), 128 pp., 1953.

STRAHLER, A. N., Equilibrium theory of erosional slopes approached by frequency distribution analysis, *Amer. J. Sci.*, **248**, 673–696, 800–814, 1950.

STRAHLER, A. N., Hypsometric (area-altitude) analysis of erosional topography, *Bul. Geol. Soc. Amer.*, **63**, 1117–1142, 1952.

STRAHLER, A. N., Quantitative geomorphology of erosional landscapes, *C.-R. 19th Intern. Geol. Cong.*, Algiers, 1952, sec. 13, pt. 3, pp. 341–354, 1954a.

STRAHLER, A. N., Statistical analysis in geomorphic research, *J. Geol.*, **62**, 1–25, 1954b.

STRAHLER, A. N., Quantitative slope analysis, *Bul. Geol. Soc. Amer.*, **67**, 571–596, 1956a.

STRAHLER, A. N., The nature of induced erosion and aggradation, pp. 621–638, Wenner-Gren Symposium Volume, *Man's role in changing the face of the Earth*, Univ. Chicago Press, Chicago, Ill., 1193 pp., 1956b.

STRAHLER, A. N., *Dimensional analysis in geomorphology*, Of. Nav. Res. Proj. NR 389-042, Tech. Rep. 7, Dept. Geol., Columbia Univ., N. Y., 43 pp., 1957.

Department of Geology, Columbia University, New York 27, N. Y.

(Manuscript received April 1, 1957; presented as part of the Symposium on Watershed Erosion and Sediment Yields at the Thirty-Seventh Annual Meeting, Washington, D.C., May 1, 1956; open for formal discussion until May 1, 1958.)

Part III

DRAINAGE BASIN CONTROLS

Editor's Comments on Papers 6 Through 9

6 **CARLSTON**
Drainage Density and Streamflow

7 **CHORLEY**
Climate and Morphometry

8 **LEOPOLD and MILLER**
Excerpt from *Ephemeral Streams—Hydraulic Factors and Their Relation to the Drainage Net*

9 **MELTON**
Correlation Structure of Morphometric Properties of Drainage Systems and Their Controlling Agents

One of the most important objectives of drainage basin investigations is to develop an understanding of the cause and effect relations that establish basin morphology. However, because of the variety of independent variables influencing the drainage basin, it is frequently difficult in one study to do more than isolate the effects of a single variable. Hence, most papers treat only the effect of one variable on the drainage basin. One reason, of course, is the considerable length of time required to obtain the needed data. For example, hydrologic data are a necessity, yet except in some experimental watersheds, hydrologic records of sufficient duration and detail are not available.

In this section a few papers have been selected that attempt to consider the effect of some important variables on drainage basin morphology. The effect of time on basin evolution has been considered in Papers 1 and 2, and the influence of geologic structure and lithology has been discussed in Paper 3. The influence of lithology and soil characteristics is reflected in the ease with which water infiltrates. The infiltration capacity of soil and the development of surface runoff, which erodes hillslopes and forms drainage channels, is discussed in part 2 (pp. 104-129) of Horton's paper (Paper 4).

Horton's emphasis on surface runoff has been criticized by Kirkby and Chorley (1967) who demonstrate that flow through soil can be important, during periods of low-intensity precipitation and for areas characterized by very high infiltration rates. Nevertheless, if a surface is relatively impermeable, most precipitation will run off, and a high drainage density is expected. Carlston (Paper 6) demonstrates that the ability of soils and rocks to transmit water significantly influences drainage density. A weakness of Carlston's model is that, although transmissibility of bedrock can be high, the infiltration capacity of the surface soil could be low. Under these conditions if the soil cover is deep, a high drainage density could result due solely to soil characteristics. Nevertheless, the relations developed by Carlston are reasonable, and they explain drainage density variations on different rock types under similar climatic conditions.

The influence of climate on landforms has long been a problem of great geomorphic interest (Tricart and Cailleux, 1965; Wilson, 1968; Derbyshire, 1973; Garner, 1974), and Chorley, in the next selection, evaluates the influence of climate on drainage systems by considering basins on similar rocks under different climates. A major difficulty in determining the influence of climate on landforms is that as climatic boundaries are crossed geology may also change. It is difficult to find similar geologic situations under very different climatic conditions. In a later paper Chorley and Morgan (1962) demonstrated the increase of drainage density with increased intensity of precipitation. The effect of climate is significant, as any student of the Pleistocene will attest.

As Carlston shows, drainage density is related to runoff. It seems probable, therefore, that stream order is a function of discharge and that discharge could be substituted for order number in a Horton analysis (Woldenberg, 1966). Of course, hydrologic data for small-order streams are not available, but Leopold and Miller (Paper 8), as part of a broader treatment of ephemeral streams, are able to demonstrate indirectly that such an assumption is correct. They show that within semiarid drainage basins, the Horton relations pertain and that, in addition, channel width is related to stream order. The hydraulic-geometry relations of Leopold and Maddock (1953) show that stream width is closely related to mean annual discharge; hence, a relation between stream order and discharge exists. This point has been discussed by Woldenberg (1966), and the relation between stream order and discharge has been demonstrated by Rzhanitsyn (1960), Hirsch (1962) and Stall and Fok (1967). This is extremely important because if stream order is a function of discharge, then the Hortonian relations between stream frequency, length, area, and order number are not simply a result of the seemingly artificial classification or subdivision of the network into streams of different orders. Instead, the relations reflect the ratio between soil erodibility and the eroding force (runoff or discharge) exerted on the soil surface.

Editor's Comments on Papers 6 Through 9

The final selection in this section is one that attempts to deal with several variables and to view the drainage basin as a system (Chorley and Kennedy, 1971). Melton shows the interrelations that exist among landform variables and how the complex system, through negative feedback, achieves an equilibrium state and adjusts to altered energy inputs. Due to the complexity of drainage basin morphology and to the numerous variables operating to modify the drainage basin, it is understandable that there are relatively few papers that take this comprehensive approach. Melton's ambitious field program of data collection, coupled with his analysis of the interrelations of the components of a drainage basin and the variables that influence basin morphology, is a model for future geomorphic studies.

REFERENCES

Chorley, R. J. and Kennedy, B. A. 1971. *Physical geography, a systems approach.* London: Prentice-Hall Internatl.

Chorley, R. J. and Morgan, M. A. 1962. Comparison of morphometric features, Unaka Mountains, Tennessee and North Carolina and Dartmoor, England. *Geol. Soc. America Bull.* 73:17-34.

Derbyshire, E. 1973. *Climatic geomorphology.* London: Macmillian and Co., Ltd.

Garner, H. F. 1974. *The origin of landscapes.* London: Oxford Univ. Press.

Hirsch, F. 1962. Méthode de prévision des debits des cours d'eau par l'analyse morphométrique des réseaux fluviatiles. *Rev. Géomorphologie Dynam.* 13:97-106.

Kirkby, M. J. and Chorley, R. J. 1967. Throughflow, overland flow and erosion. *Internat. Assoc. Sci Hydrology Bull.* 12:5-21.

Leopold, L. B. and Maddock, T. Jr. 1953. The hydraulic geometry of stream channels and some physiographic implications. *U.S. Geol. Survey Prof. Paper 252.*

Rzhanitsyn, N. A. 1960. Morphological and hydrological regularities of the structure of the river net. Leningrad: Gidrometeoizdat. Translated by D. B. Krimgold and published by U.S. Dept. Agriculture, Agricultural Research Service (1963).

Stall, J. B. and Fok, Yu-Si. 1967. Discharges as related to stream system morphology. *Internat. Assoc. Sci. Hydrology Pub.* 75:224-235.

Tricart, J. and Cailleux, A. 1965. *Introduction a la geómorphologie climatique.* Paris: Soc. D'Edition D'Enseignment Supérieur (SEDES).

Wilson, L. 1968. Morphogenetic classification. In *Encyclopaedia of geomorphology.* New York: Reinhold Book Corp. pp. 717-729.

Woldenberg, M. J. 1966. Horton's laws justified in terms of allometric growth and steady state in open systems. *Geol. Soc. America Bull.* 77:431-434.

DRAINAGE DENSITY AND STREAMFLOW

By Charles W. Carlston

ABSTRACT

Drainage density, surface-water discharge, and ground-water movement are shown to be parts of a single physical system. A mathematical model for this system, which was developed by C. E. Jacob, can be expressed in the equation $T = WD^{-2}/8h_0$, in which T is transmissibility, W is recharge, D is drainage density, and h_0 is the height of the water table at the water table divide. The rate of ground-water discharge into streams, or base flow (Q_b), is dependent on and varies directly with transmissibility of the terrane. If W and h_0 are constant, $Q_b \propto D^{-2}$. Thirteen small, almost monolithologic basins, located in a climatic area in the eastern U.S. where recharge is nearly a constant, were found to have a relation of base-flow discharge to drainage density in the form of Q_b per mi² $= 14D^{-2}$. The field relations, therefore, agree with that predicted by the Jacob model.

If transmissibility controls the relative amount of precipitation which enters the ground as contrasted with that which flows off the surface, surface-water or overland discharge should vary inversely with transmissibility. It was found that flood runoff as measured by the mean annual flood ($Q_{2.33}$) varies with drainage density in the form of $Q_{2.33}$ per mi² $= 1.3D^2$. The close relation of mean annual flood to drainage density in 15 basins was not affected by large differences among the basins in relief, valley-side slope, stream slope, or amount and intensity of precipitation. It is concluded that drainage density is adjusted to the most efficient removal of flood runoff and that the mean annual flood intensity is due predominantly to terrane transmissibility.

INTRODUCTION

This report describes the results of an investigation of some of the relations between hydrology and geomorphology. The hydrology of a stream basin involves (1) the overland runoff from the basin, (2) the ground-water recharge rate, which is dependent upon the amount of precipitation in excess of evapotranspiration losses and upon the infiltration capacity of the soil mantle, and (3) the permeability of the bedrock, which affects the rate of yield of ground water to wells, to springs, and to the streams which drain the basin. Stream-flow characteristics were examined in terms of base flow and of height of flood peaks.

The geomorphic character of drainage basins, following techniques first suggested by Horton (1945), may be measured or described in a variety of ways. For example, Strahler (1958, p. 282-283) has listed 37 such properties or parameters. In a general way these properties may be divided into four classes: (1) length or geometric properties of the drainage net, (2) shape or area of the drainage basin, (3) relation of the drainage net to the basin area, and (4) relief aspects of the basin. It has been assumed by hydrologists and geomorphologists that certain relations must exist between runoff characteristics and topographic or geomorphic characteristics of stream basins. Much effort has been devoted to statistical correlation of these relations without a clear understanding of how runoff is related to the geomorphic and geologic character of the basins. This paper advances the theory that runoff and drainage density are genetically and predictably related to the transmissibility of the underlying rock terranes.

The purpose of this research was not to develop a means of predicting stream-flow characteristics from a landform characteristic. It was to attempt to solve a complex geohydrologic problem by defining a physical system in terms of: (1) the elements of the system, (2) how it operates, and (3) why it operates in the observed fashion. The theory resulting from this type of analysis, if valid, illustrates the basic simplicity and rationality of natural processes.

During the progress of research, much effort was devoted to statistical analysis of the data. While very good correlation coefficients (0.96 to 0.98) were obtained the procedure shed no light on the physical cause and effects of the processes. The present theory, formed by inductive and deductive reasoning, is so basically simple that the correlation can be shown adequately in simple graph form and equations can be solved directly by graphical means.

The author appreciates Dr. Luna B. Leopold's penetrating and enlightening critical comments on earlier drafts of this paper. Mr. Frederick Sower, chemist and mathematician, greatly aided the author in the development of the equations.

PREVIOUS STUDIES

Early studies of the relation of topographic character of drainage basins to basin-runoff characteristics have been described by Langbein (1947, p. 128–129). The topographic parameters considered important in these studies were: area, channel slope, stream pattern, average basin width, mean length of travel (channel length), and mean relief measured from gaging station.

In 1939 and 1940, Langbein (1947) and his coworkers, through assistance from the Works Project Administration of the Federal Works Agency, compiled a large variety of topographic measurements from 340 drainage basins in the northeastern United States. The basins ranged in area from 1.6 to 7,797 square miles. The parameters determined were drainage area, length of streams, drainage density, land slope, channel slope, area-altitude distribution, and area of water bodies. No attempt was made to correlate these topographic measurements with streamflow properties. The measurements were compiled for future studies which would determine their effects on the runoff of streams.

Such studies have been carried on in the Water Resources Division of the U.S. Geological Survey by Benson (1959), who participated in the studies reported by Langbein in 1947. As a result of graphical correlation of several topographic parameters with flood flow for 90 New England drainage basins, Benson determined that basin area and channel slope were the most effective determinants of flood flow. These parameters were then statistically computed in correlation with records of 170 gaging stations.

Benson's studies were made entirely of streams in New England where the bedrock lithology, with the exception of the Triassic rocks of the Connecticut

FIGURE 1.—Map showing location of drainage basins studied and discussed in this report. Uncircled numbers and letters correspond to those in table and on graphs.

basin and the Carboniferous rocks of the southeastern New England region, is largely a hydrologically homogeneous complex of igneous and metamorphic rocks. Glacial erosion has modified the land surface of the crystalline rocks, and glacial deposits mantle much of the area. Drainage density among the basins studied ranged from 1.1 to 2.35, as measured on 1:62,500 scale maps by Langbein.

METHODS OF STUDY

The area of study comprises the central and eastern United States south of the Wisconsin glacial border. This area was chosen because the landscape is the product of normal denudational processes; landscapes shaped or modified by glacial erosion or deposition, such as those studied by Benson (1959) were thus excluded. The area is generally homogeneous in climate. Average annual precipitation ranges from about 40 to 50 inches and average annual temperature ranges from 50° to 60° F. Four criteria were used in selecting basins having a variety of bedrock geology: (1) homogeneity of bedrock, (2) availability of 1:24,000-scale topographic maps, (3) availability of discharge records, (4) lack of appreciable flow regulation of the streams. These limitations resulted in an unavoidably small sample. The basins, their geology, and their hydrologic and geomorphic parameters are listed in the following table, and their locations are shown in figure 1.

Drainage densities were obtained by sampling randomly distributed 1-square-mile plots, whose total area is generally not less than 25 percent of the total basin area. Drainage density is a measurement of the sum of the channel lengths per unit area. It is generally expressed in terms of miles of channel per square mile. Horton (1945, pp. 283-284) and Langbein (1947, p. 133) determined drainage density by measurement of the blue streamlines on topographic maps having a scale of 1:62,500. Langbein (p. 133) pointed out that the number of small headwater streams shown on these maps would vary with the season and wetness of the year in which the survey was made, as well as with the judgment of the topographer and cartographer.

The current practice, and the method used in this study, is to show drainage lines wherever a cusp or V-notch of a contour line indicates a channel. By this method, the drainage densities of the basins, on topographic maps having a scale of 1:24,000, range from 3.0 to 9.5. The measurements are believed to be of consistent accuracy because the maps used were all 1:24,000 scale maps and the measurements were made by one operator, the author. Delineation of drainage lines was generally conservative; therefore, drainage-density values, particularly in the higher values, may be lower than those ordinarily measured.

An accurate method of distinguishing the ground-water discharge component of streamflow from the surface- or overland-flow component has not yet been developed. Estimates of ground-water discharge of streams, or base flow, are usually made by assuming that all flood peaks are overland-flow discharge and

Locality (fig. 1)	Basin		Area (sq mi)	Drainage density	Average minimum monthly flow (cfs per sq mi)	Mean annual flood (cfs per sq mi)	Mean annual precipitation (inches)
	Gaging station	Bedrock lithology					
1	Sawpit Run near Oldtown, Md.	Shale	5	7.6	0.13	62	38
2	Little Tonoloway Creek near Hancock, Md.	Shale and shale interbedded with sandstone.	16.9	7.0	.29	44	38
3	South Fork Little Barren River near Edmonton, Ky.	Limestone and cherty limestone	18.1	9.5	.26	77	49
4	Crabtree Creek near Swanton, Md.	Sandstone (60-80 percent) and some shale.	16.7	4.2	.61	32	42
5	Totopotomoy Creek near Atlee, Va.	Sand and gravel	6.0	4.7	.35	23	45
6	Sawmill Creek near Glen Burnie, Md.	Sand and clay	5.1	3.0	1.22	19	45
7	Tuscarora Creek near Martinsburg, W. Va.	Limestone	11.3	3.6	.80	14	40
8	John's Creek near Meta, Ky.	Sandstone, shale, limestone, and coal.	55.7	6.3	.27	44	44
9	Fourpole Creek near Huntington, W. Va.	Sandstone, shale, limestone, and coal.	20.9	8.0	.10	86	43
10	East Fork Deep River near High Point, N.C.	Schist, slate, and greenstone	14.7	8.0	.36	102	46
11	Piney Run near Sykesville, Md.	Granite	11.4	6.0	.84	75	45
12	Ridley Creek near Moyland, Pa.	Gabbro and schist	31.9	5.0	.83	38	47
13	Beetree Creek near Swananoa, N.C.	Gneiss	5.5	5.6	.91	44	49
A	Crab Creek near Penrose, N.C.	Granite	10.9	7.4	1.86	72	62
B	Catheys Creek near Brevard, N.C.	Gneiss	11.7	8.0	2.45	67	63

that the remaining lower flow segments of the hydrograph are base flow. Many surface-water hydrologists believe that the base flow of many streams is largely derived from bank storage, rather than from discharge from the ground water. On the other hand, there are large areas of highly permeable sand formations in the Atlantic Coastal Plain where there is never any discernible overland-flow runoff and where virtually all discharge, including flood peaks, is ground-water discharge. Because of this uncertainty in the definition and computation of base-flow or ground-water discharge, the author used as a measure of base flow average minimum monthly flow, computed by averaging 6 years of monthly minimum discharge measurements.

The mean annual flood is determined by plotting annual maximum-flood discharges on modified probability graph paper in which the discharge peaks are plotted against recurrence interval. Gumbel (1958, p. 177) has pointed out that the return or recurrence interval for the mean annual flood is 2.2328 years, the median annual flood is 2 years and the most probable annual flood is 1.582 years. The U.S. Geological Survey has adopted the mean annual flood ($Q_{2.33}$) as its standard reference value. According to Wolman and Miller (1960, p. 60) most of the work of stream erosion is done by frequent flows of moderate magnitude with a recurrence interval of 1 to 2 years. The use of the $Q_{2.33}$ flood in the present correlation study is, therefore, generally in scale with the magnitude and frequency of floods which Wolman and Miller have demonstrated to be most important in geomorphic processes.

THE JACOB WATER-TABLE MODEL

Models depicting the relation of the ground-water table to ground-water drainage have been made by Horton (1936) and Jacob (1943, 1944).

Horton (1936, p. 346) considered the shape of the water table as a parabola. He stated that the elevation of the water table (h) at any point X distance from the draining stream is:

$$h=\sqrt{h_0^2+2\left(\frac{\alpha}{K_t}\right)\left(L_o x-\frac{x^2}{2}\right)}. \quad (1)$$

In this equation, h_0 is the elevation of the draining stream, K_t is the transmission capacity of the soil, L_o is the distance from the stream to the ground-water divide, and α is the rate of accretion of recharge to the aquifer, with α being equal to or less than the infiltration capacity.

In 1943 and 1944, Jacob developed a parabolic-type equation for ground-water flow in a homogeneous aquifer of large thickness and having uniform accretion from precipitation (1943, p. 566). The model for this equation is shown in figure 2. In a simplified form the equation states:

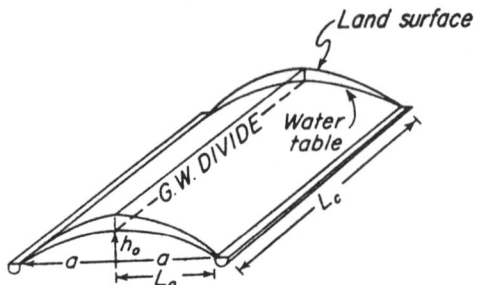

FIGURE 2.—Ground-water table base-flow model after Jacob (1943).

$$h_0=\frac{a^2W}{2T} \quad (2)$$

where h_0 is the height of the water table above the draining stream, as measured at the ground-water divide, a is the distance from the water-table divide to the stream, W is the rate of accretion to the water table (recharge), and T is the transmissibility (the volume rate of flow through a vertical strip of the aquifer of unit width under a unit hydraulic gradient). Thus, according to equation 2, the height of the water table at the water-table divide is proportional to the square of the distance of the divide from the drainage stream and to the rate of ground-water recharge. The value h_0 is also inversely proportional to transmissibility.

APPLICATION OF THE JACOB MODEL TO LANDFORM AND STREAMFLOW CHARACTERISTICS

Jacob (1944, p. 938–939) computed the transmissibility of the Magothy sand aquifer of Long Island from a water-table contour map of the island and estimated recharge as about 60 percent of the average annual precipitation. He used the following equation for his computation:

$$T=\frac{a^2W}{2h_0}. \quad (3)$$

The Jacob ground-water model can be related to a landform model by assuming that the land surface is also a parabola coincident with the water table. Then, the symbol a is equal to L_0 or length of overland flow. Thus:

$$T=\frac{L_0^2 W}{2h_0}. \quad (4)$$

Horton (1945, p. 284) pointed out that the average length of overland flow (L_0) is, in most cases, approximately half the average distance between the stream channels and is therefore approximately equal to half the reciprocal of drainage density (D), or

$$L_0 = \frac{1}{2D} \quad (5)$$

Substituting equation 5 in equation 4 the equation becomes:

$$T = \frac{W}{8D^2 h_0} \quad (6)$$

or

$$T = \frac{WD^{-2}}{8h_0}. \quad (7)$$

Jacob's water-table model as used on Long Island, simulates an aquifer having parallel boundaries at a constant head (the sea level on both sides of the island). Jacob pointed out, however, that a serious weakness of the Long Island study was the impracticability of measuring the ground-water discharge along the northern and southern shores of the island. He suggested (1944, p. 939) that the model could be tested more effectively under field conditions where the parallel shores would be replaced by streams draining the water discharged from the water table aquifer. This model is shown in figure 2. The ground-water discharge into such streams would be dependent upon the transmissibility of the aquifer, and this discharge could be measured by the gain in flow per unit length of stream.

Inasmuch as ground-water discharge into streams would vary directly with the transmissibility, ground-water discharge or base flow (Q_b) should also vary according to equation 7 or:

$$Q_b \propto \frac{WD^{-2}}{8h_0} \quad (8)$$

W and h_0 remaining constant, base flow should be related to drainage density in the form:

$$Q_b \propto D^{-2} \quad (9).$$

It may be deduced that as transmissibility decreases, the amount or rate of movement of ground water passing through the system decreases and a proportionately greater percentage of the precipitation is forced to flow directly into the streams in flow over the land surface. Streamflow, therefore, would be derived from ground-water discharge plus overland flow. Both components of discharge should vary inversely in their relative contribution to stream discharge in a regular and predictable system controlled by the transmissibility of the water-table aquifer. As T increases, ground-water discharge into streams increases and surface discharge decreases. As T decreases, there would be a corresponding decrease in base flow and increase in surface discharge.

Equation 6, which was derived from Jacob's basic equation, states that transmissibility varies inversely with drainage density squared. Thus as transmissibility increases, drainage density would decrease, and as transmissibility decreases, drainage density would increase.

This relationship may be examined from two different viewpoints. Horton (1945, p. 320) has stated that erosion will not take place on a slope until the available eroding force exceeds the resistance of the surficial materials to erosion. This eroding force increases downslope from the watershed line to the point where the eroding force becomes equal to the resistance to erosion. He named this distance the "critical distance," and the belt of land surface within the critical distance was termed the "belt of no erosion." One of the most important factors in determining the width of the belt of no erosion is the infiltration capacity of the soil. Simply stated, the greater the infiltration capacity, the less will be the amount of surface runoff. As infiltration capacity increases, the critical distance also increases because a greater slope length is required to accumulate overland flow of sufficient depth and velocity to start erosion. Thus the infiltration capacity is chiefly responsible for determining the width of the belt of no erosion and the width of the spacing of streams or channels which carry away surface runoff.

The infiltration capacity of the soil may be regarded as one part of the general capacity of a terrane to receive infiltering precipitation and to transmit it by unsaturated flow through the vadose zone above the water table and by saturated flow through the aquifer to the streams draining the ground water. This capacity is here termed "terrane transmissibility." It is recognized that this extends the strict meaning of transmissibility, which is a measure only of saturated flow.

Another way of regarding the relation of transmissibility to drainage density is to consider that as T decreases, a concurrent progressive increase in surface flow will occur. Increase in proliferation of drainage channels would provide a more efficient means of transporting the water off the land, and as the water to be so removed increases (with decreasing T) the proliferation and closeness of spacing of the drainage channels would increase. The channel spacing or geometry should operate at peak efficiency during periods of flood runoff. Since flood runoff (Q_f) of streams varies in magnitude inversely with the magnitude of their base flow runoff (Q_b), ($Q_b \propto 1/Q_f$), and according to equation 9 $Q_b \propto D^{-2}$, then flood runoff should vary with the positive second power of drainage density, or

$$Qf \propto D^2. \qquad (10)$$

Chorley and Morgan (1962) have compared the morphometry of two areas of crystalline and metamorphic rocks: the Unaka Mountains of Tennessee and North Carolina, and Dartmoor, England. Morphometric characteristics which were measured included number of stream segments, mean stream lengths, mean basin areas, mean stream-channel slopes, mean relief of fourth-order basins, mean basin slopes, and mean drainage density. Relief and drainage density were the only parameters which showed significant differences between the two areas. The difference in drainage density between the two areas (Dartmoor, $D=3.4$; Unakas, $D=11.2$) was ascribed to differences in runoff intensity caused by more intense rainfall and greater mean basin slopes. They concluded that the channel system is geared to conditions of maximum runoff. These conclusions are only partially in agreement both with the theoretical and observed relations described in the present paper.

THE RELATION OF BASE FLOW TO RECHARGE AND DRAINAGE DENSITY

Figure 3A shows the relation of drainage density to base flow per square mile for 13 basins previously described. As previously stated, average annual precipitation in the 13 basin areas ranges from about 40 to 50 inches and average temperatures range from 50° to 60° F. Recharge (W) is dependent upon the amount of precipitation less the amount of evapotranspiration losses, which is dependent largely on temperature. The variations in rainfall and temperature among the 13 basins are within a sufficiently small range that recharge can be considered to be approximately a constant for the 13 basins. The relation of base flow per unit area to drainage density can be delineated by a regression line which has a slope and intercept such that:

$$Q_b \text{ per mi}^2 = 14 D^{-2}$$

Equation 8 states that base-flow rate is also directly proportional to recharge. Two basins at the southern tip of the Blue Ridge province were examined to determine qualitatively the effect of significantly higher precipitation or recharge on base flow. In this area annual precipitation is more than 60 inches. The two basins are those of Crab Creek near Penrose, N.C. (A), and Catheys Creek near Brevard, N.C. (B). (See figs. 1, 3.) Their position on the diagram in figure 3A shows a much higher rate of base flow which may be due to the higher recharge rate. The gaging stations for both basins are, however, located in alluvium-filled

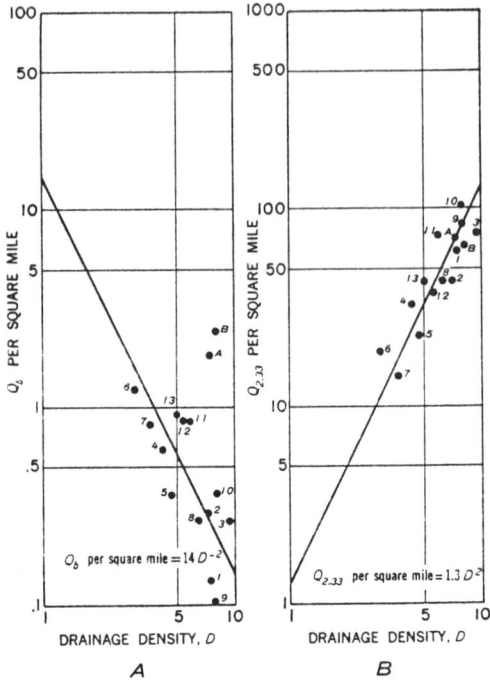

FIGURE 3.—The relation of drainage density to base flow and floods. A, Base flow (Q_b). B, Mean annual flood ($Q_{2.33}$).

valleys; therefore, the higher base flow discharges may be due in part to higher transmissibility of the alluvium.

The relation of drainage density to the 10-day base flow recession coefficient (r_{10}) was also investigated. It was found that this relation could be expressed quantitatively by the equation: Log $r_{10} = -0.008 D^2$.

THE RELATION OF THE MEAN ANNUAL FLOOD TO DRAINAGE DENSITY

In the preceding discussion of the theoretical relation of drainage density to flood runoff it was concluded that flood runoff should vary directly with the second power of drainage density. Flood runoff for the basins studied was computed in terms of the mean annual flood per square mile ($Q_{2.33}$ per mi²). A regression line having a slope of D^2 was plotted on this graph and the best fit gave an intercept with unit drainage density of about 1.3. This gives the equation:

$$Q_{2.33} \text{ per mi}^2 = 1.3 D^2.$$

The departures from this mean trend line were compared graphically with average annual precipitation and average maximum rainfall intensities as given by the U.S. Weather Bureau. There was no apparent correlation. Graphical correlation was also made between mean annual flood and total relief, local relief, the ruggedness number (the product of relief and drainage density), stream slope, and valley-side slope. If significant correlation of flood peaks with these parameters exists, it is not apparent in the graphical plots.

The apparent lack of correlation between rainfall amount or intensity and flood intensity and between relief and flood intensity may be briefly illustrated by reference to points plotted in figure 3B. Basins A and B are located in the region of highest mean annual rainfall and rainfall intensity in the eastern U.S. In addition, the relief (1,500 ft) in these basins places them among those with the highest relief in the total sample, yet they plot on and below the average for the basins. Basin 10 has the highest mean annual flood (102 cfs per mi^2), but its relief is only 200 feet. Basin 13 has a relief of 2,600 feet and a high average annual precipitation (49 in.), but departure of its mean flood intensity from the average is not significantly high.

Hydrologists have found that the delay time (and hence the attenuation) of flood peaks is composed of two parts; the inlet or overland-flow time, and the channel-transit time. The present study deals with inlet times of monolithologic terranes. The writer's analysis of flood runoff per unit area at gaging stations in the Appalachian Plateaus of eastern Kentucky (a basically monolithologic terrane) suggests that inlet time is dominant over channel transit time up to about 75 to 100 square miles in drainage area; however, because the channel transit time increases continuously, for larger basins it tends to become the dominant component of the flood-peak lag.

CONCLUSIONS

This paper has presented evidence that drainage density, surface-water runoff, and the movement of ground water are parts of a single hydrologic system controlled by the transmissibility of the bedrock and its overlying soil mantle. A mathematical model of such a system, constructed by Jacob (1943; 1944), has been adapted to show that transmissibility (T) is related to ground-water recharge (W), to drainage density (D) and to the height of the water table at the water table divide (h_0). The equation for this relation is:

$$T = \frac{WD^{-2}}{8h_0}.$$

If W and h_0 are constant, the equation may be simplified to $T = KD^{-2}$. Inasmuch as the rate of base flow (Q_b) or ground-water discharge into streams varies with and is controlled by transmissibility, $Q_b \propto D^{-2}$. A total of 13 small, basically monolithologic stream basins were selected in the eastern United States where rainfall and temperature are such that recharge can be considered to be a constant. It was found that the relation of base flow of the 13 streams to drainage density can be expressed by the equation Q_b per mi$^2 = 14D^{-2}$. The observed relation therefore is the same as that predicted by the Jacob model. Two stream basins in the southern Blue Ridge province, where rainfall is much higher than that of the 13 basins, have much higher base flows than comparable streams in the lower rainfall region.

Flood runoff as measured by mean annual flood ($Q_{2.33}$) was found to be closely related to drainage density; the equation may be expressed as $Q_{2.33}$ per mi$^2 = 1.3D^2$. It is concluded that the terrane transmissibility controls the amount of precipitation which passes through the underground system. The rejected or surface-water component increases with decreasing transmissibility. As surface-water runoff increases, an increase in the proliferation of stream channels is required for efficient removal of the runoff. The close relation of drainage density to mean annual flood per unit area indicates that the drainage network is adjusted to the mean flood runoff. Among the 15 basins in which flood runoff was correlated with drainage density, there are large and significant differences in relief, in valley-side and stream slopes, and in amounts and intensities of precipitation. These factors, however, have no discernible effect on the relation of the magnitude of the floods to drainage density. Transmissibility of the terrane appears to be the dominant factor in controlling the scale of drainage density and the magnitude of the mean annual flood for basins up to 75 to 100 square miles in area.

The research reported here should be of interest to geomorphologists in that it provides a quantitative physical model for the origin of one of the most important elements of landform characteristics, drainage density. In addition, it appears that surface-water hydrologists and ground-water hydrologists have far more in common in their hydrologic studies than is generally realized. Transmissibility or permeability is the most important aquifer characteristic in ground-water studies. The results reported in this paper indicate that the transmissibility or permeability of terranes drained by streams is also important in the study of surface-water hydrology. It is hoped that this study provides a theoretical physical basis for interdisciplinary hydrologic studies.

REFERENCES

Benson, M. A., 1959, Channel slope factor in flood frequency analysis: Am. Soc. Civil Eng. Proc., Jour. Hydraulics Div., v. 85, pt. 1, p. 1-9.

Chorley, R. J., and Morgan, M. A., 1962, Comparison of morphometric features, Unaka Mountains, Tennessee and North Carolina, and Dartmoor, England: Geol. Soc. Am. Bull., v. 73, p. 17-34.

Gumbel, E. J., 1958, Statistics of extremes: New York, Columbia Univ. Press, 347 p.

Horton, R. E., 1936, Maximum ground-water levels: Am. Geophys. Union Trans., Pt. 2, p. 344-357.

———— 1945, Erosional development of streams and their drainage basins; hydrophysical approach to quantitative morphology: Geol. Soc. America Bull., p. 275-370.

Jacob, C. E., 1943, Correlation of ground-water levels and precipitation on Long Island, N.Y: Pt. 1, Theory, p. 564-573. 1944, Pt. 2, Correlation of data: Am. Geophys. Union Trans., p. 928-939.

Langbein, W. B., 1947, Topographic characteristics of drainage basins: U.S. Geol. Survey Water-Supply Paper 968-C, p. 125-157.

Strahler, A. N., 1958, Dimensional analysis applied to fluvially eroded landforms: Geol. Soc. Am. Bull., v. 69, p. 279-300.

Wolman, M. G., and Miller, J. P., 1960, Magnitude and frequency of forces in geomorphic processes: Jour. Geology, v. 68, p. 54-74.

○

CLIMATE AND MORPHOMETRY[1]

RICHARD J. CHORLEY
Brown University

ABSTRACT

An analysis of the morphometry of three areas of similar gross lithology, structural effect, and stage of dissection shows that there is a significant difference between each of the equivalent landscape unit forms. A climate/vegetation index (I_c) was obtained for each region, employing the mean annual rainfall, the mean monthly maximum precipitation in 24 hours, and Thornthwaite's precipitation effectiveness index. It is found that the climate/vegetation index bears a remarkably consistent relationship to the mean logarithms of stream lengths, basin areas, and drainage densities.

INTRODUCTION

Theoretically, the local morphometry of fluvially dissected landscapes may be considered as a function of the following interacting variables:

1. Structure
   a) Rock type, in terms of resistance to erosion and permeability
   b) Attitude of the beds
   c) Soil and mantle characteristics—size and shape of loose load, resistance to erosion, angle of internal friction, cohesion, and infiltration capacity
2. Process
   a) Amount of precipitation
   b) Rainfall intensity
   c) Vegetational cover, being a function of precipitation, temperature, evaporation, soil, and slope conditions, and having quantitative significance in regard to the degree to which it inhibits erosion
3. Stage or time

In order that the relative effects of these variables in the production of landscape may be studied quantitatively, it is necessary, first, that they be themselves susceptible to quantitative expression and, second, that it is possible to isolate one of these variables in order to determine whether it bears a consistent relationship to empirically obtained, quantitative measures of landscape morphometry. The design of this paper, therefore, is to determine with varying conditions of process, whether the theoretically assumed relationship between process and morphometry holds empirically true under reasonably uniform conditions of structure and stage.

STRUCTURE

Three areas were selected for study, all of which have been subjected to complete fluvial dissection and all of which are completely underlain by sedimentary sequences, the predominant rock of which is sandstone.

The Exmoor region of Somerset and Devon counties, England, is an elevated tract about 30 by 12 miles in extent, rising from sea level to a maximum of 1,706 feet (Balchin, 1952), the drainage basins studied having a mean relief of 1,067 feet. It is underlain by some 20,000 feet of Devonian rocks, variously dipping to the south at angles averaging about 30° in the north (Dewey, 1935). The lithology is predominantly arenaceous, although argillites are common (see Diagram A).

It is readily apparent that neither the differences in lithology nor the dip of the beds has led to significant, mappable, morphometric variations and that the geometrical features of the landscape are broadly those which are associated with structurally and lithologically homogeneous uplands. Drainage density, drainage pattern, and relief bear little relation to major formational variations, and the main geologic contacts

[1] Manuscript received January 19, 1956.

are indistinguishable topographically. The only gross structural influence exhibited within the region is that the shapes of the three basins paralleling the east-west strike are more elongate than the rest, having a mean value of k (Chorley, Malm, and Pogorzelski, 1957) of 3.17, compared with a mean of 1.94 for the remaining ten basins.

1885; Moore, 1944; Ebright and Ingham, 1951) (see Diagram B).

The low regional dips are obscured by anticlinal structures, four of which trend northeast to southwest across the area (Wharton-Cameron, New Bergen, Rattlesnake–Kettle Creek, and Driftwood axes) and one crossing Cameron County from northwest to

DIAGRAM A

DEVONIAN	Upper	Lower and Middle Pilton beds	2,000 feet	Marine argillaceous and calcareous sandstones
		Baggy and Marwood beds	1,200 feet	Littoral shales and micaceous sandstones
		Pickwell Down beds	3,000 feet	Brown, red, and purple sandstones
	Middle	Morte slates	1,500 feet	Marine shales, becoming arenaceous at the top
		Ilfracombe beds	3,990 feet	Marine shales and sandstones; some limestones
		Hangman grits	3,600 feet	Uniform, fine-grained, red sandstones
		Lynton beds	2,000 feet	Marine and littoral bluish-gray shales
	Lower	Foreland grits	1,500 feet	Continental quartzose coarse sandstones, shales interbedded

DIAGRAM B

Pennsylvanian	Allegheny	50 feet	Coal measures and shales
	Pottsville	400 feet	Coarse sandstone, locally conglomeratic, with thin, interbedded coals, fireclays, and limestones
Mississippian	Maunch Chunk	50 feet	Discontinuous shales which, in places, may represent argillaceous facies of the lower Pottsville
	Pocono	700 feet	The upper 400 feet of massive, steel-gray sandstone are capped by thin Mississippian limestone; the basal 300 feet are similar to, and gradational with, the Catskill formation
Devonian	Catskill	500 feet	Red sandstone
	Chemung	350 feet	Dark purple and red shales and sandstones

A second group of drainage basins was selected for study in Cameron and Potter counties, north-central Pennsylvania, where some 1,500 feet of regionally horizontal Devonian and Carboniferous sedimentary rocks have been maturely dissected by a well-integrated stream system. The lithology here is even more predominantly arenaceous than that of the Exmoor region, and the following column is representative of Cameron County (Sherwood, 1880; Sheafer,

southeast (Kinzua-Emporium-Cross axis). Local dips are away from these axes and average less than 3°, although restricted dips of up to 7° occur on the north flank of the New Bergen anticline.

The third area is located in Winston, Cullman, and Lawrence counties, Alabama, where 350–500 feet of the Pottsville formation (Lower Pennsylvanian), dipping south-southwest at 2°–5°, have been dissected by the headwaters of Sipsey Fork, giving a local relief of 370–492 feet (Adams, Butts, Stephenson, and Cooke, 1926; Semmes, 1929). In this locality the Pottsville formation consists of 0–160 feet of basal, white, siliceous sandstone and conglomerate, overlain by thick to massively bedded arkosic sandstone, alternating with thinly bedded black shales.

STAGE

All three localities have been completely dissected by streams, and remains of any unconsumed, high-level, previous surfaces are not extensively evident, except in the form of some accordance of hilltops. The mean hypsometric integrals (Strahler, 1952) for the drainage basins selected in each of the areas are 55.87, 59.21, and 57.46 per cent, respectively, which would enable the basins to be classified as in the equilibrium (mature) stage of Strahler (1952). An analysis of variance (Walker and Lev, 1953) shows that there is no significant difference between the mean hyposometric integrals for the three regions, taking a critical level of significance of $F_{0.95}$. As indicated by Glock (1932), the amount of relief is a poor indicator of stage, but if one wishes to consider stage on the basis of rate of dissection rather than on the amount of bedrock removed from a basin, some measure involving relief might be perhaps more appropriate (Schumm, 1955).

MORPHOMETRY

In Exmoor, Pennsylvania, and Alabama, respectively, 13, 7, and 7 third-, fourth-, and fifth-order drainage basins were selected for study. The data for each basin were obtained from Ordnance Survey 1:25,000 (nos. 21/43, 44, 53, 54, 63, 64, 73, 74, 83, 84, 93, 94), U.S. Geological Survey 1:24,000 (First Fork, Cameron, Wharton, and Emporium quadrangles), and U.S. Geological Survey 1:62,500 (Danville and Mount Hope quadrangles) maps, the latter being photographically enlarged to 1:31,726 in order to facilitate delimitation and measurement of the landscape features. Although the accuracy of these maps may be open to question, there is no reason to suppose that the individual degrees of accuracy are of such a different order of magnitude as to prohibit the comparison of results, particularly as the drainage densities of all these regions may be classed as coarse or medium-coarse (Smith, 1950). It should be noted that, in the locality having the most suspect maps (Alabama), mean stream lengths and basin areas are the smallest and drainage density is the highest of the three regions (figs. 1–3) and that more accurate maps would tend to increase, rather than decrease, this difference with respect to the other two localities. Strahler (1950, p. 690–693) has defended the use of recent maps on a scale of 1:25,000 and 1:24,000 for morphometric work in regions of coarse texture of erosion.

For each of the twenty-seven basins, lengths of streams of various orders (L_0), areas of stream basins of various orders (A_0), drainage densities (D_d), relief (H), and hypsometric integrals were measured, according to the methods outlined by Horton (1945), as modified by Strahler (1953) and Miller (1953). The lengths of first-order streams were found to be generally too small to be measured individually with satisfaction on maps of this scale, particularly those of Alabama, and often it was found impossible to delimit areas of first-order basins with any accuracy. Consequently, they have been omitted from the following discussion.

The first step was to determine whether or not similar morphometric features are of significantly different magnitude in the three regions. The analysis of variance is based on the assumption that the individual samples are representative of normally distributed populations having equal variances, but it has been demonstrated (Miller, 1953) that the frequency distributions involving the actual values of stream length and basin area present a characteristic right-skewness. If the logarithms of the individual variates are employed, however, the resulting frequency distributions appear more nearly normal. In accordance with this observation, the frequency distributions of the logarithms of one hundred times the individual stream lengths and basin areas (to avoid negative logarithms) were constructed as histograms (figs. 1 and 2), together with

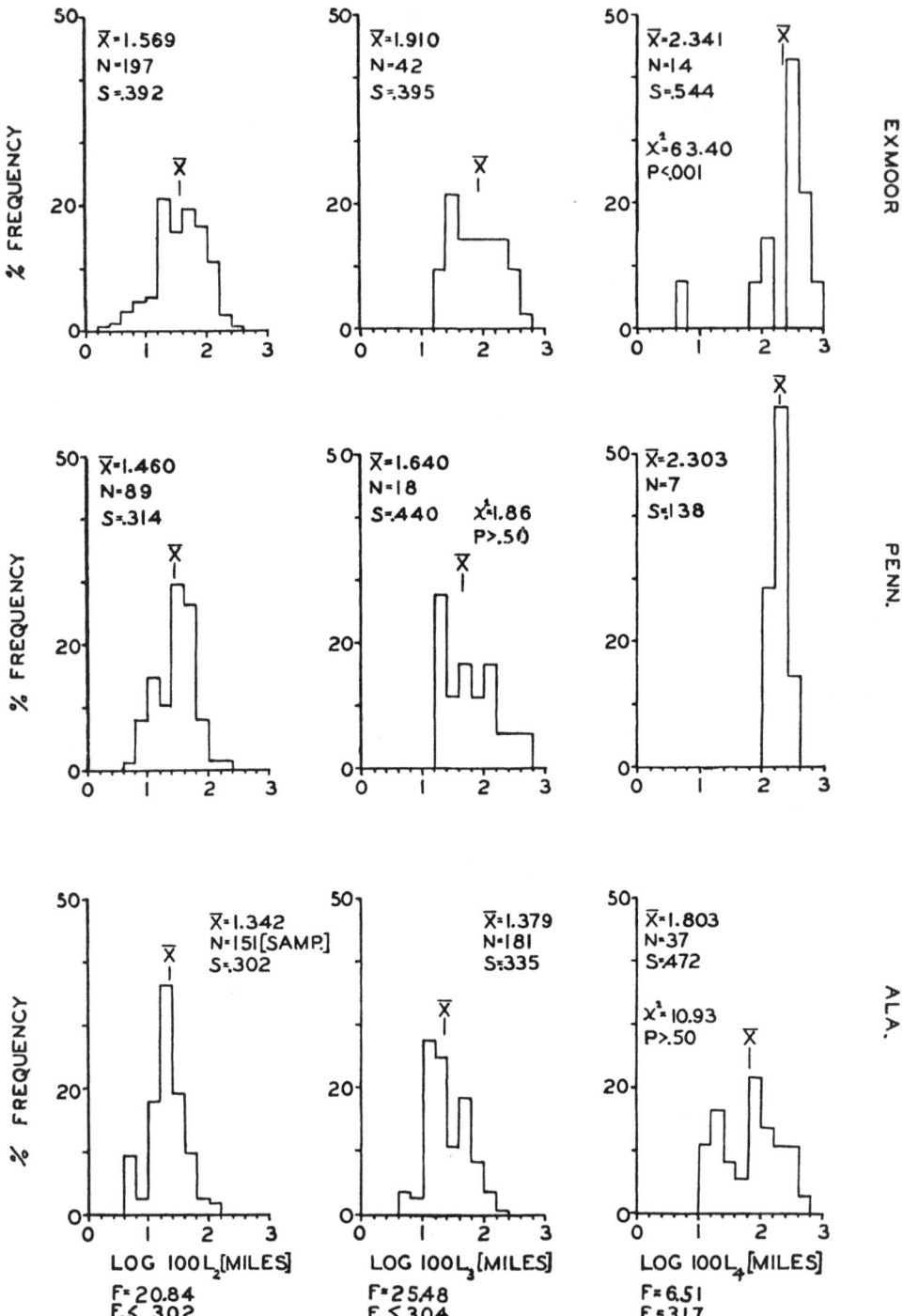

Fig. 1.—Histograms of the logarithmic values of stream lengths

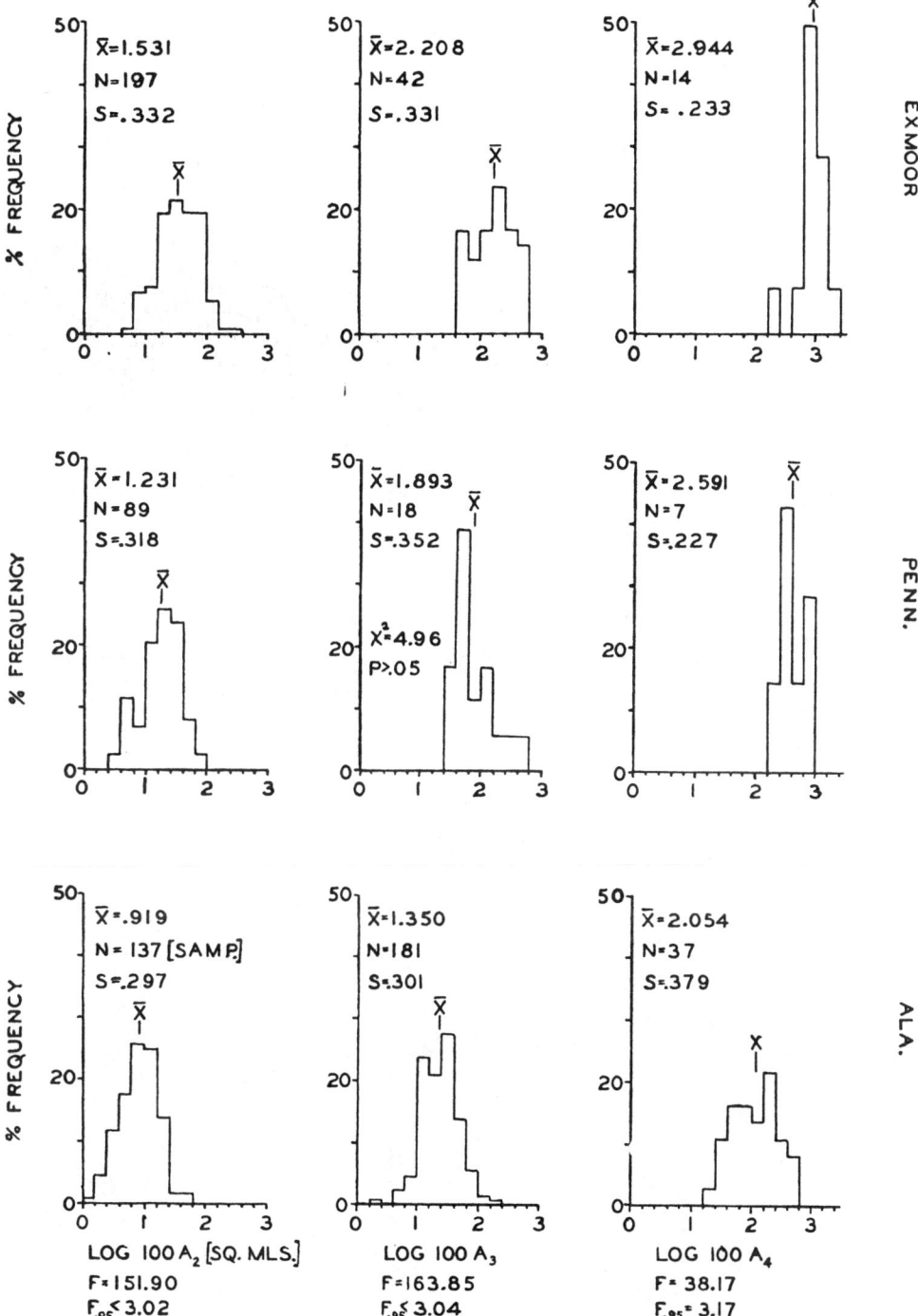

Fig. 2.—Histograms of the logarithmic values of basin areas

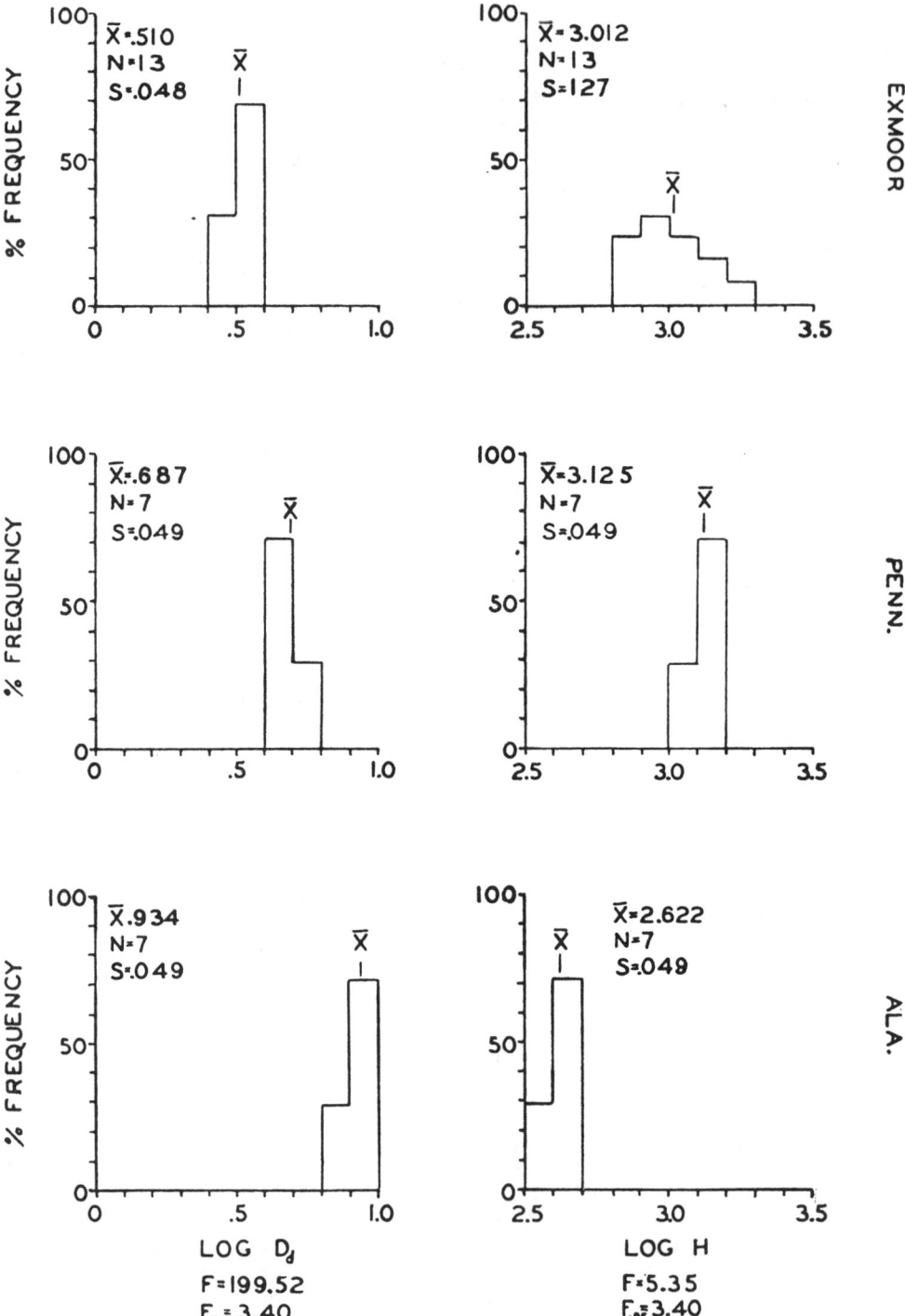

Fig. 3.—Histograms of the logarithmic values of drainage densities and reliefs

those employing the logarithm of the drainage density and relief for each of the major drainage basins (fig. 3). This resulted in more symmetrical distribution of the sample variates and in a close correspondence of the standard deviations between the three regions for equivalent morphometric features (e.g., second-order stream lengths). In order to test the normality of the distributions represented by the samples, chi-square tests were made for each of the four samples which still appeared most skewed (lengths of third-order streams for Pennsylvania, lengths of fourth-order streams for Exmoor and Alabama, and areas of third-order basins for Pennsylvania), using a critical level of significance (P) of 0.05. The results of these tests are given appropriately in figures 1 and 2, indicating that only the fourth-order stream lengths for Exmoor are non-representative of a normal distribution, probably because one variate represents a relatively large percentage of a small total sample.

Arbitrarily choosing a critical level of significance of $F_{0.95}$, analyses of variance (Walker and Lev, 1953) were made to compare the mean logarithmic values of lengths and areas of second-, third-, and fourth-order streams and basins in each of the three regions, together with the logarithms of drainage densities and reliefs, in an attempt to determine whether they present significantly different characteristics. The results of these tests are given in figures 1, 2, and 3, indicating that the means of the logarithms of equivalent morphometric measures are significantly different in each instance.

PROCESS

Process may be considered as a function of climate, acting mainly through the hydrological considerations of amount and intensity of precipitation and through the agency of vegetational cover. All these three factors are capable of quantitative expression and vary between the limited localities selected for study, for each of which meteorological stations were selected as representative (Ilfracombe and Plymouth, Devon; Emporium, Pennsylvania; and St. Bernard, Alabama).

Amount of precipitation may be conveniently indexed by the mean annual precipitation (British Rainfall, 1939; U.S. Weather Bureau, 1939) and average intensity of precipitation by the mean monthly maximum precipitation in 24 hours (Bilham, 1938; U.S. Weather Bureau, 1952). These data are given in table 1, together with mean monthly temperatures.

As an index of vegetative growth, Thornthwaite's (1931) precipitation effectiveness

TABLE 1

CLIMATIC DATA

	Mean Monthly Precipitation (Inches)			Maximum 24-Hour Precipitation (Inches)			Mean Monthly Temperatures (°F.)		
	Ilfracombe	Emporium	St. Bernard	Plymouth	Emporium	St. Bernard	Ilfracombe	Emporium	St. Bernard
Years of record	30+	20	23	30+	61	42	30+	20	23
January	3.17	3.23	5.53	1.5	1.66	4.47	43.3	26.2	43.0
February	2.65	2.98	4.81	1.4	2.06	4.05	43.3	24.8	45.9
March	2.77	3.98	6.00	1.5	2.26	3.80	44.4	35.5	52.4
April	2.01	3.13	5.61	1.1	2.56	5.30	48.2	47.3	60.8
May	1.98	4.65	5.62	1.6	5.85	4.25	54.0	59.2	67.6
June	2.08	4.47	4.64	2.2	2.38	4.49	58.7	67.2	75.4
July	2.45	4.92	4.97	1.8	7.67	5.51	60.9	70.7	78.0
August	3.46	3.70	4.35	1.9	2.51	5.84	61.3	68.2	77.0
September	2.59	3.38	2.93	1.5	2.41	3.67	58.7	62.2	73.4
October	4.40	3.10	3.48	2.2	2.43	6.15	53.3	50.1	61.4
November	3.76	2.88	3.95	2.2	1.80	5.22	48.5	39.4	50.6
December	4.65	3.53	5.74	1.7	1.81	5.02	45.3	29.9	43.7
Mean annual	35.97	43.95	57.63	....	....	....	....	....	....
Mean	....	....	....	1.72	2.95	4.81	51.3	48.4	60.8

index (P-E index) (I) has been employed, where

$$I = \sum_{n=1}^{12} 115 \left(\frac{P}{T-10}\right)_n^{10/9}.$$

Here P and T are the mean monthly precipitation and temperature figures, respectively. That this index may be used in this connection was inferred by Thornthwaite (1931, p. 641) when he showed the close relationship which it bears to Weaver's (1924) plant growth index, which is a quantitative expression of the dry weight of vegetation per unit area.

Using the mean monthly precipitations and temperatures given in table 1 for the Exmoor, Pennsylvania, and Alabama localities, P-E indexes of 78, 126, and 111, respectively, were obtained.

TABLE 2

Locality	I_c	$1/I_c$
Exmoor	1.2607	0.7932
Pennsylvania	0.9718	1.0290
Alabama	0.4004	2.4975

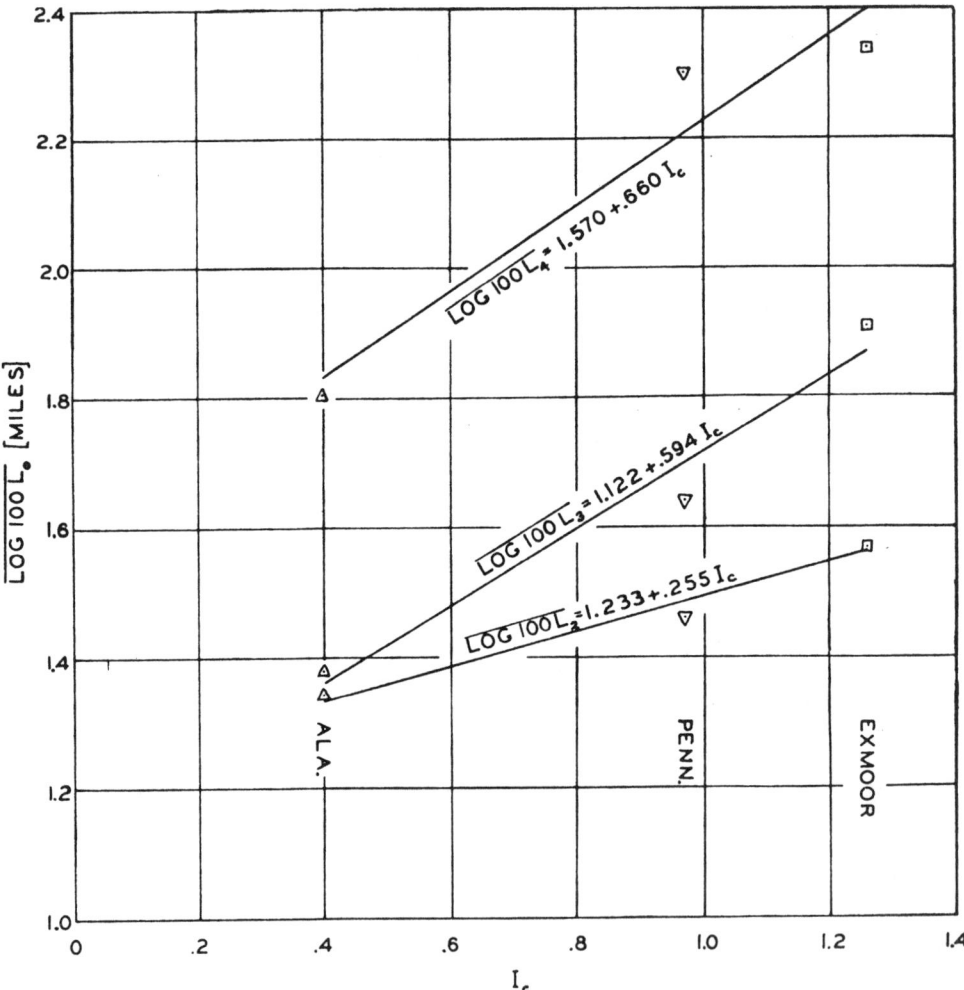

Fig. 4.—Plots of the arithmetic mean values of the logarithms of stream length versus the climate/vegetation index.

THE CLIMATE/VEGETATION INDEX

Assuming, then, a reasonable constancy of lithology and stage between the three regions, one might expect that quantitative variations in process, expressed in terms of precipitation characteristics and vegetative cover, should theoretically find sympathetic and consistent variations in the corresponding morphometric features between the localities.

Theoretically, linear and areal aspects of landscape morphometry are directly related in magnitude to the amount of vegetative cover and inversely to the amount (P) and intensity (Q_s) of precipitation, whereas drainage density is directly related to the amount and intensity of precipitation and inversely to the amount of vegetative cover. An index of climate and vegetation (I_c) can therefore be derived, in which

$$I_c = \frac{\text{Vegetation amount}}{\text{Precipitation} \times \text{precipitation intensity}}$$

$$= \frac{I}{P \times Q_s}.$$

Thus mean logarithms of lengths of streams and areas of stream basins of various orders should, theoretically, be directly related to I_c, whereas mean logarithms of drainage density should be inversely related to I_c.

With the P-E indexes and the values of P and Q_s (table 1), I_c and $1/I_c$ were calculated for each of the three localities (table 2).

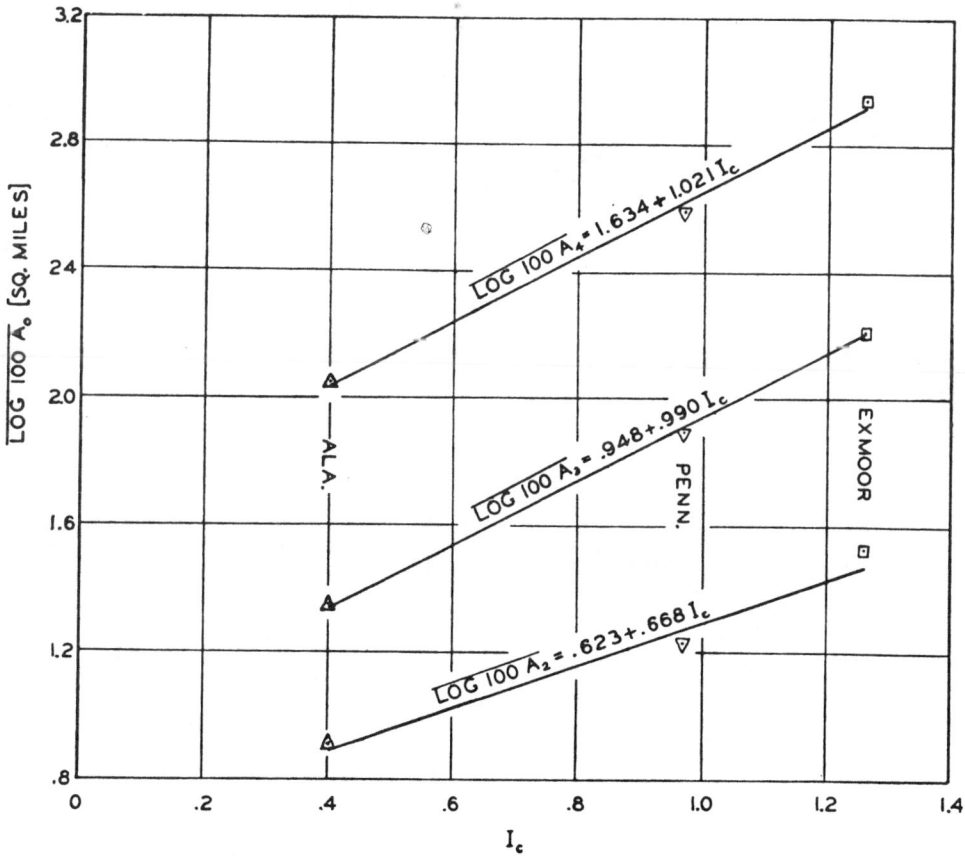

Fig. 5.—Plots of the arithmetic mean values of the logarithms of basin area versus the climate/vegetation index.

The plots of $\overline{\log 100\,L_0}$ and $\overline{\log 100\,A_0}$ versus $\bar{I}_c$ (figs. 4 and 5) and of $\overline{\log D_d}$ versus $1/\bar{I}_c$ (fig. 6) yield remarkably consistent results, which support, in every instance, the assumed theoretical relationships. It is of note that the respective increases (drainage density) and decreases (stream lengths and basin areas) of the morphometric measures which, as has been suggested, might be expected to appear with the use of more accurate maps of Alabama would, in every instance except one, tend to increase the degree of correlation.

More striking than the general constancy between the theoretical and quantified relationships is the fact that analyses of covariance (Walker and Lev, 1953) show that, at the $F_{0.95}$ level of significance, there is no significant difference between the regression coefficients in figure 4 and a coefficient of 3.18 (i.e., $\log^{-1} 0.503$) ($F = 1.08 < F_{0.95}$) and no significant difference between those in figure 5 and a coefficient of 7.94 (i.e., $\log^{-1} 0.900$) ($F = 4.88 < F_{0.95}$). It is therefore possible to derive the following general relationships for these three regions:

$$\overline{\log 100\,L_0} = K_L + 0.503\,\bar{I}_c,$$

$$\overline{\log 100\,A_0} = K_A + 0.900\,\bar{I}_c,$$

where K_L and K_A are constants, depending on the stream order.

CONCLUSION

Because of the controversial nature of several basic assumptions, the questionable accuracy of two of the maps used, the arbitrary selection of the precipitation factors, and the generalized nature of the vegetative index employed, the value of the foregoing work in absolute terms is obviously limited. It is for precisely the same reasons, however, that the consistency of the results obtained and the harmony between theoretical considerations and empirical measurements are all the more noteworthy. This reasoning assumes, of course, that the means about

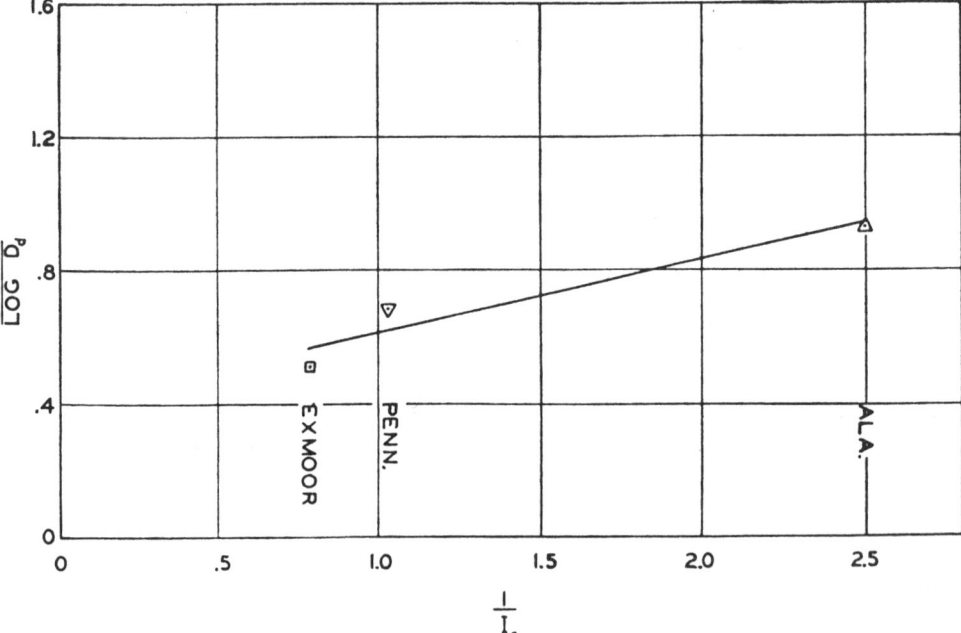

Fig. 6.—Plots of the arithmetic mean values of the logarithms of drainage density versus the reciprocal of the climate/vegetation index.

which the climates of the three areas have been deviating bear a relationship to one another not significantly different from that relationship expressed by the modern climatic data employed here. What this study does provide, perhaps, is stimulus for further work along the same lines of reasoning, together with some indication that the quantitative aspects of landscape morphometry may have such an ultimately rational basis that they may soon be placed within the realm of prediction.

ACKNOWLEDGMENTS.—This investigation formed part of a quantitative study of erosional landforms sponsored by the Geography Branch of the Office of Naval Research as Project no. NR 389-042, Technical Report no. 14, under Contract N6 ONR 271, Task Order 30. The writer wishes to thank Professor A. N. Strahler, of Columbia University, project director, for his help and advice.

REFERENCES CITED

ADAMS, G. F., BUTTS, C., STEPHENSON, J. W., and COOKE, W., 1926, Geology of Alabama: Ala. Geol. Survey, Special Rept. 14, p. 25–27 and 40–230.

BALCHIN, W. G. V., 1952, The erosion surfaces of Exmoor and adjacent areas: Geog. Jour., v. 118, p. 453–476.

BILHAM, E. G., 1938, The climate of the British Isles: New York, Macmillan Co.

British rainfall: London, H.M. Stationery Office, 1939.

CHORLEY, R. J., MALM, D., and POGORZELSKI, H., 1957, A new standard for estimating drainage basin shape: Am. Jour. Sci., in press.

DEWEY, H., 1948, South-west England: British Regional Geology, 2d ed., London, H.M. Stationery Office.

EBRIGHT, J. R., and INGHAM, A. I., 1951, Geology of the Leidy gas field and adjacent areas, Clinton County: Pa. Geol. Survey, 4th ser., Bull. M34.

GLOCK, W. S., 1932, Available relief as a factor of control in the profile of a land form: Jour. Geology, v. 40, p. 74–83.

HORTON, R. E., 1945, Erosional development of streams and their drinage basins: Geol. Soc. America Bull., v. 56, p. 275–370.

MILLER, V. C., 1953, A quantitative geomorphic study of drainage basin characteristics in the Clinch Mountain area, Virginia and Tennessee: Tech. Rept. no. 3, Dept. of Geology, Columbia University, New York.

MOORE, R. C., 1944, Correlation of Pennsylvanian formations of North America: Geol. Soc. America Bull., v. 55, p. 657–706.

SCHUMM, S., 1955, The relation of drainage basin relief to sediment loss: Pub. no. 36 de l'Association Internationale d'Hydrologie (Assemblée générale de Rome, v. 1), extrait.

SEMMES, D. R., 1929, Oil and gas in Alabama: Ala. Geol. Survey, Special Rept. 15, p. 34–68, 76, and 126–138.

SHEAFER, A. W., 1885, The township geology of Cameron County, Pa.: Pa. 2d. Geol. Survey, Rept. R2.

SHERWOOD, A., 1880, The geology of Potter County, Pa.: Pa. 2d Geol. Survey, Rept. G3.

SMITH, K. G., 1950, Standards for grading texture of erosional topography: Am. Jour. Sci., v. 248, p. 655–668.

STRAHLER, A.N., 1950, Equilibrium theory of erosional slopes approached by frequency distribution analysis: Am. Jour. Sci., v. 248, p. 673–696 and 800–814.

——— 1952, Hypsometric (area-altitude) analysis of erosional topography: Geol. Soc. America Bull., v. 63, p. 1117–1142.

——— 1953, Revisions of Horton's quantitative factors in erosional terrain: paper read before Hydrology Section of American Geophysical Union, Washington, D.C., May, 1953.

THORNTHWAITE, C. W., 1931, The climates of North America according to a new classification: Geog. Rev., v. 21, p. 633–655.

U.S. Dept. of Commerce, Weather Bureau, 1939, Climatological data, Pennsylvania and Alabama sections.

——— 1952, Maximum 24 hour precipitation in the U.S.: Tech. Paper no. 16.

WALKER, H. M., and LEV, J., 1953, Statistical inference: New York, Henry Holt & Co.

WEAVER, J. E., 1924, Plant production as a measure of environment: Jour. Ecology, v. 12, p. 205–237.

EPHEMERAL STREAMS—HYDRAULIC FACTORS AND THEIR RELATION TO THE DRAINAGE NET

Luna B. Leopold
John P. Miller

[*Editor's Note:* The introductory material and a discussion of the hydraulic characteristics of ephemeral streams have been deleted.]

INTERRELATION OF DRAINAGE NET AND HYDRAULIC FACTORS

RELATION OF STREAM ORDER TO STREAM NUMBER, STREAM LENGTH, AND DRAINAGE AREA

The quantitative description of drainage nets developed by Horton (1945) related stream order to the number, average length, and average slope of streams in a drainage basin. Our purpose in this section is to show how this useful tool may be extended to include the hydraulic as well as drainage-net characteristics.

All the data required for the Horton type of drainage-net description can be obtained from maps. As maps of several different scales were required for our own analysis, some explanation of the procedure is in order.

Figure 12 presents planimetric maps of a sample area near the city of Santa Fe. The map on the right, which shows the drainage net in a typical watershed about 9 miles long by 2 miles wide, was compiled from planimetric maps made by the Soil Conservation Service from aerial photographs at an original scale of 2 inches to the mile. The left map shows in more detail the drainage net in one small tributary which for purposes of this report we will refer to as Arroyo Caliente. This map was made by pace and compass after a planetable traverse had been run for control. Each tributary rill was paced out to its farthest upstream extension in order that the map would include all recognizable channels.

The orders of various channels in the basin of Arroyo Caliente are indicated by numbers appearing in the upper part of the left map of figure 12. A small unbranched tributary is labeled "order 1," and the stream receiving that tributary is labeled "order 2." All streams of orders 3, 4, and 5 are labeled with appropriate numbers near their respective mouths.

On the right map of figure 12, the little basin called Arroyo Caliente is one of the minor tributaries which even on this small-scale map appears to be unbranched like the tributary just west of it. If only this map were available, one would conclude that Arroyo Caliente is a first-order stream. This points up an important qualification to the Horton scheme of stream-order classification; namely, that the definition of a first-order stream depends on the scale of the map used. The first-order stream, by definition, should be the smallest unbranched channel on the ground. The designation of which stream is master and which is tributary is somewhat arbitrary, but we have followed the guide suggested by Horton (1945, p. 281).

The largest drainage basin which is included in the present analysis is that of the Rio Galisteo (fig. 1). At its mouth this basin drains about 670 square miles. Such an area contains a very large number of small tributaries. It was desired to estimate the number of tributaries of various sizes, their lengths, and other characteristics. The task of counting and measuring each individually would be inordinately great but approximate answers could be obtained by a sampling process. Arroyo Caliente is one of the samples used.

The detailed map of Arroyo Caliente was used to determine the number, the average length, and average drainage area of each order of stream in its basin. At its mouth, Arroyo Caliente is of fifth order.

The Arroyo de los Frijoles basin shown in small scale at the right in figure 12 was used as another sample and similar measurements were made. The small, unbranched tributaries on this 1 mile to 2 inch map, of which Arroyo Caliente is one, would be designated order 1, in accordance with the definition of stream order. The detailed study of Arroyo Caliente showed, however, that on the ground this tributary which had appeared unbranched was, in reality, composed of a drainage network of still smaller tributaries. Arroyo Caliente and other channels which appeared as order 1 on the right-hand map are, on the average, actually of 5th order. Thus the true order of any stream determined from the right-hand map of figure 12 is increased by adding 4, so that an order 1 stream on that map becomes order $1+4$, or 5.

This provides a way of combining maps of different scales to carry the numbering of stream order from one map to the other. It can be seen in the plot of stream length against stream order (left graph of figure 13) that this relation in stream orders 1 to 5 (average values of Arroyo Caliente) fits well with data from the small-scale maps after the values of stream order were adjusted as described above.

Similarly, this principle was used to obtain the estimate of the number of streams of each order in an 11th order basin in the area studied, as shown in the right diagram in figure 13. An actual count of the number of streams of highest order was made. From the small-scale maps the order of the Rio Galisteo at its mouth (fig. 1) was determined to be 7, which when adjusted by 4, indicates the true order of 11. Because this is the only stream of order 11 in the area studied, the graph must go through the value of 1 on the ordinate scale at an abscissa value of 11. The mean relation was drawn for the numbers of streams of highest orders and extrapolated to determine the number of streams of order 5, the smallest tributaries shown on the small-scale map. The graph of number of streams of orders 1 to 5 in the Arroyo Caliente basin which included only one stream of order 5 was superimposed on the graph determined from the 2 inches to 1 mile map and placed so that the points representing order 5 coincided. By extending the graph to order 1, the number of 1st order tributaries in the 11th order basin could be estimated.

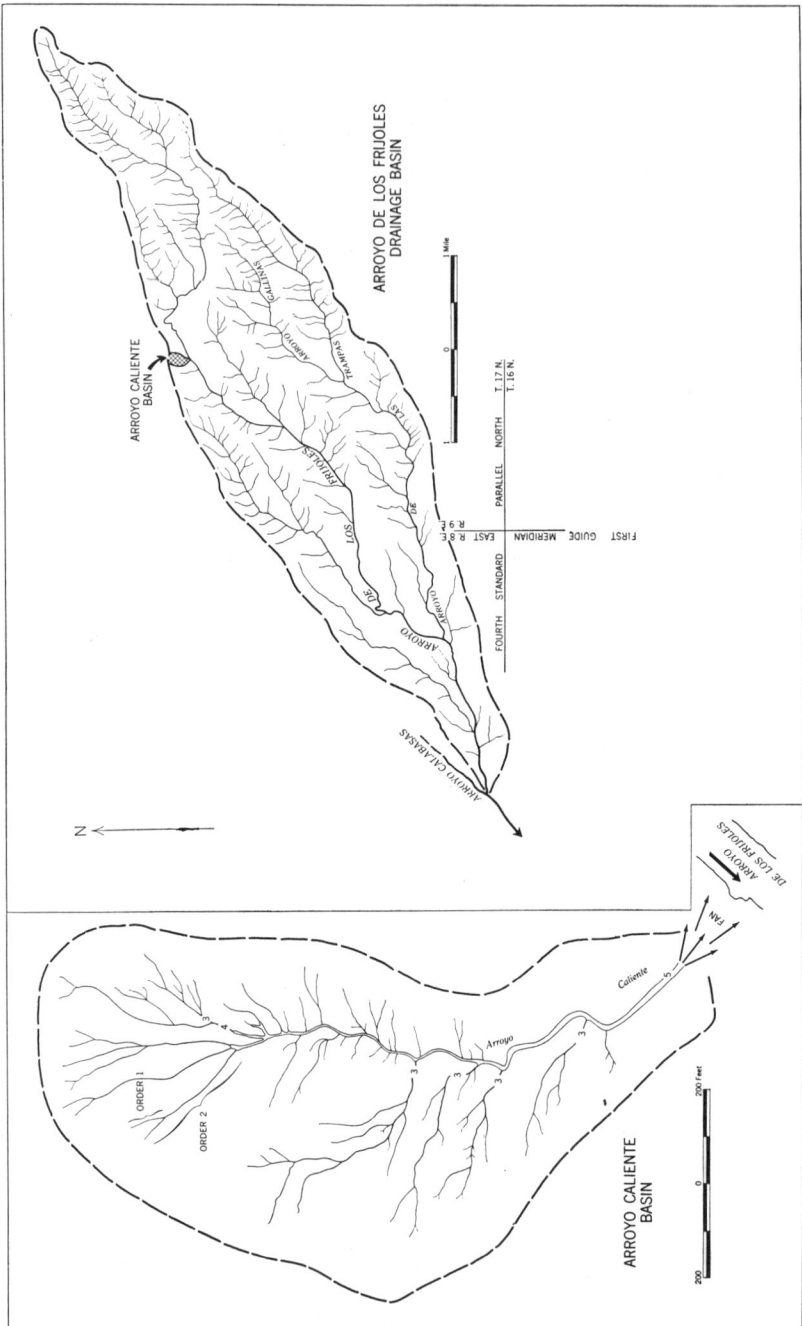

FIGURE 12.—Drainage basin of typical ephemeral arroyo near Santa Fe, N. Mex. Left, basin of Arroyo Caliente, a tributary to Arroyo de los Frijoles; right, basin of Arroyo de los Frijoles showing location of the tributary Arroyo Caliente.

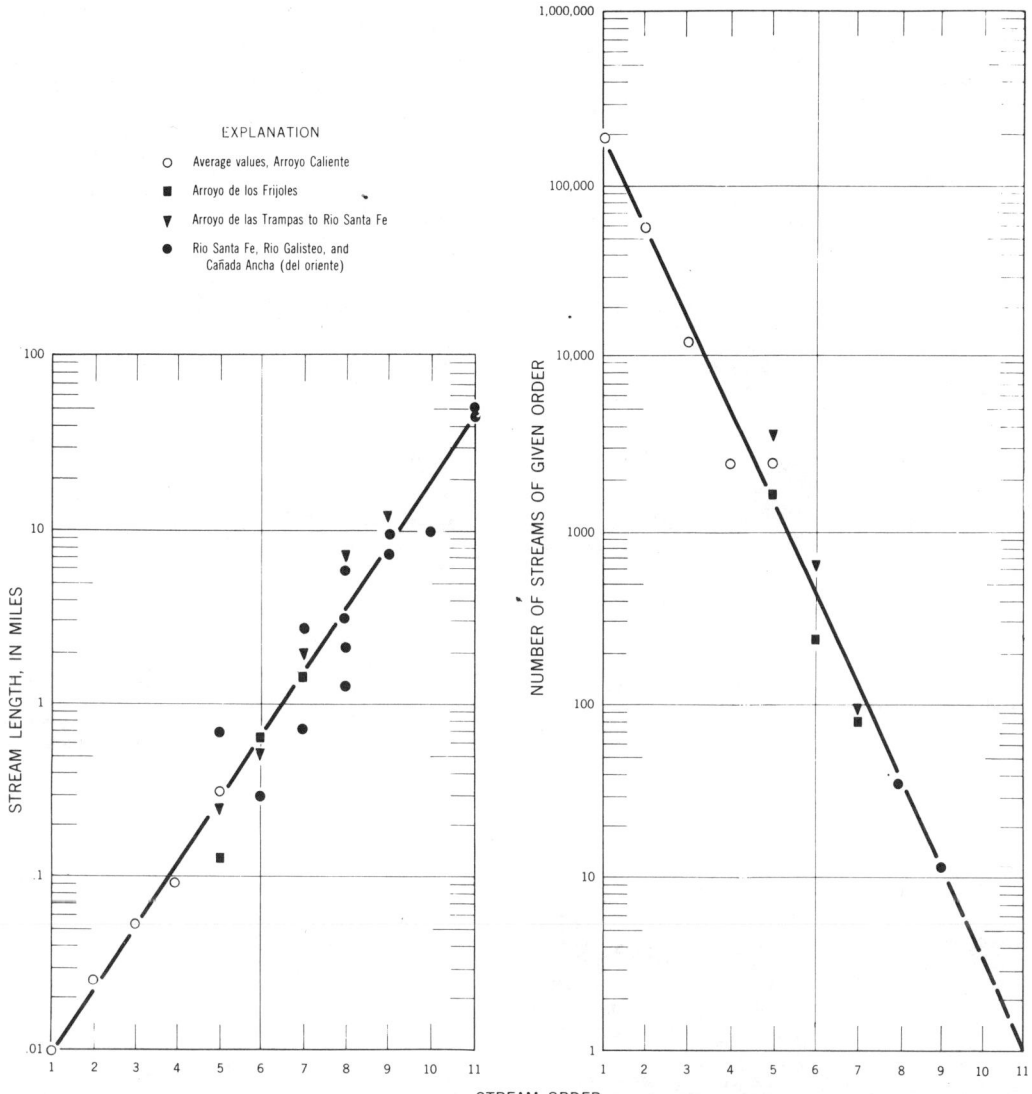

FIGURE 13.—Relation of stream length and number of streams to stream order in basins of 11th order in central New Mexico. Left, stream length plotted against stream order; right, number of individual streams of a given order plotted against order.

This same general procedure was followed for several other sample drainage basins. Order 11 was the highest found, this being for the main stem of Rio Galisteo. The number and average length of streams in each order were determined and the results are included in figure 13.

It will be noted that the plots of stream order against stream length and stream order against number of streams are straight lines on semilogarithmic paper, as Horton discovered. Lengths range from about 50 feet for the 1st order tributaries to 54 miles for the Rio Galisteo.

In order to visualize better the types of channels studied and their relation to stream order, the reader is referred to the photographs in figure 14, which show first-order tributaries in the basin of Arroyo Caliente,

FIGURE 14.—First-order tributaries in basin of Arroyo Caliente. These show the most headward extensions of the smallest tributary rills in the area.

and figure 15 which pictures Arroyo de los Frijoles at a place where its size is typical of an 8th order stream. Channels of order 5 and order 10 can be seen in figures 2 and 3, respectively.

Because maximum stream length is a function of drainage-basin area, it is not unexpected that the relation of drainage area to stream order is also a straight line on semilogarithmic paper, as can be seen in figure 16. The smallest unbranched tributaries, which are rills about 8 inches wide and 1 to 4 inches deep, drain on the average about .00006 square miles or .04 acre. In the 670-square-mile basin of Rio Galisteo there are roughly 190,000 such first-order tributaries, as estimated from figure 13.

EQUATIONS RELATING TO HYDRAULIC AND PHYSIOGRAPHIC FACTORS

From the previous work of Horton or from our data plotted in figures 13 and 16, it is apparent that stream order, O, bears a relation to number of streams, N, in the form

$$O = k \log N \text{ or } O \propto \log N \quad (1)$$

and a similar relation to stream length, l, slope, s, and drainage area, A_d,

$$O \propto \log l \quad (2)$$

$$O \propto \log s \quad (3)$$

$$O \propto \log A_d \quad (4)$$

FIGURE 15.—Arroyo de los Frijoles at place where it typifies a stream of eighth order.

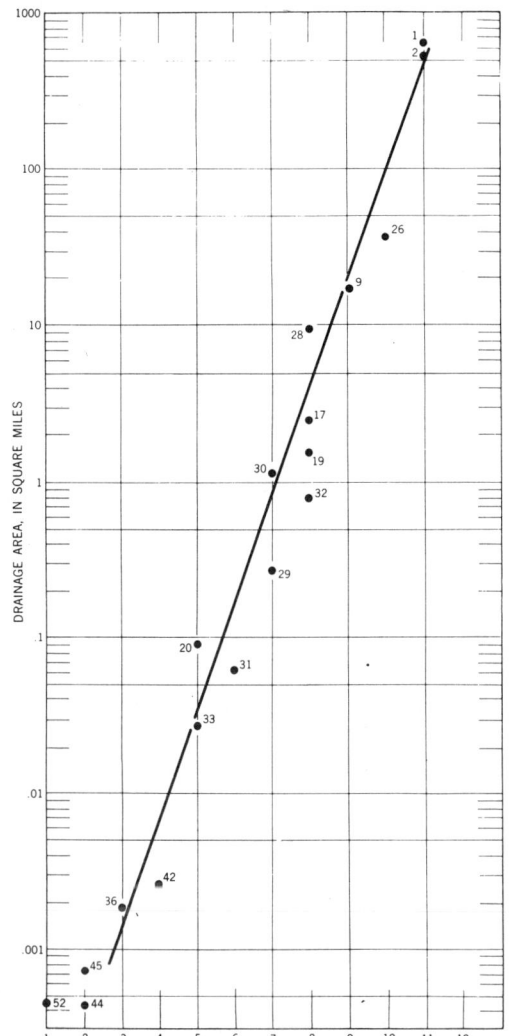

FIGURE 16.—Relation of drainage area to stream order for ephemeral channels in central New Mexico. Number beside point is serial identification keyed to data in appendix.

It follows, then, that there must exist a power-function relation between each two of these variables other than order, in the form

$$l \propto A_d^k,\ s \propto A_d^k,\ s \propto l^k$$

where k is some exponent having a particular value in each equation. Moreover, because it is known that discharge, Q, bears a relation to drainage area of the form

$$Q \propto A_d^k \tag{5}$$

then discharge must be related to stream order according to

$$O \propto \log Q \tag{6}$$

Reiterating some of the hydraulic relations discussed earlier:

$$w \propto Q^b \tag{7}$$

$$d \propto Q^f \tag{8}$$

$$v \propto Q^m \tag{9}$$

$$L \propto Q^j \tag{10}$$

$$s \propto Q^z \tag{11}$$

$$n' \propto Q^y \tag{12}$$

Where Q = discharge
w = width
d = depth
v = velocity
L = suspended load
s = slope
n' = a roughness parameter
$b, f, m, j, z,$ and y are numerical exponents

It follows from equation (6) and equations (7) to (12) that there definitely is a relation between stream order and width, depth, velocity, suspended load, slope, and roughness of the form

$$O \propto \log (\text{width, depth, etc.})$$

Also, there are several interrelations among variables in equations (1) to (4), and among those in equations (7) to (12). An example of such a relation is that which would exist between sediment load, L, and drainage area, A_d.

Because $L \propto Q^j$
and $Q \propto A_d^k$
then $L \propto A_d^{jk}$

Thus, it can be seen that a whole series of hydraulic and drainage-network factors are interrelated in the form of power or exponential functions. These equations can be added to the several others derived by

Horton (1945, p. 291) which, in his words, "supplement Playfair's law and make it more definite and more quantitative. They also show that the nice adjustment goes far beyond the matter of declivities."

The equations listed above merely state the condition of proportionality; for them to be definitive the constants involved must be determined. In addition to the values of certain exponents already discussed, data collected during this investigation established the relations among order, slope, and width, as will now be described.

Although channel depth, velocity, and discharge at a particular cross section can be measured only when the stream is flowing, the important hydraulic variables, channel width and channel slope, can be estimated even in a dry stream bed. During the many days when no thunderstorms were occurring in the area studied, our field work included measurement of width and slope in the dry channels. The procedure used will be described briefly.

The width was defined as width near bankfull stage. In Eastern United States where the development of the river flood plain is the rule rather than the exception, the bankfull width is relatively easy to define and measure. It is the width which the water surface would reach when at a stage equal to the level of the flood plain. In dry arroyos where flood plains are the exception and alluvial terraces exceedingly common, it is difficult to point specifically at an elevation which might be called bankfull. Nevertheless, field inspection allows fairly consistent estimates of width corresponding to an effective or dominant discharge, even by different observers. The positions on the stream banks representing the two ends of the cross section were chosen independently by each of us, and the recorded widths represent a compromise between our individual judgments.

Channel slope is somewhat easier to determine. Although there are local dry pools or deeps resulting from both definite patterns of flow and random channel irregularities, a smooth profile of channel bed drawn through several points in a reach provides an estimate of channel slope which is reproducible by successive measurements. Our procedure was to place the planetable in the center of the dry stream bed and take a series of sights both upstream and down as far as the stadia rod could be read with a leveled instrument. In most instances this gave a measurement of slope through a reach of about 600 feet.

Width and slope were measured at more than 100 channel cross sections. Because of the widespread geographic distribution of measured cross sections and because of the lack of adequate maps, we have not determined the order of the streams on which many of these are located. The graphs of figures 17 and 19

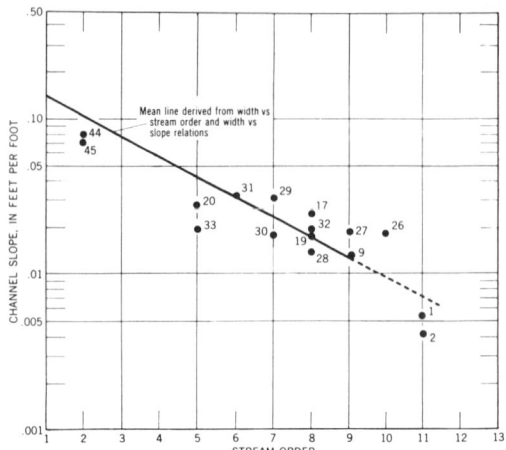

FIGURE 17.—Relation of channel slope to stream order, showing the individual measurements and the mean relation (solid line) derived from other parameters. Number beside point is serial identification keyed to data in appendix.

were plotted from data on cross sections of streams of known order. As expected, order is related to width and slope by equations of the type.

$$O = k \log w \text{ and } O = k \log s$$

The relation between width and slope can be determined by equating the relations above or by direct plotting of the data as in figure 18. To represent the relation in larger streams, a straight line on logarithmic paper has been drawn through the numbered points (fig. 18). The equation of this line is

$$s = 0.12 w^{-0.5}$$

This means that for the area studied, channel slope decreases downstream approximately as the reciprocal of the square root of channel width. This expresses an interrelation and does not imply a direct dependence between these two parameters.

Establishing the relations among width, slope, and order merely requires collecting and plotting the appropriate data. It should be possible to combine those factors which are most easily and definitively measured, and thereby arrive at approximations of other factors which are more difficult to measure. Such a scheme would expand the usefulness of the techniques now available. An example to illustrate this possibility will now be cited.

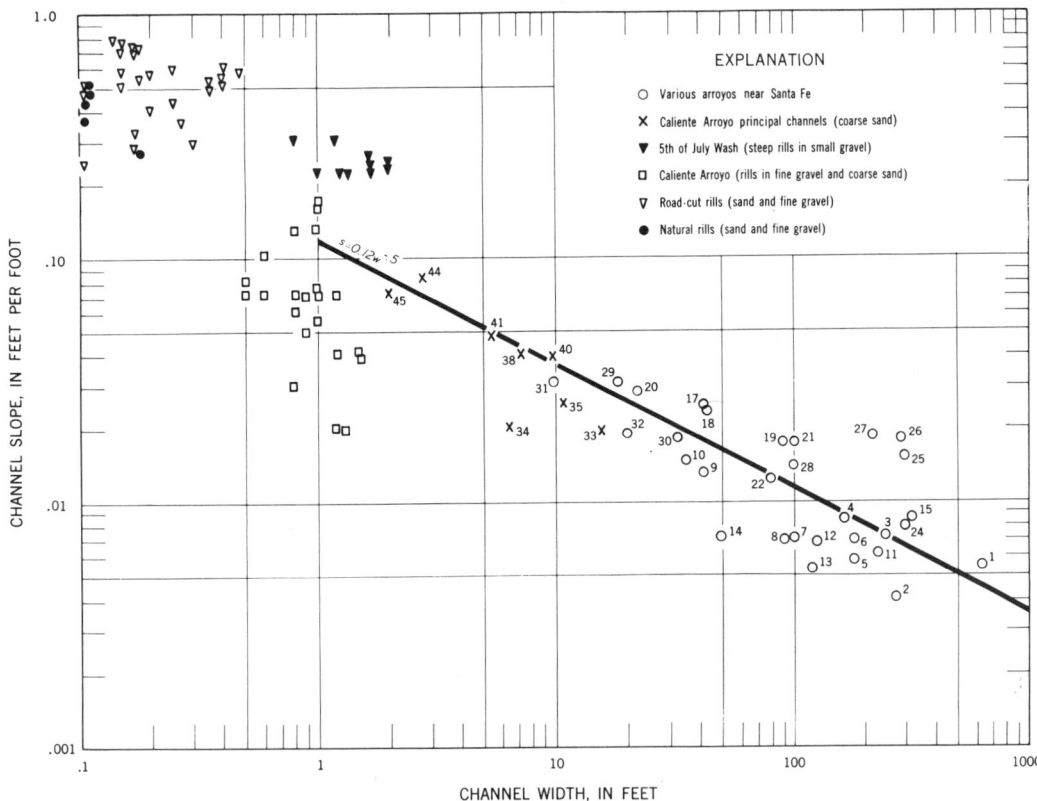

FIGURE 18.—Relation of channel slope to channel width. Data from moderate and large arroyos and differentiated from rills by the different symbols. Number beside point is serial identification keyed to data in appendix.

The manner in which discharge increases with stream order is of particular concern, as it provides the link between the Horton analysis and the hydraulic geometry. For perennial streams, gaging-station data are available in quantity for the larger stream orders. The determination of stream order at a gaging station is laborious but can be made, and order can be plotted directly against discharge of a given frequency derived from gaging-station data. In arid regions there are relatively few gaging stations, and for the most part they are located on the larger streams. To obtain a relation between order and discharge for ephemeral streams requires discharge measurements of the smaller streams. But even these measurements do not provide at a given cross section a specific value of discharge which can be plotted against stream order, for it must be remembered that discharge at any given location may fluctuate through a wide range and that sufficient record is required to allow some kind of frequency analysis.

If the discharge chosen has some specific relation to the position of the cross section along the length of the stream, an approximation to constant frequency might result. It is proposed to use that discharge which corresponds to the average width representative of the stream order at the point in question. To obtain this value, the relations already demonstrated can be utilized. Specifically, the relation between width and order indicated on figure 19 will be combined with the relation of downstream increase of width with increasing discharge. The latter is, fortunately, one of the most consistent of the graphs representing the hydraulic variables. From figure 19 the stream width for each order can be obtained and, by use of those values of width, the corresponding discharges can be read from the width-discharge graph of figure 11. In such a manner the relation between stream order and discharge may be derived, and it is plotted as the unbroken line on figure 20.

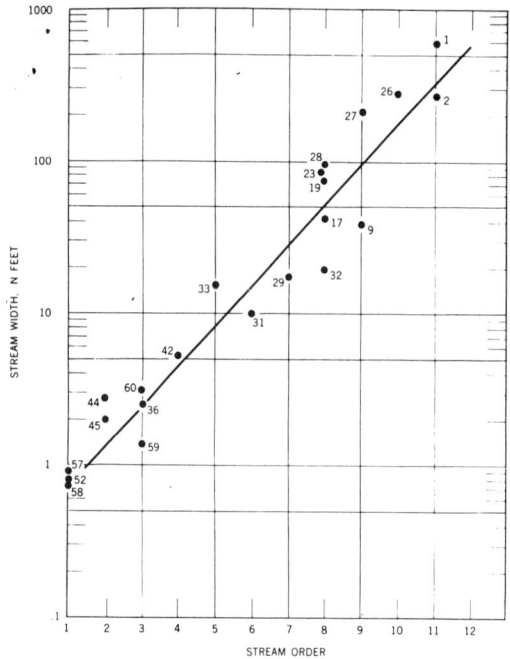

FIGURE 19.—Relation of stream width to stream order in arroyos. Number beside point is serial number in appendix.

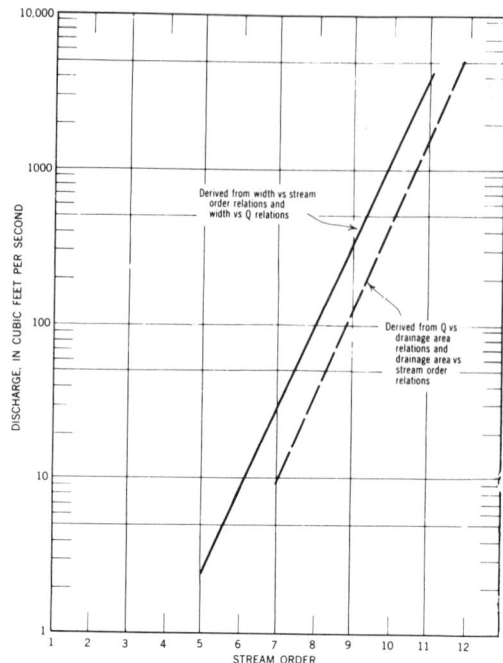

FIGURE 20.—Relation of discharge to stream order derived by two separate types of analyses.

A second approach to the discharge-order relation makes use of flood-frequency analysis. On streams of small and moderate size in central New Mexico there are only a few gaging stations that have a relatively long period of record. Despite the paucity of data, flood-frequency curves for all gaging stations considered applicable to the area being studied were obtained from an unpublished study by H. H. Hudson. Also, the records of some additional stations were analyzed by the authors. In eastern streams the bankfull stage is attained about once a year. A conservative quantity little affected by the length of record is the average of the highest flood each year of record. This is called the mean annual flood. It has a recurrence interval of 2.3 years, and thus we assume roughly approximates the discharge at bankfull stage. The relation between 2.3-year flood discharge and drainage area for gaging stations in central New Mexico is presented on figure 21.

It is known by hydrologists that the discharge of a flood of a given frequency increases somewhat less rapidly than drainage basin size. It is typical for the relation of flood discharge to drainage area to plot as a straight line on logarithmic paper, expressed by the equation

$$Q \propto A_d^k$$

where k is a constant which for the mean annual flood has a value of between 0.7 and 0.8. The slope of line on figure 21 is consistent with values known from other areas. In drainage basins of the same size, those studied in the West produce floods of about one-fifth the magnitude of the typical one in the East.

By combining the discharge-area graph with the order-area graph, discharge can be related to order. The drainage area corresponding to each stream order is read from figure 16, and the value of discharge for an equal drainage area is determined from figure 21. The resulting plot is shown as the dashed line in figure 20.

Thus, the relation between stream order and discharge, which is the link between the Horton analysis and hydraulic geometry, has been derived for a central New Mexico area in two ways which are at least somewhat independent. Comparison between the results is indicated by the two lines in figure 20. Only the slopes

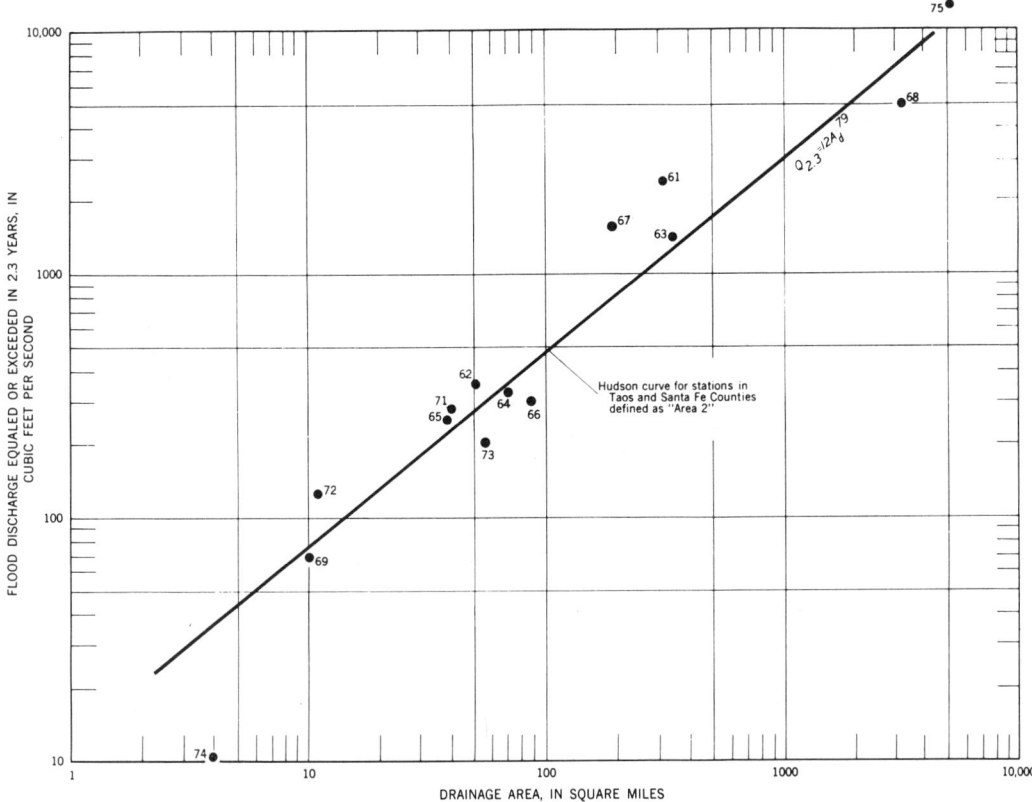

FIGURE 21.—Relation of mean annual flood discharge (equaled or exceeded in 2.3 years) to drainage area. Number beside point is serial number in appendix.

of the two lines should be alike, and indeed, they are similar. The intercepts should not be expected to be exactly the same because the discharge determined from flood-frequency data was chosen to represent a frequency of 2.3 years. The frequency of that discharge which corresponds to full channel width is unknown and need not be identical to the discharge having a 2.3-year recurrence.

It should be recognized that each of the graphical relations presented has considerable scatter owing to the nature of the measurements and the inherent variability of these factors in the field. It should not be inferred that a relation such as that presented between discharge and stream order is considered precise. Rather we are concerned with explaining a methodology by which generalized relations can be obtained, and with demonstrating the nature of the interrelations between a variety of hydraulic and physiographic factors.

In summary, it has been shown that the ephemeral streams in New Mexico are characterized by a uniform downstream increase of width, depth, and velocity with stream order, and also with drainage-basin size. Increasing size of drainage basins is accompanied by a downstream increase in discharge. Furthermore, the interrelations among all of these factors, both hydraulic and physiographic, may be expressed in simple terms, either as exponential or power functions. The method of combining the hydraulic variables with factors measured on a map or obtained in the field from a dry stream bed allows a simple means of obtaining interrelations which cannot be measured directly.

[Editor's Note: A discussion of the longitudinal profile and equilibrium in ephemeral streams has been deleted.]

REFERENCE

Horton, R. E. 1945 Erosional development of streams and their drainage basins; hydrophysical approach to quantitative morphology. Geol. Soc. America Bull. 56:275-370.

9

Copyright © 1958 by the University of Chicago

Reprinted from *Jour. Geol.* 66(4):442–460 (1958)

CORRELATION STRUCTURE OF MORPHOMETRIC PROPERTIES OF DRAINAGE SYSTEMS AND THEIR CONTROLLING AGENTS[1]

MARK A. MELTON
University of Chicago

ABSTRACT

To facilitate analysis of quantitative data obtained by measurements of many elements in nature, variables are mapped onto plane sets of points to form variable systems, within which (1) each variable is fairly highly correlated with every other, (2) the direction of causality, if any, is given, and (3) one or more variables is correlated with a variable not contained in the system. Variable systems are classified by the presence and direction of feedback among the variables. Variable systems may be infinite, and constitute hypotheses about the contained variables.

Fifteen variables of geomorphic, surficial, and climatic elements are arranged into two related variable systems on the basis of correlation coefficients for every possible pair in the study. One system contains variables that measure and determine scale of topography; the other contains variables that measure and determine scale-free elements of topography, i.e., the appearance of the landscape. On the basis of the feedback that possibly exists in nature, there is some tendency for geomorphic elements to control the processes of erosion that shape them, but the control does not extend to the external elements of climate, geology, etc., and therefore the condition of maturity (which is perhaps a shifting equilibrium) is not stable but depends on constancy of climate and other governing factors. A change in some governing factor results in changes in the topography that neither return the governing factor to its original condition nor affect the processes of erosion in the particular way necessary to restore the original form of the land. Drainage density and valley-side slope angle, in particular, are by-products of the rates of energy expenditure on the land surface by running water and the rates of restoration of soil and vegetal cover in relation to the resistance of the surface to erosion. Valley-side slope, however, may exert control over internal erosive and depositional processes to maintain a steady angle consistent with external controlling factors.

1. RATIONALE OF STUDY—THE THEORY OF VARIABLE SYSTEMS

The words "factor," "element," and "variable," as used throughout this paper, can be defined as follows: (1) an *element* is a particular feature of the landscape, climate, flora, etc., that possesses recognized individuality and that can be measured on a simple linear scale; (2) a *variable* is a mathematical object that assumes values obtained by a particular method of measuring a corresponding element; (3) a *factor* is a set of many elements that form a complex unity in nature. For example, the availability of moisture to plants is an element of a precipitation factor; it can be measured in a variety of ways, defining as many variables, one of which is the *P-E* index (Thornthwaite, 1931). The maximum steepness of a valley side is an element of a surface-slope factor and can be measured by finding the angle in degrees which a contour orthogonal makes with the horizontal plane; a variable, θ, is defined that can assume all such values.[2]

Topographic texture, average maximum valley-side slope, basin circularity, and other morphometric elements of mature drainage basins in fluvially controlled landscapes are related to causative and determinative factors of climate, lithology, and geologic structure. However, it is misleading and frequently meaningless to study isolated elements of the environment of a drainage basin in connection with a few morphometric properties because of the very great complexity of the drainage basin–environment system as a whole. Many elements are neither dependent nor completely independent but are intermediate between

[1] Manuscript received March 10, 1958.

[2] A complex element of a certain type can be considered a "vectorial element" and can be measured by a two- or three-vector. For example, a valley-side slope at the steepest point could be represented by a two-vector on a map, pointing in the downslope direction perpendicular to the contour line, whose length is proportional to a function of the angle of slope. No vectorial elements will be treated in this study.

more independent, environmental elements and more dependent, morphometric elements. For example, the percentage of vegetal cover, infiltration capacity, soil strength, surface roughness, etc., depend on regional climatic elements; in turn, they influence the surface form.

Clearly, a method of handling data obtained on the many elements pertinent to a general study is needed that will furnish insight into the relations of each of these elements to all others. Conventional multiple regression analysis is helpful, provided that the means are at hand for doing the arithmetic and every pair of variables is linearly related. But the results must still be interpreted, and with fifteen, twenty, or more variables this may not be easy.

Other researchers have found the concept of a "cluster of variables" to be useful in handling many variables of a comprehensive study (Miller and Kahn, 1958). A cluster is a set of variables, each of which is "linked" with every other variable of the set. Following Miller and Kahn, two variables, x_1 and x_2, are linked if the probability

$$P\{|\rho_{1,2}| \geq k\} = 1 - \alpha,$$

where k is a particular level of correlation between 0 and 1, selected arbitrarily or to suit the purposes of the study; α is the probability of erroneously rejecting a null hypothesis; and $\rho_{1,2}$ is the population correlation between x_1 and x_2. The value of k determines the content of a cluster and also the variables contained in common with other clusters. An interesting application of cluster analysis is the consideration of the changes in a cluster's content and its intersections as k assumes successive values (*ibid.*).

A variation of cluster analysis is used in this study that seems somewhat easier to apply, does not depend on arbitrarily chosen levels of correlation, and assures that the intersections of sets of variables that are obtained are empty. The following definitions will be used:

1. For a particular collection of variables under consideration, the "correlation set,"
$\mathcal{G}_k$, contains a variable x_1 if and only if x_1 is more highly correlated with some variable $x_{ak} \in \mathcal{G}_k (x_{ak} \neq x_1)$ than with any other variable y, whether y is contained in $\mathcal{G}_k$ or not.

2. For a collection of variables in a particular study, the "isolated correlation set," $\mathcal{F}_j$, contains variable x_1 if and only if (in that sample) x_1 is more highly correlated with every variable $x_{ij} \in \mathcal{F}_j$ ($x_{ij} \neq x_1$) than with any variable z not contained in $\mathcal{F}_j$.

Any finite correlation set must contain at least two variables that are mutually most highly correlated, i.e., an isolated correlation set of two or more variables.[3] It is not necessary for every pair of variables in a $\mathcal{G}$ to be highly, or even significantly, correlated. If all correlation coefficients are calculated from equal sample sizes, it would be simple, however, to make a probability statement about the lowest population correlation level between any pair of variables in $\mathcal{F}$. Because in any sample each variable has only one highest correlation coefficient, ties being removed by some suitable method, it is obvious that no variable can be contained in two different $\mathcal{G}$ sets in the same study. It is easily shown that a variable cannot be contained in two $\mathcal{F}$ sets.[4] Finally, placing x_1 in $\mathcal{G}$ or $\mathcal{F}$ is a factual statement, relevant to the particular sample at hand, and says nothing about population correlations between the variables.

3. A "variable system," $\mathcal{V}$, is an abstract set of variables such that (*a*) each is, in reality, rather highly correlated with every other one; (*b*) the direction of causality (if any) between each pair of variables is stated; and (*c*) one or more variables in $\mathcal{V}$ may be correlated with variables not in $\mathcal{V}$.

A variable system might be thought of as a "mapping" of variables onto plane sets of

---

[3] The isolated correlation set is a generalization of the basic-pair concept (Olson and Miller, 1958, p. 47–55).

[4] Let x_m be a member of two isolated correlation sets, $\mathcal{G}_1$ and $\mathcal{G}_2$, if that is possible. Then, for $x_i \in \mathcal{G}_1$, $y_i \in \mathcal{G}_2$, $r_{\min}(x_m, x_i) > r_{\max}(x_m, y_i)$ by definition, since $x_m \in \mathcal{G}_1$. Likewise, $r_{\min}(x_m, y_i) > r_{\max}(x_m, x_i)$ by definition, since $x_m \in \mathcal{G}_2$. Of course, $r_{\max}(x_m, x_i) \geq r_{\min}(x_m, x_i)$, from which is deduced the contradiction $r_{\min}(x_m, y_i) > r_{\max}(x_m, y_i)$.

points. Such sets may be infinite but probably of enumerable cardinality. To treat a point set representing a variable system as a metric space, a metric, μ, must be defined such that, for three arbitrary points, x, y, z,

$$\mu(x, y) = \mu(y, x), \qquad (1)$$

$$\mu(x, y) = 0 \text{ if, and only if, } x = y, \qquad (2)$$

$$\mu(x, y) + \mu(y, z) \geq \mu(x, z) \qquad (3)$$

(Newman, 1954, p. 17). If we grant that $|\rho(x, z)| \geq |\rho(x, y) \cdot \rho(y, z)|$, ρ being the population correlation, then $\mu = -\log \rho$ satisfies the three conditions and might be a suitable metric. However, in this paper the structural properties only of variable sys-

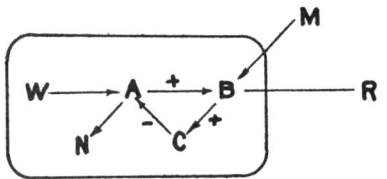

FIG. 1.—A simple negative-feedback system; variables W and M are independent, N is dependent, A, B, and C are intermediate, and R is simply correlated and perhaps varies *pari passu* with B.

tems will be investigated, those that do not depend on any particular metric.

Stating that certain variables can be placed together in a variable system is a hypothetical statement about those variables.[5] If the direction of causality is from x_1 to x_2, it should be understood that this means only that, all other variables being constant, changes in x_2 will always and only follow changes in x_1. In any interesting physical system, specifically in drainage systems, the majority of elements and all the variables associated with each of those elements cannot change independently and without involving changes in other elements

---

[5] This may be less true for all those variables that measure the same element in different ways. If an element A is measured by variables $A_1, A_2, \ldots$, all the A_i's could be placed in the same variable system with complete confidence if A is a fairly simple geometrical element. There would not be much point in doing this, however.

also; and, of course, all degrees of association are found. In these cases, metaphysical statements about causes and effects are largely meaningless, and, because statistical methods cannot determine the direction of causality or whether causality exists, a practical solution is simply to make hypothetical statements about temporal sequences of changes in values of variables that can be inferred from the nature of the elements.

Any variable system could contain a very large, perhaps infinite, number of variables, each being correlated with all others. However, only a very small fraction of these would necessarily be connected by arrows representing a time lag or temporal sequence. Large groups of variables could be connected with lines indicating correlation only, because they vary *pari passu* with one another. This is the best explanation of the meaning of "element." Any other would depend ultimately on one's point of view, whether macroscopic or microscopic, and one person's "element" would be another's "factor." Even so, the distinction between *factor* and *element* will not always be clear.

Three types of variable systems can be distinguished on the basis of the presence and effect of feedback among the variables of the system:

1. *Negative-feedback variable system.*—This type is characterized by the presence of at least one circular path or loop in the directions of causality, by which a variable, A, governs other variables, B, C, etc., which in turn govern A in such a way that a change in the value of A induces changes in B, C, etc., that return A to its prior value (fig. 1). In the closed loop, an odd number of negative correlations must be present.

2. *Positive-feedback variable system.*—This type similarly contains a closed loop in the directions of causality by which a variable governing other variables is in turn governed by them, so that changes in the value of the first variable result in changes in the others that produce changes in the first variable *in the same direction* as its original variation (fig. 2).

3. *No-feedback variable system.*—This

type contains two or more variables, one of which may govern the others, or there may be no unique direction of causality between any pair of variables. An example is the set of variables defined by different methods of measuring the same element (fig. 3).

The "environment" of a variable system is the set of variables governing, or correlated with, at least one variable in the system.[6] If a variable in the environment governs some variable in $\mathcal{V}$, it is, by definition, correlated with, or governs indirectly, all other intermediate and dependent variables but not the independent variable in $\mathcal{V}$. Two truly independent variables cannot be in the same variable system. If a variable is not constant in time (steady), other members of the system may or may not be steady, depending on whether the unsteady variable

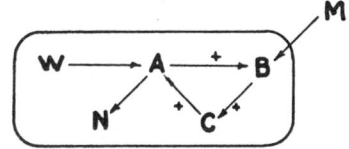

FIG. 2.—A simple positive-feedback variable system.

FIG. 3.—Simple no-feedback variable systems

is governed by the environment and whether it governs other variables in the system (fig. 4). Suppose a variable w fluctuates about a mean value μ but that other members of a variable system containing w are steady, even though some of them are governed by w. In other words, the mean value of w is

[6] A variable X_2 governed by a variable X_1 is ideally correlated with every variable that is correlated with X_1 and might be contained in the same variable system as X_1. If many other variables intervene between X_1 and X_2 and between X_1 and some X_3 in the system containing X_1, the correlation between X_2 and X_3 may be so close to zero that whether X_2 is placed in the same variable system or not is largely immaterial.

significant in the physical system corresponding to this hypothetical variable system, but minor fluctuations in w have no effect on other elements because of their short duration, threshold effects, etc. This situation can be handled formally by treating w as the sum of two variables: a variable, μ, which, by hypothesis, is constant for long periods of time, and a random variable, ϵ, whose mean is zero and has some distribution about 0. Then $w = \mu + \epsilon$. Now μ can be placed in a variable system in which all members are steady except w, and some or all of the other variables may be governed

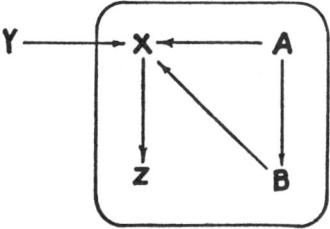

FIG. 4.—A partially steady, no-feedback variable system; fluctuations in Y produce fluctuations in X and Z, but not in A or B.

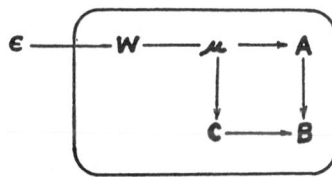

FIG. 5.—Formal treatment of a randomly fluctuating variable, $w = \mu + \epsilon$, in an otherwise steady system.

by μ; w is correlated with μ and ϵ (fig. 5). As ϵ is a random variable, it is not correlated with any non-random variable except w and cannot be a member of any $\mathcal{V}$.[7]

[7] In a tidy, determinative, Cartesian universe ϵ might be a member of another complex negative-feedback variable system rather than a measurement of an element that actually has a random or completely unpredictable behavior. The variability in any natural environment is the product of the happenings in many geologic periods and of the unknown habits of some grotesque animals. To argue that this variability could ever be entirely explained is absurd. Geology differs from physics and other exact sciences in the kind of variability encountered,

Variable systems, as they are abstract objects, do not contain energy or entropy and are otherwise different from physical systems. However, some of the terminology of physical open systems may be borrowed for analogous characteristics of variable systems. Variable systems are "open" because well-defined relationships exist between their members and their "environments." Variable systems may be steady, fluctuating, or monotonically varying, depending on whether the absolute values of the contained variables are constant, increasing and decreasing at different times, or increasing or decreasing continually.

A class of physical systems, such as all mature drainage basins, will have a number of variable systems associated with it. For each particular member of the class, i.e., for each particular drainage basin, the variables will have the values applicable to that member (basin). If the form, content, energy, entropy, and other characteristics of the physical system change, the corresponding variables in the associated variable systems will assume different values. Thus, the symbolic representation of a variable system gives an abstract picture of the interrelations among the elements of members of a class of physical systems; and particularization of the values of the variables gives a picture of one example of all the possible physical systems of that class. This abstract picture of the physical systems of a class will be of greatest value for the most complex systems, in which the variables are not simple mechanical quantities such as force, energy, power, stress, etc., but are quantities whose mechanical relations are not known.

The objective of the remainder of this paper is to consider a body of data obtained[8] by measurements of elements of climate, topography, and surface features in mature drainage basins in terms of the rationale development above. Theoretical variable systems that are consistent with the data and with what is known of erosional processes will be constructed by assuming that isolated correlation sets, to be determined, form the nuclei of variable systems, and the (non-isolated) correlation sets contain part of a variable system and its environment. Hopefully, this will be of some use in visualizing the interrelations of elements of drainage-basin and environmental factors on a structural basis.

II. VARIABLES TREATED, FIELD AREAS EXAMINED, AND DESCRIPTION OF DATA

The fifteen variables treated in this study can be grouped into three categories: morphometric, climatic, and surficial. Brief definitions of each and the symbolism used are given below, and the data sources are mentioned. More complete descriptions and discussions may be found in other publications (Strahler, 1954; Melton, 1957).

A. Morphometric variables

Valley-side slope (θ) is the angle of the steepest part of a graded valley side; measured largely in the field with an Abney hand level.

Drainage density (D) is the ratio of channel length to area drained, in miles; measured from maps.

Circularity (C) is a measure of the resemblance of a drainage basin's outline to a circle, given by the ratio of the basin's area to the area of a circle with the same perimeter; measured from maps.

Ruggedness number (H) is a scale-free measure of the relative relief of a basin, given by rD, where r is the total basin relief; measured from maps.

Ratio of total channel length to basin perimeter (L/P) is a scale-free measure of the relative development of the channel net within a basin's outline; measured from maps.

Relative density (F/D^2) is a scale-free measure of the completeness with which the

---

as well as the amount. It should be acknowledged that the concept of the universe applicable to geology and physics is accordingly quite different; i.e., that there is not just a single scientific "world view," but two or perhaps many.

[8] The raw data, which are not presented here, were taken largely from an earlier work (Melton, 1957) with a few minor additions. As the previous treatment was based on the regression and correlation analysis of only a few of the variables, the present study is believed to be a more adequate investigation.

channel net fills the basin outline for a given number of channel segments, given by the ratio of channel frequency to the square of drainage density; measured from maps.

Map area of basins of a given order (A_u); mainly fourth-order basins in this study; measured from maps.

B. Climatic variables

Precipitation-effectiveness index (Thornthwaite) (P-E) measures the availability of moisture to plants and is an estimate of the ratio of precipitation to evaporation for an area; obtained from climatic records and field estimates based on flora and elevation.

Relative January precipitation intensity (J) is given by the ratio of the mean January precipitation to $\frac{1}{12}$ mean annual precipitation; data obtained from climatic records.

Runoff intensity (q) is a measure of the excess precipitation over infiltration capacity for 5-year, 1-hour storms; obtained from Weather Bureau compilations and field infiltration-capacity measurements.

C. Surficial variables (all are field measurements)

Infiltration capacity (f) is the stable, usually minimum, rate at which rain can enter the A_0 soil layer; measured with a sprinkle-plot ring infiltrometer.

Wet and dry soil strength (S_w and S_d) is the strength of the soil when thoroughly soaked and when dry, under the impact of a 12-lb. shot dropped a standardized distance.

Per cent bare area (b) is the percentage of ground surface not covered by vegetation, plant litter, cobbles, bare rock, etc.; measured by counting the number of bare spots under foot-marks of a tape measure.

Roughness number (M) is the mean total length of rock-fragment diameters $\geq \frac{1}{2}$ inch in circles of 1-foot radius, within a drainage basin.

A limitation to the validity of interpretations of correlations among these variables is the omission of one or more possibly very important elements from the study. Three elements that might have added significant information are (1) the degree of departure of a basin from a mature state of development, such as the hypsometric integral (Strahler, 1952, p. 1130); (2) some channel characteristic, such as gradient or width at the basin's mouth; and (3) mean annual, or mean annual peak, discharge from the basins. However, a great deal of implied information about the basins is contained in these fifteen variables. Channel length is involved in D and can be found by multiplying D and A_u; number of channel segments can be found from D, A_u, and F/D^2; basin order can be estimated from D and A_u plus F/D^2 or θ (Melton, 1958); total basin relief is known from D and H; H supplies some knowledge about the diastrophic history of the basin; the caliber of bed load is probably related to M; general type and extent of vegetation are implied by P-E and b; the mean intensity of 5-year, 1-hour storms is known from f and q. There is some reason to believe that degree of departure from maturity is reflected in values of F/D^2 and θ (Melton, 1958); for basins of a given (low) order, there is very likely a relation between θ and main-channel gradient (Strahler, 1954, p. 350), and gradient varies with caliber of bed load and inversely with area drained (Hack, 1957, p. 55). Finally, an estimate of the mean annual runoff or peak runoff could be obtained from q and C.

Thus, out of a very large number of possible variables, it is necessary to have only one variable for each element, avoiding measuring essentially the same thing in different ways. And for an exploratory study such as this, it might suffice to represent each major factor by a single element or omit a factor entirely if something is known about its relationships to other factors.

The drainage basins that were studied were selected on the basis of their map and field appearance. Basins with cliffs, well-developed floodplains for the main channels, obvious compound slopes, extensively gullied slopes, or trenched channels were avoided. Each basin lies, for the most part, within a single lithologic and climatic environment, but many lithologic and climatic types are represented. All the basins approach the "mature" stage of development. Fifty-nine major basins were studied, and 23 of these were subjected to a full field study. In each basin so studied, 50–75 slope-angle measurements, 3–10 infiltration measurements, and 6–20 soil-strength, roughness, and bare-area measurements were taken.

The basins studied are covered by recent 1:24,000 USGS topographic maps or special plane-table maps. The areas are in Arizona near Cameron, Prescott, Tucson, Nogales, and Sonoita; in Colorado near Nederland, Golden, and Madrid; in New Mexico near Capulin, Arroyo del Agua, Los Alamos, Tijeras, Santa Fe, Grants, and Silver City; and in Utah near Springville, Lehi, and Spanish Forks. Basin areas range from 0.0004 to 5.5 square miles; most are fourth order, with a few third- and fifth-order basins; climate types range from quite arid to humid.

III. STATISTICAL ANALYSIS—PRESENTATION OF CORRELATION COEFFICIENTS AND EXAMINATION OF CORRELATION SETS

For the purposes of finding correlation sets and constructing variable systems from them, it is necessary to find a measure of correlation for every combination of the fifteen variables, taken two at a time. Excepting valley-side slope angles, the population distributions of the variables are generally unknown (Strahler, 1954, p. 349). For this reason, the standard parametric method of correlation analysis could lead to erroneous interpretations of tests of significance. To use the parametric measure of correlation, it would further have been desirable to transform the scale of each variable to give a linear scatter plot against every other variable, a very tedious job with fifteen variables. A nonparametric measure of correlation, Kendall's tau (τ), was therefore selected (Siegel, 1956, p. 213 ff.). To find τ, it is necessary only that the data be subject to ordering, and, as order and rank are invariant under the logarithmic and square-root transformations commonly used to linearize scatter plots, the difficulty just mentioned is avoided.

To speed computation of τ for each pair of variables, a set of fifty-nine edge-punched cards was prepared, one per basin. For each variable, the rank of the basin in the ordered sequence of sample values of the variable was recorded by punching the appropriate positions. If a certain basin has the greatest area of the fifty-nine, "59" was punched in the field allotted to basin area. If the same basin ranked tenth in order of increasing drainage density, "10" was punched in the drainage-density field. The set of cards could be ordered in any variable very quickly, and the values of τ for that variable with all others were obtained by observing the orders of ranks of the other variables (Siegel, 1956, p. 213). The complete matrix of values of τ is displayed in figure 6. The small numbers in the lower right-hand corners are the sample sizes.

For a one-tailed significance test, $\alpha = 0.075$, the limit of the critical regions for each sample size encountered are given in table 1. Because the power of Kendall's τ to detect true correlations is somewhat less than that of the product-moment correlation coefficient, using $\alpha = 0.075$ reduces the probability of making a type II error, and the results are roughly comparable to those obtained by using the parametric measure of correlation and a t test for significance of difference from zero, with $\alpha = 0.05$. At this stage of the analysis, it is preferable to have a low value of β than a particularly low value of α.

Comparing the list of significant correlations obtained with the τ measure of correlation, presented in figure 7, with the significant correlations obtained by parametric methods (Melton, 1957), the method using τ failed to detect possible true correlations between the following variables: S_w and θ, S_w and M, θ and b, M and D, M and q, and perhaps others.

For purposes of comparing the correlation sets obtained according to definitions presented in section I with clusters of variables obtained by the methods presented by Miller and Kahn, the following particular definition of a link will be used:

$$P\{\,|\,\tau_p|\geq 0.40\,\} = 0.925 = 1 - 0.075\,.$$

That is, two variables are linked if the probability is 0.925 that the absolute value of the true τ for those variables is equal to or greater than 0.40. The lowest permissible values of τ for linkage between two variables for the sample sizes encountered are given in table 2.

The first step in finding the correlation sets was to examine the matrix of significant values of τ (fig. 7) for possible basic pairs and isolated correlation sets. Four basic pairs were found ($P\text{-}E\text{—}b_k$ $f\text{—}q$; $F/D^2\text{—}H$; $S_w\text{—}S_d$); further inspection showed that no other variable was correlated more closely with both members of one of these groups than with any other variable of the study.

Each of these isolated correlation sets is then considered to be the nucleus of a (non-isolated) correlation set; the remainder of the set can be found by looking through the values of τ for each variable not contained in one of the above basic pairs and adding any variable whose maximum correlation is with one of the pair to the set. The process is cumulative, and, after adding a variable, the list must be looked over again to see whether some other variable has its highest correlation with the variable just added. Table 3 shows four correlation sets and gives a descriptive name to each. The variables are attached by arrows, indicating highest correlations, to the encircled pair comprising

X\Y	θ	D	C	H	L/P	F/D²	f	P-E	S_w	S_D	b	M	q	J	A_u
θ	1	+.0229	-.0079	+.495	-.0662	-.294	+.302	+.0709	-.158	-.0953	-.1106	0.00	-.236	+.391	-.0455
D		1	+.00709	+.00234	-.0339	-.0280	-.273	-.376	-.0905	+.0381	+.649	+.1905	+.428	+.285	-.426
C			1	-.160	-.0160	+.0975	-.177	+.112	-.116	+.0577	+.169	+.212	+.275	-.0702	-.216
H				1	+.253	-.560	+.362	+.133	-.167	-.0381	-.149	+.110	-.308	+.348	+.178
L/P					1	-.322	+.0451	+.0853	+.0905	-.0857	+.101	+.352	-.0336	+.0386	+.506
F/D²						1	-.311	-.225	+.224	-.0857	-.0622	+.0286	+.254	-.197	-.231
f							1	+.341	-.0239	-.0957	-.584	-.153	-.737	+.0693	+.276
P-E								1	+.0962	+.0863	-.681	-.307	-.536	+.0408	+.366
S_w									1	+.496	-.168	+.210	+.0625	-.178	+.0286
S_D										1	-.120	+.200	+.0911	+.0719	-.133
b											1	+.278	+.488	+.198	-.499
M												1	+.187	+.0624	-.114
q													1	+.111	-.518
J														1	-.0878
A_u															1

FIG. 6.—Matrix of all τ values

the nucleus, in the order in which they were added to the set.

For the most part, the contents of the sets make sense in terms of geomorphic theory. (1) The precipitation-effectiveness index was proposed in the first place as a climatic indicator of the type of vegetation that can be expected in a region; hence it is not surprising that it appears in the same isolated set with a measure of the proportion of bare, uncovered area. The high correlation of D with b fits immediately with the Horton theory of drainage density as a function of the resistivity of the surface to erosive forces,

TABLE 1

LIMITS OF CRITICAL REGIONS FOR FOUR SAMPLE SIZES

Sample Size	$\|\tau_c\|$	Sample Size	$\|\tau_c\|$
59	0.1286	29	0.1889
56	0.1322	21	0.2269

TABLE 2

MINIMUM VALUES OF τ FOR LINKAGE AT 0.40 LEVEL

Sample Size	Minimum $\|\tau\|$	Sample Size	Minimum $\|\tau\|$
21	0.227	56	0.268
29	0.211	59	0.271

X\Y	θ	D	C	H	L/p	F/D^2	f	P-E	S_w	S_o	b	M	q	J	A_u
θ	1			+.195		−.294	+.302						−.236	+.391	
D		1				−.273	−.376				+.649		+.428	+.285	−.426
C			1	−.160									+.275		−.216
H				1	+.253	−.560	+.362	+.133					−.308	+.348	+.178
L/p					1	−.322							+.352		+.506
F/D^2						1	−.311	−.225					+.254	−.197	−.231
f							1	+.341			−.584		−.737		+.276
P-E								1			−.681	−.307	−.536		+.366
S_w									1	+.496					
S_o										1					
b											1	+.278	+.488		−.499
M												1			
q													1		−.518
J														1	
A_u															1

FIG. 7.—Matrix of τ values significantly different from zero

determined by vegetation elements (among others), which, in turn, determines the mean length of overland flow (Horton, 1945, p. 320; Strahler, 1956, p. 625). (2) The appearance of f and q in the same isolated set needs no explanation, as q was obtained from values of f. The high correlation of C with q is rather surprising, though (very elongate) gullies would form most readily where rainfall and runoff intensity were highest. Like D, A_u is a measure of the drainage texture, and it might as well be most closely associated with D or b. However, its present high correlation with q emphasizes that length of overland flow and therefore drainage texture depend on erosive forces as well as the resistance to erosion (*ibid.*). Within a homogeneous area, L/P will increase as larger and larger basins are considered, because the length of channels will tend to increase linearly with the area, and the perimeter linearly with the square root of the area. The high correlation of M with L/P is almost certainly a type I error, as roughness is rationally more closely related to the proportion of bare area or to the precipitation-effectiveness index. Correlation sets 1 and 2 contain variables pertaining to the scale of topography and to climatic and surficial elements that determine or affect the scale of topography and for that reason will be classed together in the treatment below. Further, this is reasonable because the six variables, $P-E$, b, D, f, q, and A_u, are mutually correlated fairly highly and could also be placed together in a Miller-Kahn type cluster.

3. The appearance of F/D^2 and H in the same isolated set is very interesting and puzzling. The relative channel frequency and the ruggedness number are both dimensionless measurements applicable to a specific basin. F/D^2 is algebraically identical with NA/L^2; for a constant N (number of channel segments), relatively longer channel segments mean smaller values of F/D^2 for a certain basin area. The negative correlation with H means that in rugged areas F/D^2 is low and the basin outline is filled with relatively longer channels than are basins of the same value of N in less rugged areas (Melton, 1958). This is probably the result of low-order channels extending much closer to the divides in rugged areas, with steep slopes, than in less rugged areas. The high correlation of θ with H is understandable because the maximum valley-side slope is an essential adjunct to the concept of ruggedness. The influence of relative intensity of winter precipitation on θ may be the result of greater susceptibility to erosion at that time. Or it may merely be the result of the higher mountain areas of the sample having both more rugged terrain and higher winter precipitation rates because of greater elevations. A careful sampling program is needed to clarify this point.

4. The two soil-strength variables comprising the last correlation set, while not measuring exactly the same thing, are closely related to the same determining factors, viz., soil texture and mineral content. The fact that no other variable is most closely correlated with soil strength surely means that the method of measurement and/or sampling did not detect the relation rather than that the relation does not exist.

TABLE 3

BASIC PAIRS AND DESCRIPTION OF CORRELATION SETS OF WHICH THEY ARE THE NUCLEI

	Correlation Set	Descriptive Name
(1)	$(P\text{-}E \leftrightarrow b) \leftarrow D$	Moisture-availability—vegetation set
(2)	$(f \leftrightarrow q \leftrightarrow C)$ $A_u \leftarrow L/P \leftarrow M$	Infiltration—runoff set
(3)	$(F/D^2 \leftrightarrow H) \leftarrow \theta \leftarrow J$	Relative channel-density—ruggedness set
(4)	$(S_w \leftrightarrow S_d)$	Soil-strength set

The four correlation sets discussed above are related by intercorrelations of some of their members. These relations can be displayed as in figure 8. The moisture availability-vegetation set is combined with the infiltration-runoff set into a single, larger correlation set that is shown in the largest enclosure. Solid lines connect variables whose τ correlations are significant: dashed lines connect variables whose parametric correlations are significant but whose τ correlations are not significantly different from zero. The improbable correlation between M and L/P is not shown.

To relate figure 8 to the cluster-analysis techniques of Miller and Kahn, table 4 shows the total clusters (first col.), the non-contained clusters (second col.), and the clusters arranged around basic pairs (third col.). Basic pairs, which here are identical with the isolated correlation sets, are circled. The arrangement of variables in the third column is based on placing more highly correlated variables nearer to each other and on allowing each basic pair to appear in only one cluster.

The clusters in the third column of table 4 can be combined to give an integrated picture of the correlation structure, similar to that for the correlation sets (fig. 9). Each enclosure contains mutually correlated variables (in contrast to fig. 8, where enclosures contain variables whose highest correlation is with some other variable in the enclosure), and lines represent links between variables, as defined above. The resemblance between the two diagrams is pronounced, and they

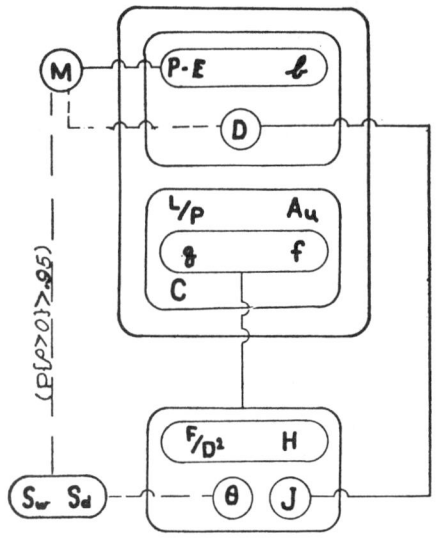

FIG. 8.—Diagram of correlation sets

TABLE 4

CLUSTERS OBTAINED AT 0.40 LEVEL OF CORRELATION

θ, (H, F/D^2), (f, q)	(F/D² H)—θ (f q)	(F/D² H)—θ
θ, H, J	A_u	
D, (P-E, b), (f, q), A_u	(P-E b) (q, f)	(P-E b)—D
D, J	D	
C, q		(q f)
L/P, F/D^2		A_u
L/P, M		
L/P, A_u		
(P-E, b), M	(b P-E)—M	
(S_w, S_d)	(S_w S_d)	(S_w S_d)

convey essentially the same picture of the correlation structure. A basic difference between the two does exist, however, in that the diagram of correlation sets is purely a statement of fact about the particular sample at hand, whereas the cluster diagram constitutes a hypothesis about the true correlations among the variables. In some cases, when many variables are considered, it is conceivable that the cluster diagram and correlation-set diagram might differ greatly in appearance. This could probably always be remedied by choosing a different level of correlation to use in the definition of linkage between variables. The only advantage in using the correlation-set method would be in

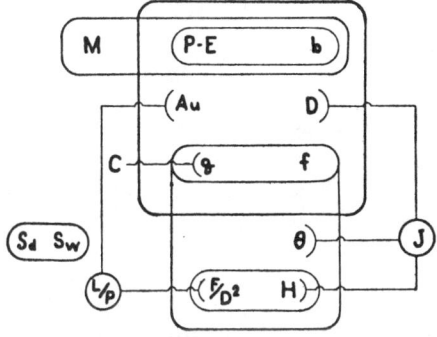

Fig. 9.—Cluster diagram

the simplicity of preparing the diagrams, the lack of intersection of sets, and the lack of an arbitrary level of correlation. The cluster method, on the other hand, may be somewhat more flexible in allowing examination of the intercorrelations at any chosen level, from very low to very high.

IV. INTERPRETATION OF STATISTICAL ANALYSIS BY VARIABLE-SYSTEMS THEORY

Simply confronting the reader with the correlation sets or clusters as arranged in figures 8 and 9 is not an adequate way of explaining the effects of various elements on one another, though it is preferable to listing all correlation coefficients among the variables. It is desirable to interpret the correlation sets in terms of the variable-systems theory already developed, as that should provide a much greater insight into the workings of drainage systems. This is largely an inductive process, so there is no reason to restrict the variables considered to those for which correlation measures have been obtained, so long as no known natural laws, statistical or otherwise, are violated.

We shall begin by constructing the smallest variable systems, of two and three variables, then fit these together into larger sys-

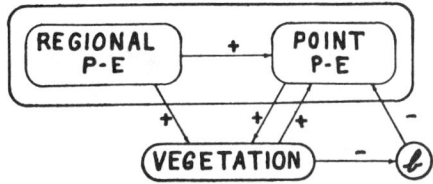

Fig. 10.—P-E—vegetation positive-feedback system.

tems. First, the scale-determining elements:

1. Precipitation-effectiveness can be analyzed into two variables: a regional-climatic variable plus a microclimatic variable. The regional P-E is determined by major wind patterns, elevation, latitude, situation with respect to coasts, mountain chains, etc., all very well known. Superimposed on this are the effects of location on a hill, orientation, steepness of hillside, amount of shade at the point in question, albedo of vegetation and soil, etc., also well known (Hursh and Connaughton, 1938). Thus, if for a point on the ground, the regional P-E index increases for some unspecified reason, the increased vegetation density that results will reduce insolation at the point, modify the temperature, reduce evaporation, intensify soil-forming processes—and so the point P-E index will be increased further because of microclimatic effects. This positive-feedback variable system can be diagrammed as in figure 10.

2. The infiltration capacity of the soil is decreased by erosion of the surface, by removing the organic fractions of the soil that are most important to permeability. Follow-

ing lowered infiltration capacity, runoff will be greater in amount and rate, increasing erosion and further lowering the infiltration capacity. A positive-feedback system can be constructed with the variables f and q, as in figure 11. Once a change has started, the system should continue to vary monotonically until at least the A soil zones are removed or until further erosion either does

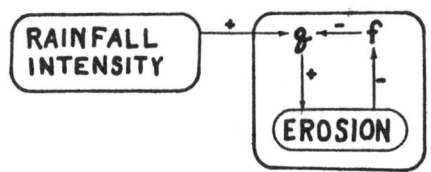

Fig. 11.—Infiltration capacity—runoff intensity positive-feedback system.

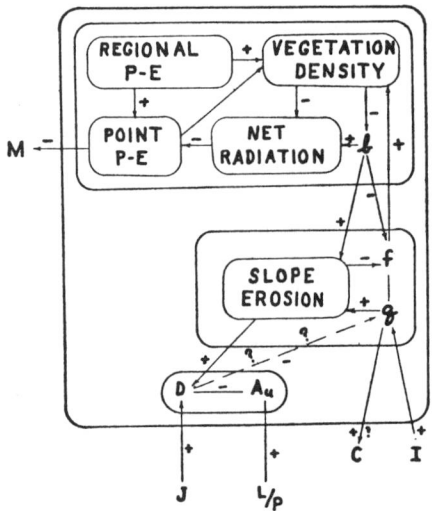

Fig. 12.—Composite variable system containing climatic and surficial controls of scale of topography.

not change the infiltration capacity or would increase it. This would explain in part the rapid erosion of a slope, once a sufficient change in one of these variables occurs, and also why the slope is not completely destroyed. A hypothetical rate-of-erosion element and the rainfall intensity-duration element (I), used to define q, have been introduced for clarity and completeness; it is assumed that they can be measured in a satisfactory way.

3. The two variables measuring topographic texture, D and A_u, can be placed together in the same no-feedback system, with negative correlation indicated.

Reference to figure 8 or figure 9 prompts us to include these three variable systems in one major one, whose environment is largely contained in the corresponding correlation set of figure 8. The erosion-rate element is the connecting link among the three subordinate systems (fig. 12). The arrows drawn in this system indicate probable direct influences (on this level of analysis); e.g., regional P-E is undoubtedly a controlling variable of D, but, as it probably acts entirely through vegetation and infiltration, no separate arrow need be drawn. Wherever two variables of the study would be connected with a double arrow indicating feedback, additional elements are inserted to clarify the mechanisms involved. Such is the case with "vegetation density," "net radiation," and "erosion rate."

The most controversial and interesting arrow in the diagram (fig. 12) is that going from D to q, indicated as questionable. The only possible negative-feedback system in the diagram is dependent on the existence of that arrow. If it represents an actual natural process, then it is apparent that the topographic texture of an area would be more than just the by-product of the rates of energy expenditure in erosion (running water —work done on the surface in entrainment, transportation of debris, etc.) balanced against resistance of the surface to erosive forces and rates of "repair of damage" by erosion. If, perhaps, through a mechanism involving an element not shown here, D could control the slope-runoff rate or intensity, then anomalously low values of D would cause an increase in q, increasing the erosion rate and finally increasing D. Then q would decrease, etc., until D became adjusted to all the other variables. The commonly rather low variability in D over a region of homogeneous lithology, relief, and climate seems to confirm that such a mechanism of "self-control" does exist in D. However, until more data are available and such a mechanism is explained, the writer believes that the fairly constant macroscopic value of D throughout a region is the result of a frequently realized mean value of runoff intensity capable of performing a certain amount of work on the surface in erosion and that, because the region has a certain mean surface resistance to erosion, a definite length of slope is needed before the accumulated runoff is sufficient to initiate a channel (Horton, 1945, p. 319; Strahler, 1956, p. 625). Once formed, a channel requires the expenditure of energy at a particular mean rate, depending on rates of soil creep and slope-wash erosion, simply to maintain its form by removing the debris added by these agencies. Thus the regularity in the drainage density over a region is an expression of the regularity in climate, soil-forming processes, rates of plant growth, lithology, and structure rather than an expression of a hypothetical tendency for a particular drainage density to develop because it controls the regional elements that in turn control it.

In detail, however, the length of channel in a first-order basin is regulated in part by a method employing negative feedback. With a constant total basin-drainage area, the area drained by the head of the channel is an inverse function of the length of channel, and for a particular slope-runoff intensity the discharge at the channel head is a function of the area above the channel head and therefore of the channel length. Channel length need have no effect on the intensity of slope discharge (q_s), this being determined by regional and surficial factors, as in figure 12, but it can control the channel discharge by varying the area drained. A variable system describing this is presented in figure 13, L being channel length, A_{hc} the area drained by the channel head, and q_{hc} the channel discharge at its head. This negative-feedback system is important only on a small scale, but for very many local areas in any fluvially controlled region.

The relative January precipitation intensity, J, is shown in figures 8 and 9 as correlated with D, and it is shown as governing D directly in figure 12. A good case could be argued for combining it with I, the precipitation intensity-duration variable, and forming a more complex precipitation-intensity distribution factor that would ultimately be related in some manner with the regional $P\text{-}E$. However, it will become evident in the next section that the mechanism by which J governs D, H, and θ is not too well understood, and combining it with I would lead to a sign paradox.

Passing to the variables contained in the smaller correlation set in figure 8, these determine and measure the scale-free elements

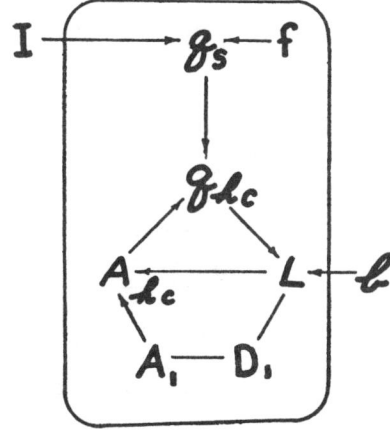

FIG. 13.—Detailed mechanism of control of channel length by slope runoff intensity (q_s) and proportion of bare area (b), in a first-order basin.

of a fluvial landscape. This means that the "appearance" of a region is not so much related to size of geomorphic elements as to these ratios and angular measurements (Strahler, 1958). A subordinate no-feedback variable system can be built around the isolated correlation set (or basic pair) H, F/D^2 by adding θ, J, and L/P (fig. 14). Partial correlation coefficients show that the arrows are probably properly placed. The questionable return influence of H on J, mentioned above, has been treated by introducing another factor, called here "relief factor," which might include such elements as recency and amount of orogenic uplift or recency and amount of lowering of a controlling base level. Almost certainly the only effect on J would be from regional uplift, if even that. It is not the ruggedness (a scale-free element) per se that would affect J but rather the altitude, which certainly affects climate. Thus the questionable double arrow between H and J is removed and with it the possibility of a closed loop in the directions of causality within this limited system.

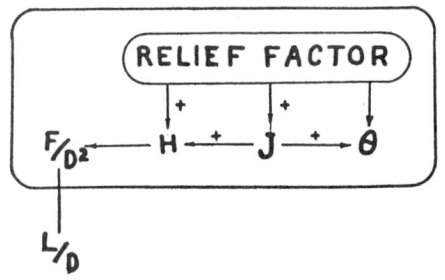

FIG. 14.—Diagram of variable system containing scale-free morphometric elements.

Some doubt may exist as to whether an arrow between H and θ, is not needed, but the partial correlation of H with θ, relief held constant, is simply the correlation of D with θ, which is not significantly different from zero (fig. 7).

From figures 8 and 9 we see that q and f can be placed in the same variable system with H and θ, or in its environment. The connection with H and θ is apparently through the agency of slope erosion; greater rates of slope erosion will reduce the relief, the slope angle, and therefore the ruggedness. However, reducing the relief will not in itself alter J, so the "relief factor" of figure 14 must be separated into at least two components, which in figure 15 are "altitude" and "immediate relief." Because the effect of J on H and θ is opposite the effect of slope erosion, it seems apparent that the mechanism through which J must act is something else;

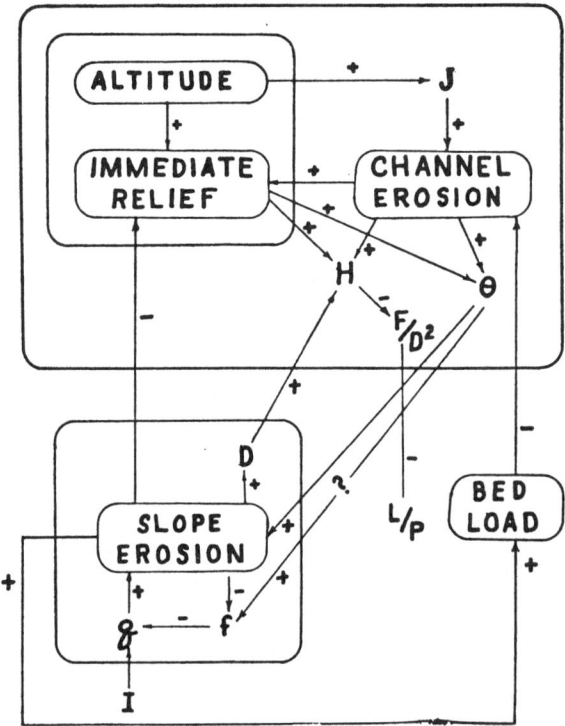

FIG. 15.—Diagram of correlation structure of morphometric, climatic, and surficial variables as interpreted by variable-system theory.

channel erosion (down-cutting) seems most likely, as this acting alone would increase relief and therefore H and θ. Increased slope angle, augmenting the downslope gravity vector, will increase the rate of slope erosion and should decrease f somewhat; however, θ is positively correlated with f, and an additional arrow from θ to f may be drawn because steeper slopes present a greater surface area to a particular depth of rainfall, and the effective infiltration capacity is therefore greater. Still, most of the positive load and through θ, must certainly be related in a more complex manner. Both "channel erosion" and "slope erosion" could be analyzed into a number of components that interlock in a complex way with one another and with their environment. The mechanism for linking the two here assumes that increased slope runoff and erosion present greater debris to the subjacent channel; channel discharge is not appreciably increased because the increment of slope runoff is merely a diversion from interflow; and

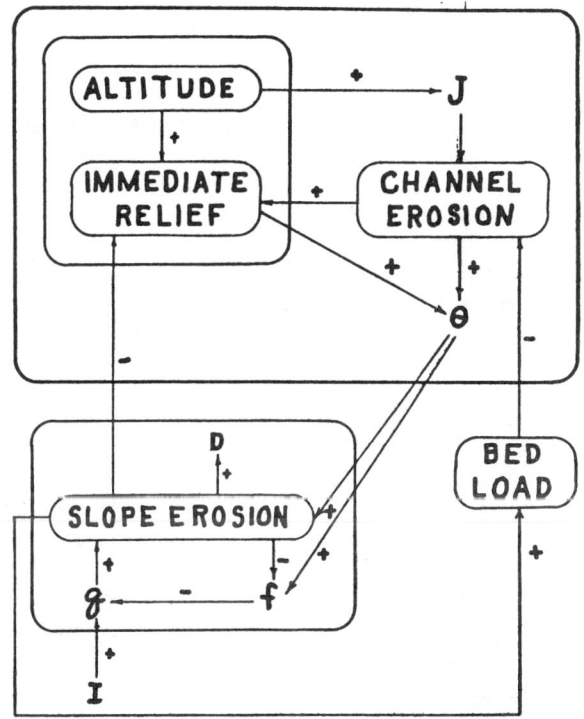

Fig. 16.—Simplified correlation structure, emphasizing negative-feedback loop

correlation between θ and f may be explained by the path from f through q, slope erosion, and relief (fig. 16), which results from a presumably higher ratio of channel discharge to slope runoff when f is higher, because of the greater interflow (shallow subsurface flow) rate (Melton, 1957, p. 32).

The two elements (or factors, perhaps), "slope erosion" and "channel erosion," joined in figure 15 by paths through bed

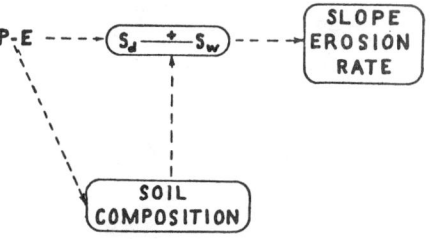

Fig. 17.—Soil-strength system

the channel is aggraded with the increased bed load. However, because of negative feedback among these variables, the aggradation does not continue indefinitely, as the slopes will be lowered, the angles reduced, erosion slowed (even though a side effect tends to reduce the effective f, increasing q), and channel erosion allowed to resume until the variable system is steady once again, if that is possible. This particular negative-feedback operation must be very common in semiarid and arid regions, with a sufficient supply of coarse debris that must be transported as bed load, and provides at least a partial explanation for the commonly low variability of maximum slope angles in a mature region. It has been proposed as a fairly general law (Strahler, 1956, p. 635; Melton, 1957, p. 32). However, there must be many combinations of conditions for which this does not hold, depending on the reasons for increased or decreased slope runoff and erosion, character of the soil, caliber of bed load imported from upstream, etc. Further, both "slope erosion" and "channel erosion" are shown as related positively to D, so they are probably not always antithetical in effect. A more detailed variable system should be constructed, perhaps contained wholly within one or the other of those presented here (fig. 12 or 15), that would clarify the conditions for the various types of behavior not explained here by considering many more variables, such as S_d and S_w (see fig. 17), grazing intensity, type of soil structure, frost action, soil creep and other mass-wastage processes, curvature of slope profile, and many others.

Without further data, the above is about as detailed an analysis of these elements as should be attempted. Very much has been inferred, both from the statistical analysis of the data collected for this study and from a general knowledge of the erosional processes that has been developed by geologists and hydraulic engineers. It should be apparent that this method can be extended into very detailed studies and that the number of variables potentially included in almost any variable system is very large. At about the stage of complexity reached in figure 15, the utility of the method declines as more variables are added or as some of the variables already present are subdivided into variables defined by restricted measurements of detailed aspects of the element represented. Then, as the scope of the study narrows and a more intensive investigation of a few variables is pursued, subordinate variable systems can be constructed whose environments are the larger systems containing them. Following construction of a variable system, detailed study of a few of the more interesting variables by standard multiple regression analysis should be profitable, particularly in instances where the effect of variable A on B is opposite in sign if two different paths in the directions of causality are followed (as in the effect of θ on f, fig. 16).[9]

The variable-systems method, using directed-line segments between variables, not only is a simple and graphic way of presenting the complete correlation structure of many variables but is an aid in synthesis. Each combination of two variables must be considered when putting several subordinate variable systems together into one, the direction of causality determined (if any), and the sign of correlations explained where conflicts occur. Possibilities for circular paths in the directions of causality and the need for inserting additional variables to explain the mechanisms by which one variable affects another will become apparent. Lastly, a variable system provides a mathematical model

[9] After variable systems have been constructed, the method of path coefficients (Wright, 1934) might be used with advantage, provided that (1) parametric correlation coefficients are used and the functional relations among the variables are known; (2) only a very few variables have been introduced for which data are not available; and (3) path coefficients could be found in feedback systems. The method of path coefficients could elucidate how effective the feedback is; i.e., whether all changes in a variable A are eliminated by negative feedback, or only a small amount of the actual change is eliminated (e.g., as in $\theta \rightarrow f \rightarrow \ldots \theta$). Also if A controls D through variables B and C separately, and the control is opposite in sign, then the more important could be determined. Why this could not be done with partial correlation or, preferably, regression analysis is not clearly explained in Wright's paper.

for the correlation structure of a large number of variables when only a few of the total possible variables are represented in the study and when several possible ways of measuring a particular element in nature could be used. It is a model based on set theory and topology rather than on analysis and can be made sufficiently complex to suit the purpose at hand.

V. CONCLUSIONS

A rigorous, or even barely satisfactory, discussion of the pertinence of the variable systems presented above to the theory of drainage basins as open systems tending toward steady states cannot be given without clear definitions of system limits, environments, contents, and steady features. It is, of course, clear that mature drainage systems are somehow adjusted to the regional climatic and geologic environment and that this environment is not affected by man-induced alterations in the form of the basin surface. It was shown that negative feedback probably operates (1) to keep channel length constant *if* basin area, slope-runoff rate, slope-erosion rate, and perhaps soil-creep rate are constant; (2) to keep valley-side slope constant *if* mean rainfall intensity and (likely) infiltration capacity are constant. By "constant" is meant "constant mean value." The negative feedback operates to remove anomalous variations in any variable that is part of a loop or circular path in the directions of causality. Obviously, if the regional P-E index, J, or I were to change, there would be no tendency for the resultant morphological changes in drainage basins to return the climatic elements to their original values, and so the morphological changes would be permanent. If the proportion of bare area (b) were changed artificially, the resulting changes in morphology would not tend to return b to its original value, and so the changes in morphology would last as long as b was not allowed to return to the natural value dependent on the climate (Strahler, 1956, p. 630). Artificial changes in f will induce changes in basin morphology; the changes in channel length would probably not tend to return f to its former value (fig. 13), but a change in θ would modify f at least to a degree, probably in the direction of its former value (fig. 16). Depending on the circumstances, it seems unlikely that f could be returned to its original value by θ alone, however. An artificial change in the rate of slope erosion, if this could be effected without altering f, would induce changes in the basin morphology (particularly θ) that probably would eventually return the erosion rate to its former value and, with it, all basin morphological elements. Thus, though negative feedback among some of the morphological and surficial elements surely exists, the "self-control" paths of the morphology do not extend very far into the regional and climatic elements that determine morphology and may not hold for more than minor changes in these determining, surficial elements. By and large, any tendency for drainage basins to remain steady over a fairly long period of time must be the result of a constant environment and the absence of internal, artificial changes by man.

ACKNOWLEDGMENTS.—I am indebted to Professor A. N. Strahler, of Columbia University, for guidance in planning and carrying out field and map research, and to Professor R. L. Miller, of the University of Chicago, for kindly letting me read portions of the manuscripts of the books of which he is co-author, for many helpful discussions on statistical problems, and for critically reading the manuscript. As is commonly the case, the field work would never have been completed without the help of my wife, Sara.

REFERENCES CITED

Hack, J. T., 1957, Studies of longitudinal stream profiles in Virginia and Maryland: U.S. Geol. Survey Prof. Paper 294-B, p. 45-97.

Hursh, C. R., and Connaughton, C. A., 1938, Effects of forests on local climate: Jour. Forestry, v. 36, p. 864-866.

MELTON, MARK A., 1957, An analysis of the relations among elements of climate, surface properties, and geomorphology: Tech. Rept. 11, Project NR 389-042, Office of Naval Research, Department of Geology, Columbia University, New York.

——— 1958, Geometric properties of mature drainage basins and their representation in a E_4 phase space: Jour. Geology, v. 66, p. 35–54.

MILLER, R. L., and KAHN, J. S., 1958, Statistical problems in geology: New York: John Wiley & Sons, Inc. (in press).

NEWMAN, M. H. A., 1954, Elements of the topology of plane sets of points: 2d ed., Cambridge Univ. Press.

OLSON, E. C., and MILLER, R. L., 1958, Morphological integration: Chicago, Univ. of Chicago Press.

SIEGEL, SIDNEY, 1956, Nonparametric statistics: New York, McGraw-Hill Book Co.

STRAHLER, A. N., 1952, Hypsometric (area-altitude) analysis of erosional topography: Geol. Soc. America Bull., v. 63, p. 1117–1142.

——— 1954, Quantitative geomorphology of erosional landscapes: Comptes Rendus, 19th Session, International Geological Congress, Algiers, Sec. XIII, p. 341–354.

——— 1956, The nature of induced erosion and aggradation; *in* W. L. THOMAS (ed.), Man's role in changing the face of the earth, p. 621–637: Chicago, Univ. of Chicago Press.

——— 1958, Dimensional analysis applied to fluvially eroded landforms: Geol. Soc. America Bull., v. 69, p. 279–300.

WRIGHT, SEWALL, 1934, The method of path coefficients: Ann. Math. Statistics, v. 5, p. 161–215.

ERRATA

On page 443, in the first column, the probability statement $P\{|\rho_{1,2}| \geq k\} = 1-a$ is meaningless, although the operation that is followed to identify linked variables is valid. The author is indebted to Dr. William Kruskal, Department of Statistics, University of Chicago, for pointing this out and suggesting the following restatement of the definition of linkage:

Two variables x_i and x_j are linked if r_{ij} is such that either (1) $r_{max} \geq k$ and $r_{min} > 0$ or (2) $r_{min} \leq -k$ and $r_{max} < 0$ is true for arbitrary k, $0 < k < 1$, and for the confidence interval $r_{ij} \pm \epsilon$ depending on n.

Page 443, in footnote 4, all Gs should be Fs.

Page 449, first column, line 12 should read: "pairs were found (P-E—b;f—q;..."

Part IV
DRAINAGE BASIN EVOLUTION

Editor's Comments on Papers 10 Through 13

10   JAGGAR
           Experiments Illustrating Erosion and Sedimentation

11   GLOCK
           The Development of Drainage Systems: A Synoptic View

12   SCHUMM
           Excerpts from *Evolution of Drainage Systems and Slopes in Badlands at Perth Amboy, New Jersey*

13   LEOPOLD and LANGBEIN
           Excerpt from *The Concept of Entropy in Landscape Evolution*

What happens when a submerged surface is raised above sea level, or when a surface is uplifted and the poorly developed drainage systems on that surface are rejuvenated? How does a drainage pattern develop and how does the pattern reflect the underlying geology? As the system evolves, how does the quantity and character of sediment exported from the drainage basin vary through time? These questions are of real concern to geologists who must interpret drainage patterns and to the geomorphologist who is interested in drainage basin evolution. They should also be of interest to soil conservationists because the headward extension of many small gullies and drainage patterns and the rejuvenation of small drainage basins appears to be a major erosional problem in many areas of the world. Such questions were considered by de la Noe and de Margerie in 1888, and Davis (Paper 1), and they are still the subject of controversy and discussion today. The answers can be obtained with certainty only by documentation of the slow erosional evolution of a drainage system. This may be done in part by study of the alluvial deposits resulting from erosional reduction of the sediment source area, by careful study of small rapidly eroding areas, or by experimental studies.

The required information is commonly obtained by arranging landforms in a sequence that appears to represent developmental stages. That is, it is assumed, and reasonably so, that space or location can often be substituted for time (ergodic hypothesis). Therefore, a series of slope profiles or a series of small drainage basins, when arranged in a sequence that is assumed to represent increasing age, will provide information on the erosional evolution of drainage basins.

Perhaps because of scale problems and the need for a large experimental area, the experimental approach is only now providing data of significance on drainage basin evolution and dynamics (Schumm and Parker, 1973; Schumm, 1977). Nevertheless, during the early part of this century several pioneering experimental studies were attempted. Jaggar's is reprinted because it is an early, serious attempt to study drainage-network development experimentally. Also, Jaggar considered an aspect of drainage pattern evolution that is usually ignored, the effect of groundwater on pattern evolution (see also Paper 6). He describes the competition within growing drainage systems and the complexity of their growth, and the reader is left with a feeling that drainage system growth is deterministic, and that with adequate information each detail of the system can be explained. This is in marked contrast to the approaches taken by Shreve (1967), Smart (1972), and Scheidegger (1970) in later work.

Geologists are little interested in drainage patterns developed on essentially uniform material and simple structure because the emphasis in drainage pattern interpretation is to determine how a pattern is influenced by structure and lithologic variations (Papers 1 and 3). Hence, the development of drainage patterns in homogeneous materials tended to be disregarded, but Glock (Paper 11), by comparing drainage networks of different stages of development, outlined a model of the growth and eventual decay of a dendritic drainage pattern. However, Glock was subjected to severe criticism by Douglas Johnson (1933) who showed that the blue drainage lines on the topographic maps used by Glock represent only a very small part of the total drainage pattern. This is a criticism that can be made of many studies of drainage patterns. Nevertheless, Glock's conclusions are correct, and they have been supported by recent experimental work (Schumm, 1977) and field studies in Australia (Abrahams, 1972). Glock's paper is reprinted here because it is an important attempt to understand the erosional development of the drainage pattern without regard for structural or lithologic variations.

The qualitative description of drainage patterns was satisfactory from the earth scientist's point of view, but obviously a quantitative description is required if landform characteristics are to be related to hydrologic, climatic or other quantitative data. As early as 1932 Horton described ways of describing drainage systems quantitatively for hydrologic purposes. Geologists remained unaware of this approach until his 1945 paper was published. The reader is here referred to the balance of Horton's classic (Paper 4, pp. **129-167**) that presents a model of drainage system evolution. Horton envisioned a rapid development of parallel drainage lines (rills) that are then integrated by capture to form a dendritic pattern. In nature this is only likely when a very steep slope is dissected, and on gentler slopes a dendritic pattern forms directly. In addition, Horton assumed that there is a critical length of overland flow from a divide over which channelized flow does not occur. Unfortunately he termed this the "belt of no erosion"; however, in this "belt of no erosion" the importance of raindrop impact, overland flow, and sheet erosion is obvious, and erosion does occur there. Horton used the term to mean that there was no erosion by channelized flow in this upper few meters of the drainage basin, and it is channelized flow that forms the drainage network that concerned him.

An obvious weakness of the discussions of drainage basin evolution is that there is not adequate time for an investigator to document changes in the field. Models can be proposed but rarely can they be tested (Ruhe, 1952). However, where erosion progresses at extremely high rates, documentation of drainage basin evolution is possible. Such areas are the badlands of the western United States or any unvegetated surface susceptible to rapid rates or erosion (Morisawa, 1964). One badland area is described in Paper 12, where a drainage pattern that began to develop in about 1929 was mapped by A. N. Strahler in 1948, and then was studied in detail by Schumm in 1952 and 1953. Changes through time were documented, and drainage network growth was related to a unit drainage area that was required to maintain a unit of stream channel. Thus, in contrast to Horton's length-control of pattern growth, Schumm related network growth to an area control. The relation between stream order and discharge, as developed by Leopold and Miller (Paper 8) and others, is very important in this context.

In marked contrast to the preceding papers are those of theoretical geomorphologists who study drainage systems that are simulated by random-walk techniques. One should note, however, that in all of these models a unit area is the building block of the random system. The Leopold and Langbein paper (Paper 13) stimulated a number of theoretical studies of this type that are summarized by Smart (1972) and Howard (1971). Howard (1971) concludes that in the future the most satisfying models of drainage network development will be those in which most of the morphologic characteristics of the system can be explained and in which randomness plays the smallest part.

REFERENCES

Abrahams, A. D. 1972. Drainage densities and sediment yields in eastern Australia. *Australian Geog. Studies* **10**:19-41.

de la Noe, G. and de Margerie, E. 1888. *Les formes du terrain, Paris.* Imprimerie Nationale, Service Geographique de l'Armee.

Howard, A. D. 1971. Problems of interpretation of simulation models of geologic processes. In *Quantitative geomorphology: Some aspects and applications,* ed. M. Morisawa, pp. 61-82. Publications in Geomorphology. Binghamton, N.Y.: SUNY.

Johnson, D. 1933. Development of drainage systems and the dynamic cycle. *Geog. Rev.* **23**:114-121.

Morisawa, M. 1964. Development of drainage systems on an upraised lake floor. *Am. Jour. Sci.* **262**:340-354.

Ruhe, R. V. 1952. Topographic discontinuities of the Des Moines lobe. *Amer. Jour. Sci.* **250**:46-56.

Scheidegger, A. E. 1970. *Theoretical geomorphology,* 2nd edition. New York: Springer-Verlag.

Schumm, S. A. 1977. *The fluvial system.* New York: Wiley Interscience, in press.

——. and Parker, R. S. 1973. Implications of complex response of drainage systems for Quaternary alluvial stratigraphy. *Nature Phys. Sci.* **243**:99-100.

Shreve, R. L. 1967. Infinite topologically random channel networks. *Jour. Geol.* **75**:178-186.

Smart, J. S. 1972. Channel networks. *Advances in Hydroscience* **8**:305-346.

10

Copyright © 1908 by the Museum of Comparative Zoology

Reprinted from *Harvard Univ. Mus. Comp. Zoology Bull.* 49:285–304 (1908)

EXPERIMENTS ILLUSTRATING EROSION AND SEDIMENTATION

By Thomas Augustus Jaggar, Jr.

CONTENTS.

	PAGE.
Introduction	285
Catchment basins and water-bearing strata	285
Imitative character of rills	286
Purpose of experiments	286
Previous studies of rill drainage	287
The Laboratory of Experimental Geology (Harvard University): preliminary experiments	287
Seepage rills	287
Underground drainage areas and stream shadow	289
Flood rills	290
"Grand Canyon" model	290
Trickle pattern in clay	292
Models sprayed with atomizers	293
Stream-robbery model	299
Discussion of principles	300
Bibliography	304
Explanation of plates	305

INTRODUCTION. The fundamental conceptions of river development in current physical geography are as follows:— (1) young rivers cut down their channels where the slope is steep enough to give them an active current, but where the slope is faint, the streams lay down their load of detritus and build up the surface. (2) A branch stream of a larger river has the advantage of the greater depth to which the main valley is worn. In competition with like branches of smaller nearby rivers its divide will shift faster. This doctrine is commonly applied to "belted coastal plains." (Davis.)

CATCHMENT BASINS AND WATER-BEARING STRATA. It is of course recognized that velocity of current increases with increase of volume, and where two streams have like slope, the more voluminous will have greater corrasive power. There are two very important ele-

Note: — This research was carried on in the Laboratory of Experimental Geology, Harvard University, while the author was there a teacher of Geology, and was aided by Grant 101 Elizabeth Thompson Science Fund. The author wishes to express to the Trustees his grateful appreciation of this grant.

ments in the problem, however, which have been given less attention by geomorphologists:— initial catchment basins, and initial attitude of aquiferous strata. The whole surface of an uplifted sea-bottom, for example, does not immediately become covered with streams. Rivers are extended from the old land and a limited number of new tributaries and longshore streams develop, apportioned in number to the rainfall and to the underground water supply. Their loci and spacing are dependent first upon extremely faint irregularities dividing the surface into a few flat catchment areas. After some gorges are cut, strata relatively impermeable or water-bearing are opened, and underground reservoir areas take control. When this happens if there be the faintest possible longitudinal tilt to the beds, local or general, there will be unsymmetry in the development of subsequent branches. A flat syncline of deposition with axis normal to the shore line would give mastery to any stream, small or great, cutting into aquiferous strata along its axial line. A much greater stream from the old land might cut a deep gorge along the adjacent faint anticlinal axis. It would have no power to send young branch ravines "gnawing headward" right and left. The underground pores would tend rather to drain it than to feed it, if the river cut into a permeable stratum. Hence extremely slight initial surface slopes and the underground slopes of impermeable water-bottom strata must be taken into account in any argument which deals with river piracy, divide migration, and "consequent," "subsequent," or "obsequent" streams.

IMITATIVE CHARACTER OF RILLS. The subject of drainage modifications is not without experimental possibilities. The delicate rill patterns seen on beaches are but one step removed from "bad-land" valleys; the latter again are analogous to all land drainage, when due allowance is made for the effects of geological structure. Such allowance is too often neglected on the one hand, or imagined structural influence is too much accentuated on the other, as in those cases where river pattern is attributed indiscriminately to the influence of joints and faults, and residuals are believed, because of their saliency, to possess specially resistant rocks.

PURPOSE OF EXPERIMENTS. The present study deals with a series of experiments designed to perfect apparatus for reproducing rill patterns in the laboratory. In the later of these experiments some success was attained, and meanders, piracy, digitation, corrasion, aggradation, and delta building have been reproduced in miniature. Some account is here given of the history of these experiments, which have been carried on from time to time in the Harvard laboratory for eight years. In addition, the author's last models are described in detail, as they illustrate the argument outlined above.

PREVIOUS STUDIES OF RILL DRAINAGE. The principal studies of rill pattern which have been made by geologists (Chamberlin and Salisbury, Nathorst, W. C. Williamson, Meunier) have been directed to their imitative character when preserved as fossil markings. In such a relation they frequently resemble organic forms. Daubrée (1879) believed many river systems to be controlled in development by faults and joints, and he has been followed in the United States by Hobbs (1901, 1905) and Iddings (1904) who seem to have a similar belief. Dodge (1894), from the physiographic standpoint, has pointed out the similarity of beach rills to continental drainage.

THE LABORATORY OF EXPERIMENTAL GEOLOGY (HARVARD UNIVERSITY): PRELIMINARY EXPERIMENTS. The mechanical difficulties of realizing even beach conditions in a laboratory are great. Ground breaking studies of importance were made in 1899 in the Harvard laboratory by Dr. Ernest Howe, now geologist to the Panama Canal Commission. In connection with experimental laccoliths Howe (1901) eroded domed surfaces of sand and marble dust with a coarse spray, and obtained radial drainage and infacing hog-backs.

SEEPAGE RILLS. In 1900-1901 the method was changed and instead of etching the surface of a model with a spray, various devices were used to produce seepage through a bank of sand. It was hoped that the phenomena observed on beaches might be duplicated.

In Plate 1, fig. 1, there is shown a meandering stream produced by seepage through a sand bank. A beach of sand was built on a suitable metal pan and a trench was dug along the upper margin. The trench was kept full of water which soaked the model. The oozing water gathered in two principal rills, which meandered from the time of their inception, and never showed any tendency to develop tributaries. The streams corraded profoundly even while meandering, and the complex terrace and distributary phenomena are part of a process of planation (Gilbert) uninfluenced by change of tilt.

In Plate 1, fig. 2, are shown the beginnings of digitation, coördinate with a meandering tendency. This stream, as in Plate 1, fig. 1, was produced by seepage through an inclined sand beach. The model differed from that of Plate 1, fig. 1, however, in that it was built of alternate sedimentary layers of marble dust, clay and sand, whereas No. 1 a was homogeneous sand. The oozing waters of the main stream bed undermined water-bearing strata on the walls of its gorge. The waters so released into the valley undermined superjacent beds, starting lateral channels. The loci of these channels are interdependent, because each tributary has a potency of position dependent on an initial underground drainage area. The moment a stream flows freely it discharges the waters of a certain upslope district and that district thereafter slowly enlarges until it is delimited by boundaries which are underground divides from other drainage. Such divides are not topographic elevations in any sense; they are the boundaries of what might be called the "sphere of influence" of any stream. No lower tributary in the same up-and-down zone of flow can use from the same aquiferous stratum any of the underground waters of an area above a higher tributary; hence a lower tributary, to gain a drainage area of its own, must eat laterally until it is beyond the zone of higher tributaries, or make use of aquiferous strata stratigraphically below their level. Moreover a higher tributary is always apt to take off the underground waters of a lower, and therefore rob it in fact, though none of the ordinary superficial evidences of piracy may appear. Initial rhythm occurs where uniform undermining affects uniformly a cliff of uniform height for some distance; instance the rhythmic spacing of the four lower tributaries on the right[1] bank of the main stream in Plate 1, fig. 2. A similar notching, having a tendency to rhythm, may be seen on both cliffs which bound the widening flood plain farther down stream. The four dextral tributaries mentioned are graduated in length, the highest being the longest. The reason for this is that the highest controls the headward drainage area underground, the next lower stream is thereby impoverished of supply, and so on to the smallest. As a stream gains width it increases the number of orifices of exit of water, its volume and its underground drainage area. This process has reached its maximum in the wide flood-plain area of the main stream, which by both scour and lateral planation has opened a broad water channel beneath the flood plain. That channel satisfies the discharge of a large drainage area on either side, hence there are no lower tributaries in the cliffs bordering the plain. Wells dug in the flood plain would find abundant water anywhere representing a

---

[1] "Right" and "left" in this paper always refer to an observer facing downstream.

flowing sheet from under the lateral cliffs. The river bed itself is a locus of seepage wherever the stream corrades.

UNDERGROUND DRAINAGE AREAS AND STREAM SHADOW. The foregoing analysis brings out two principles which may be defined before going farther. Each tributary in this model of inclined strata, has an underground drainage area. So has every natural orifice of exit of seeping waters in nature. Every spring controls such drainage area. Every brook is apt to be a line of loci of springs, and collectively they control a certain underground drainage area. Every forking system of tributaries, on larger scales, controls a similar underground area. The supply of every river system is dependent on this underground area (King, F. H., 1899), on the one hand, and a surface catchment basin, on the other. If, as in this model (Plate 1, fig. 2), strata variously aquiferous slope with the drainage, then each impervious stratum is an underground *surface* on which are separated the underground drainage areas of different orifices. Hence there may exist underground divides dependent on initial attitude of strata, and quite independent of initial surface topography. Apparently the presence of drainage *surfaces* as distinct from *pores* favors digitation. This is shown by the contrast between figs. 1 and 2, Plate 1. This is borne out by the fact that the spray models hereinafter described also develop arborescent drainage, the drainage surface in such case being the superficial topography. In contrast to these, in a model supplied wholly by seepage (Plate 1, fig. 1), there is no initial drainage surface.

The control by a tributary of a certain drainage or accumulation area of its own prevents that area from supplying water to any orifice farther down the general slope. Whether this be a superficial or an underground area, the effect is the same. Accordingly any such lower orifice may be said to be in the *shadow* of the higher stream and its drainage area. In Plate 1, fig. 2, the lowest of the four rhythmic tributaries on the right bank is thus shadowed by the next higher, and so on. Headward development by undermining will go on by a multiple ramification of those streams which can free themselves from shadow. The development of tributaries is more difficult mouthward than headward, because the mouthward region of a straight river is always in the shadow of the headward tributaries. If a river can for any reason bend laterally, new tributaries may develop wherever a curve undermines a portion of the general slope not in shadow. This principle is illustrated in Plate 1, fig. 2. If a lower tributary can eat laterally far enough to go beyond the shadow of higher branches, it may acquire a drainage area of its own. The outer bank of a river's main curves is the one which will naturally develop tributaries.

FLOOD RILLS. In the model illustrated by Plate 2 an accidental development of arborescent rills on the surface was produced by flooding. The model consisted of stratified marble dust, coal dust and sand, the strata overlapping. Two faint constructional terraces had been left in the topography. It was planned to erode with seepage, but the trough at the upper edge of the model accidentally overflowed. Almost instantly, as the rush of water flowed off the surface, the drainage pattern developed as shown in the plate. The three steeper slopes show arborescent drainage, while flood plains occupy the terraces. An interesting feature of this model is the development of arborescence not from baselevel headward, but all over the surface flooded, with its maximum on the steeper slopes and in the medial region.

"GRAND CANYON" MODEL. A considerable advance beyond earlier experiments was made in 1901 by R. W. Stone, now of the U. S. Geological Survey, and the author. A tank five feet (1.52 m) square and ten inches (.254 m) deep was used. The water was maintained at a constant level by a flood-gate. A rectangular island was built by sedimentation, 25 × 36 inches (.63 × .91 m) in area and 3 inches (.076 m) thick. Separate layers were sifted into a frame, the whole being kept moist. The model consisted of 61 very thin layers of marble dust, coal dust, clay and red lead, with three massive layers of sand each $\frac{1}{2}$ inch (.013 m) thick, (Plate 3). A fine spray was produced by means of two special nozzles. These were constructed so that a fine direct jet was broken up by impact against the ragged surface of a finely punctured tin plate, fastened at 45° inclination to the line of the jet. A mist-like spray was thus thrown at an oblique angle, while the excess of water was allowed to run off. These two nozzles were arranged on opposite sides of the model so as to produce a uniform rainfall over its surface. The model was tilted, sprayed daily for some weeks, and photographed frequently. Arborescent drainage developed, deep canyons with esplanades and waterfalls were cut, and a flood plain was formed along the lower reaches of the confluent main streams which terminated in a lobate delta built out into the lagoon. A cross-section of this model, after $718\frac{1}{2}$ hours of erosion, is shown in Plate 3. On the right is shown the delta in section. In the middle of the section may be seen profiles of two cascades and a perspective view of the canyons. On the left is shown the upper margin of the model and the back slope. The three light bands in the cross-section consist of about twenty thin layers each of marble dust, coal, etc.; the darker bands are sand. During the final days of spraying the model was tilted 10 degrees.

The important part played by volume of water, as contrasted with slope, in producing erosion, is well illustrated in this model. The back-slope, shown on the left in Plate 3, had a grade of some 45 degrees, while the surface of the model sloped only 10 degrees. All the stream development was on the main surface, however, because nearly all the water which fell on the model flowed down that surface. Moreover, the underground structure sloped the same way, producing seepage in that direction, and away from the backslope. The high divide at the back margin of the model remained practically uneroded, and what water trickled down the backslope never acquired volume or load enough to do any considerable trenching. The backslope would be the equivalent of "obsequent" or infacing slopes in a coastal plain escarpment. Such slopes are commonly supposed, by reason of their steepness, to cause or give evidence of a rapid retreat of the escarpment. No sign of such retreat was observed in this model, and all of the hydrostatic conditions indicate that such a slope would absorb moisture and carry the water down the dip of the strata. It would yield no springs for the development of "obsequent" streams.

It is of interest to note in the structure of the sloping frontal delta beds the following sequence:—

>Thick lower white marble dust layer.
>Sand layer.
>Second marble dust layer.
>Upper sandy layers.

These correspond to the materials of the model in the order in which they were reached by the eroding streams, viz:—

>Thick upper marble dust layers.
>Sand layer.
>Second marble dust layers.
>Sand layer.

The lowest group of light colored layers in the original model forms the bottom of the deepest canyon sectioned. Its effect is shown in the lighter color of the highest delta beds seen farthest to the right. On a steep submarine continental slope receiving the detritus of many rivers, some such corresponding inverse sequence of beds eroded and beds deposited might be looked for.

TRICKLE PATTERN IN CLAY. The experiment last described was coarse. The model was bulky and very laboriously made. The coarse sand layers were made of material the equivalent of gravel in proportion to the size of the streams developed. Hence it seemed desirable, if in any way possible, to work with finer material, so as to reduce the scale of the phenomena studied and likewise diminish the labor by making less bulky models. Plate 4 shows a pattern developed in liquid modelling clay. The phenomenon is closely analogous to the trickle of raindrops on a windowpane. A glass plate 15 × 20 inches (.38 × .50 m) rectangular, was flooded with smooth liquid clay of the consistency of syrup. The clay was "flowed" over the surface, held horizontally until it was evenly covered. The plate was then allowed to rest on one of its longer edges and raised until the surface dipped 45 degrees. It was supported in this position and left to drain. The greater part of the clay ran off and formed a pool at the lower edge of the plate. A portion, however, clung to the glass, settled, and its water tended to separate and run down the slope in drops, making clear spaces along the streams and leaving the divides opaque. Plate 4 is a portion of the upper margin of the plate, printed after it was dry by direct contact with solar paper. (The right and left are therefore reversed). For a width of five inches (.127 m) the upper margin showed an arborescent tracery on so fine a scale that from 20 to 25 streams are crossed in a distance of 7 inches (.178 m) (Plate 4). Lower down the slope distributary phenomena interfere with the arborescence, and the pattern is a complex of streaks with V-shaped accumulations of clay pointing up the slope.

A number of experiments were made with this process. In one series of plates, made under similar conditions, the duration of draining was systematically varied, so as to show all stages from the first initiation of arborescence to its completion. After the first sheet-flood run-off, the arborescence always develops near the upper margin of the plate in a definite zone about two inches from the margin. It seems to develop all at once and to grow very little thereafter. This is to be expected, as there is no source of added water, and nothing to erode when the glass is reached. There is interference of adjacent streams in many places, but the control of initial drainage basins and of "shadow" are clearly shown. The trickle pattern is complicated with capillarity and much of it is mere drop-trickling. It shows, however, that the tendency to digitation may be studied on a very small scale, and these experiments led to an effort to produce something intermediate between these too quickly drained clay films and the coarse spray model illustrated by Plate 3. The desideratum would be a model of some very fine material like clay, with an impalpable but continuous moistening of the surface, and sufficient thickness to permit the development of a distinct relief under erosion from the resulting rills.

MODELS SPRAYED WITH ATOMIZERS. Such conditions were finally realized by using the finest kind of crushed rock — slimes from a stamp mill — and spraying with atomizers. The atomizers used were the ordinary bottle style used in a barber-shop. Air pressure was maintained at about 10 pounds (ca. 4.5 kg) per square inch (625

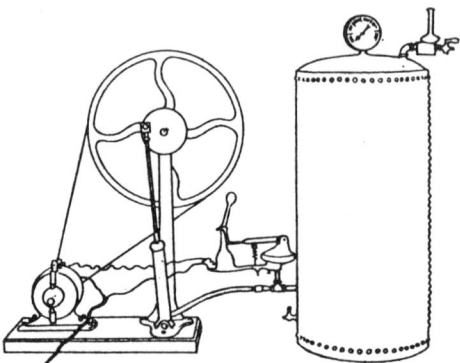

Fig. 1. Electric Air Compressor.

sq. mm.) while two atomizers were in operation. The apparatus for compressing the air is shown in Fig. 1. A ⅛ h. p. electric motor operates a single-cylinder air-pump. The air is compressed in a five-foot tank provided with pressure gauge and an automatic shut-off valve which may be adjusted to stop the motor at any required pressure and start again when pressure is reduced. A rubber tube from the tank carries the air to the atomizers which are each provided with independent cocks. This apparatus may be obtained of dealers in barber's supplies.

A model of slimes was built [1] as follows:— a plate of glass 6½ × 8½ inches (.165 × .216 m) rectangular was placed on the bottom of a deep ash-pan. A quantity of the slimes was stirred in water and the

[1] The laboratory work of this and the last experiment was done by Mr. H. G. Ferguson, now government geologist in the Mining Division, Bureau of Science, Manila, P. I.

turbid mixture poured into the pan on the glass and left overnight. The slimes settled on the glass in a layer about ½ inch (.013 m) deep. The slime-covered plate was removed and supported at a tilt of 20 degrees, one end resting on a brick, the other on a sheet of slate. A portion of the lower edge of the model fell off abruptly, the remainder rested on the slate, which was horizontal. The sprays from the atomizers were now turned on and in two hours the drainage pattern

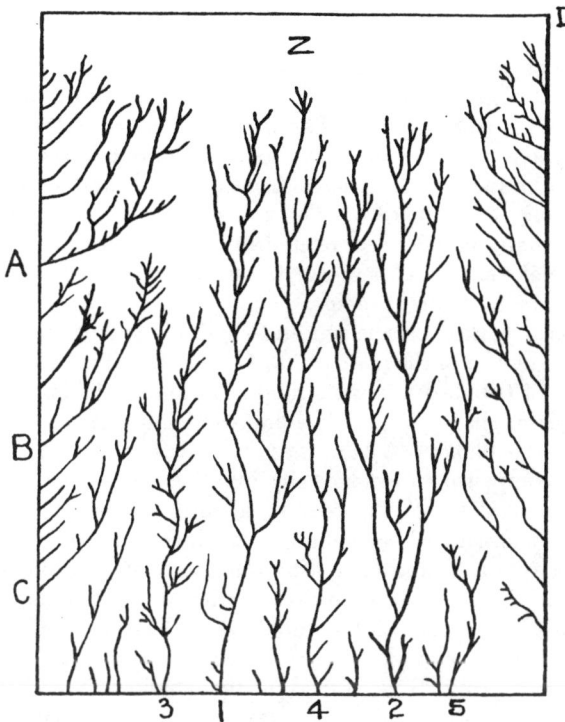

Fig. 2. Slime model. First stage. Sprayed with atomizers. Tracing of Plate 5, fig. 1.

shown in Plate 5, fig. 1, was produced. Continued spraying produced more mature topography, with deeper canyons on the side of the steep fall-off and flood-plains at the mouths of the rills on the opposite side. (Plate 5, fig. 2.)

An analysis of the drainage in Plate 5 shows a number of rhythmic features (Fig. 2). The model differs from those of Plate 1 in that the eroding waters were all supplied from the surface. The homogeneous model was soaked and the excess of superficial water flowed off. Under these conditions the controlling drainage areas for individual rills are true catchment basins. The upper margin of the model (D, Fig. 2) is free from corrasion; here the waters accumulated, flowing down the slope until individual streams gained velocity and load sufficient to trench. Each trench then became, by reason of its depression below the general level, the medial line of an elongate drainage area, delimited by superficial divides. Lateral tributaries have tendency to parallelism and rhythmic spacing according as the general slope, initial sheet-flood, and valley-slopes were uniform. The angle of junction, in plan, of tributary and main stream in this model varies from 25 to 28 degrees. Parallel sets of spurs and streams occur in many places marking local rhythms; these are best seen by holding the plate obliquely sloping down from the eyes, and sighting downstream along the main rills. The spurs are thus seen to be equally spaced in many parallel sets related to the general slope rather than to individual valleys. There are frequently parallel features of this sort tributary to different mains.

The most marked parallelism and rhythmic arrangement is shown at the right border of the plate (A, B, C, Fig. 2). This may be viewed as a series of tributaries to the straight fall-line at the border. Three sets of streams shaped in plan like half candelabra are seen, one above the other, the highest (A) the largest. For each group, the lowest stream is compelled to eat laterally into the model farther than the next higher, and so on, in order to acquire its own drainage area. When such a stream is beyond the area which is *in the shadow* of the drainage basin next above, it wins the water of an oblique upslope district, and the drainage from that district unites in a channel which makes a distinct bend with the initial lateral channel. The bends up the slope, for the same group, are progressively nearer the fall line. The whole of group B is clearly in the shadow of group A next above. The highest stream of a lower group is prevented from development by the proximity upslope of the group next above, which catches all the upslope water. The stunted growth of the highest stream gives an advantage to the next below which acquires a longer drainage area, and the third still longer. In this respect each group is like the uppermost group, the form of which is clearly determined by the margin of accumulating rainfall Z at the upper edge of the plate. The larger rhythm of the three groups does not follow the same law, for there the upper group A is the largest. The reason for this is that the upper group, from the time of inception of trenching, had the waters of the upper margin to draw on, whereas the lower streams stood in its shadow. Its own capacity for trenching on across the plate was delimited by competition with slopeward streams, Nos. 1 and 2 of the medial zone, of like grade with itself. In spite of the lateral stream's advantage in fall, the increasing obliquity of its course rapidly reduced its slope. Therefore, with like volume of upper sheet flood, its divide finally became adjusted to medial streams which possessed the advantage of direct flow down the slope. The larger rhythm of the three groups, A, B, and C, appears to have been propagated *down*, not up, the slope. The rhythm of streams flowing off the opposite or left-hand side of the plate is of a different sort, with five major streams of increasing length from top to bottom of the plate.

The main medial streams, 1 and 2 (Fig. 2), and the side streams, show signs of having either shadowed or beheaded the three subordinate medial streams 3, 4, and 5. Nos. 3 and 5 appear to be shadowed by the right and left side streams, respectively, No. 4 appears to have had its right fork beheaded by the left tributary of No. 1, and its left fork by the right tributary of No. 2. Considering the side streams as tributaries of the right and left fall lines as though the latter were two master stream courses, the two fall lines may be considered to have propagated across the plate the rhythmic arrangement observed. Shadowing 3 and 5, they left space for 1 and 2: 4, however, in the middle, was overshadowed by unequal competition with its two rivals.

The process of shadowing begins with the first determination of drainage areas before distinct channels have been eroded. The process of beheading or capture takes place after channels have determined local competitive base levels. The result of competition is determined for the same general zone by relative volumes of water. Beheading is accomplished by undermining headward. Shadowing is accomplished from above mouthward in a very early stage of the development of drainage. Probably on a mathematically uniform surface with uniform conditions, shadowing is a rhythmic process involving many miniature or embryonic beheadings. The theory of arborescent stream development is dependent on such a rhythmic process. The writer believes a complete statement of the mechanism of this process has still to be worked out. Like the many processes of mottling, rippling, wave motion, and bilateral, concentric, and radial symmetry in nature, the development of digitate drainage is a simple group of rhythms in its ideal form, but probably never occurs simply in natural examples. Nevertheless, probably the cases so often described of parallel tributaries in nature, supposedly caused by joint systems, are generally an original or superposed rhythmic tributary system controlled by the law of drainage rhythm. The

Fig. 3. Stream-robbery model, ultimate rill pattern. Tracing of Plate 6.

object of such experiments as those here described should be to reduce each element in the problem to its simplest form, and to vary one

298 BULLETIN: MUSEUM OF COMPARATIVE ZOÖLOGY.

condition at a time, until the primitive types have been discovered. To this end future experiments must be directed — not toward the imitation of complex landscape conditions. The just interpretation

Fig. 4. Condition at 10.50 A. M.

of problems in river drainage can be rigorously attacked only after the primitive types have been worked out experimentally and some careful study given to the influence of each variable in the process. The experiments hitherto described in this paper show results which are still far too complex, but they point the way.

STREAM-ROBBERY MODEL. The last experiment, the results of which are shown in Plate 6, produced an interesting illustration of stream robbery. The size, material, and method of this model were like those of the next preceding (Plate 5), the thickness slightly greater. The cracks shown in the photograph were the result of drying and should be disregarded in the discussion. Just as in the model of Plate 5, the drainage which developed left a clear zone of accumulation above, lateral streams formed on the two sides, and long straight slopeward streams formed medially. The ultimate stream pattern is shown in Fig. 3.

At the end of the first day's spraying the right side streams were about like the left ones shown between R and the left edge of the model. That portion of stream E above the letter E was the headward portion of stream C. There were thus three principal medial streams initially. There was an initial difference of baselevel assumed whereby stream R fell off abruptly at its mouth F, whereas M, B, and C flowed out onto a flat plain P. As a result R became entrenched after the same fashion as the side streams. M, B, and C were depositing their loads and forming floodplains from about the zone of the letter M downwards. The right side stream E, having an abrupt fall-off at the edge of the plate, became entrenched and so possessed a greater fall than the sluggish and relatively overloaded C. Consequently E undermined its way headward rapidly.

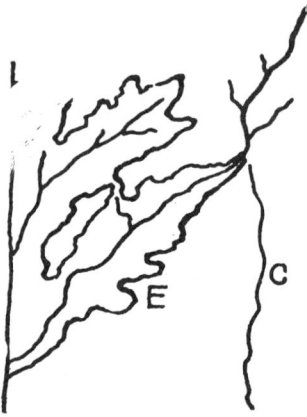

Fig. 5. Enlarged sketch of stream E at moment of capture 11.45 A. M.

On the second morning of the experiment (the spray was turned on for three or four hours each day) E, pushing headward, captured C. Fig. 4 shows a rough sketch of the model at 10.50 A. M., and Fig. 5 shows the effect at the instant of capture. The captured district of C at once revived and entrenched itself and the swollen stream E ate its way across the old floodplain of C. The lower part of C became quite inactive, being effectually shadowed by its captor. The latter (E), however, made no great headway in extending its new headwater basin, because it in turn was shadowed by the next side stream above. The latter controlled a broad portion of the accumulation zone, and being at no disadvantage as to fall, yielded nothing to E. Fig. 3 and Plate 6 show the final result.

The other features of this model which deserve some notice are the symmetry of M (Fig. 3) in contrast to the unsymmetry of R. M is a typical medial stream which has developed freely with a delicate pattern of arborescence. R has had its left bank shadowed by the left side-streams from the first inception of drainage. The left side-streams possessed an advantage in quicker fall to baselevel, consequently they appropriated the waters of the whole area to the left of R, and nothing was provided whereby R could develop left-bank tributaries. R was not so shadowed to the right, where it controlled the accumulation area A over the space between M and the left side-streams. Its right tributaries dominate the space between R and M in a fashion similar to that whereby the left side-streams control the space between the left edge of the model and R. Probably there was a faint initial slope from M left-ward in the original construction of the surface. R was not captured by the left side streams because it was enabled by a low baselevel to trench and maintain a divide on its left bank even though it could not gather water from a wide enough lateral area to develop important tributaries.

DISCUSSION OF PRINCIPLES. The foregoing experiments suggest many questions and answer few. They are based on the assumption that the extraordinary similarity of the rill pattern to the mapped pattern of rivers is due to government in both cases by similar laws. The writer recognizes the fact that in drawing analogies, only the mechanism of falling, running, and seeping waters is imitated, and not the erosion mechanism resulting from degeneration, winds, vegetation, rock-joints, and other phenomena which complicate the problem in nature. He believes, nevertheless, that river, creek, brook, rill, spring, and underground water coöperate in an orderly system of land sculpturing related to structure. This system implies a mechanism hydrostatic, corrasive, and depositive. Physical geography has produced valuable studies of form, and has classified forms in accordance with inferred processes of corrasion and deposition. But physical geography has neglected hydrostatics, and has not quantitatively nor experimentally investigated the processes inferred. This paper will have accomplished its purpose if it starts certain fundamental lines of investigation connected with land drainage. These are suggestive and qualitative, as the experiments have not yet progressed to the quantitative stage.

Drainage areas and stream shadow have been discussed on p. 289; the presence of drainage surfaces essential to arborescence on p. 289; the importance of volume as contrasted with slope on p. 291; the inverse relation of deposits to beds eroded on p. 291; parallelism and rhythm on p. 295 and thereafter; the comparison of capture and shadow on p. 296; and in several places reference has been made to the undermining of surfaces whereby tributaries arise and whereby a propagation both upslope and downslope of certain rhythms may start from an intermediate region. In the trickle-pattern on glass, V-shaped deposits pointing upstream form in a mouthward (p. 292) zone of overload. In most of the experiments there is an upper zone of water accumulation without channelling. Parallelism and rhythm in distribution of tributaries and spurs may by the mechanism of arborescent drainage be satisfactorily accounted for, without any influence of parallel rock joints (p. 297).

In Plate 5, figs. 1 and 2, and Plate 6, a marked characteristic is the presence of much larger portions of the original surface uneroded near the lower edge of the models than in the middle right-and-left zone of many tributaries. The general surface is more lowered in the middle zone than in the upper or lower zones. The middle zone is the region of maximum corrasion, maximum removal of material, and maximum maturity of topography. The interstream divides have been lowered below the original surface level, whereas the original surface level is still preserved in the flat interstream uplands of the lower part of the models. Hence the general upland surface is bevelled headward for a certain distance from the region of the mouths of the streams. This would give it a flat catenary profile from the headward district mouthward.

This profile would be uniform wherever measured. In Plate 5 nineteen or twenty stream depressions would be crossed from right to left in the middle zone and about twelve in the lower zone. Interstream divides in each case rise to a common level. The relief is greater in the lower zone. This is contrary to the prevalent conception, that would expect least relief and greatest maturity in the lower zone (Tarr, 1898). Bevelling and uniformity of crests in the same zone have been discussed by Tangier Smith (1899). He has developed a law of slopes as follows: summits follow slopes, and slopes are dependent on the rate of cutting of the streams at their foot. "If the altitude and rate of cutting in neighboring streams of the same class are approximately the same, then the adjacent divides should approximate equal altitudes" (Tangier Smith, loc. cit., p. 164). "If at the beginning of the cycle of erosion the upland sky-line is markedly irregular, * * * it will depend on circumstances whether or not the uplands will tend to approach uniformity of altitude after graded slopes have been attained" (p. 170). "Adjusted slopes are graded slopes" (p. 165). That slopes tend to become graded and equalized, and that summits or crestlines follow the lead of adjacent slopes is a fact of observation clearly stated by Tangier Smith and further illustrated by the models under discussion here. But Tangier Smith throughout his paper has failed to mention the controlling feature of the process, namely, underground water. The underground water-surface varies with the topography, rising higher under the higher divides and approaching the valleys to form springs. (King, 1899, p. 97–99.) On the slopes adjacent to higher water-table, there will be higher hydrostatic spring pressure and consequently more undermining. Other things being equal, therefore, a high water-table tends to pull down adjacent slopes and consequently, by Tangier Smith's law, lowers superjacent summits. When the water-table level, and with it the summit level, equal those of neighboring divides, the spring-pressure is equalized and the opposing slopes of a valley become adjusted. Accordance of summit levels is probably largely controlled by this process (see also R. A. Daly, 1905, p. 105), with the tree-line and snow-line as correlated levels. A corollary of this statement of the levelling controlled by the water-table is that in a zone of overloaded streams depositing on flood-plains, the water-table may be dammed back into a relatively high position in the low interstream divides. If this be true the profile of the water-table from head-zone to mouth-zone along an interstream divide line should be concave upward with the two ends relatively higher than the middle portion above the catenary curve of the adjacent stream-beds. The water-table will thus have a headward bevel like that of the general surface mentioned above (p. 301) and shown in Plate 5. It should be noted, however, that a headward bevel is different from a headward slope. While still sloping mouthward, the underground water-surface is bevelled headward to the zone of maximum number of tributaries, because the divides in that direction are lower relatively to the original upland surface.

No attempt is made in this paper to discuss "peneplanation," meandering, or the mechanism of flood-plain aggradation. It is believed by the writer that bevelling of divide surfaces to produce even sky-lines has been frequently confused with planation in works on geomorphology. Planation, where it is not marine, requires a flood-plain. (See Gilbert, G. K., 1880, p. 120.) True fluviatile planation is due to lateral corrasion and the test for it is the presence of facetted surfaces across diverse structure, where glaciation or marine planation are positively absent. It is to be hoped that the complex mechanism of planation may some day be subject for the experimental method.

BIBLIOGRAPHY.

Chamberlin, T. C. and Salisbury, R. D.
Geology. New York, 1904, vol. 1, figs. 325–326.
Daly, R. A.
The accordance of summit levels etc. Journ. Geol., 1905, vol. 13, p. 105.
Daubrée, A.
Études synthétiques de la géologie expérimentale. Paris, 1879, p. 352–375, 4 pls.
Davis, W. M.
Physical geography. Boston, 1899.
Dodge, R. E.
Continental phenomena illustrated by ripple marks. Science, 1894, vol. 23, p. 38.
Gilbert, G. K.
Geology of the Henry Mountains. Washington, 1880, p. 120.
Hobbs, W. H.
The river system of Connecticut. Journ. Géol., 1901, vol. 9, p. 469–485, 2 pls., 2 figs.
Examples of joint-controlled drainage from Wisconsin and New York. Journ. Geol., 1905, vol. 13, p. 363–374, 7 figs.
Howe, E. See Jaggar, T. A. Jr.
Iddings, J. P.
A fracture valley system. Journ. Geol., 1904, vol. 12, p. 94–105, 1 pl.
Jaggar, T. A. Jr.
The laccoliths of the Black Hills. With a chapter on Experiments in intrusion and erosion by E. Howe. U. S. G. S. 21st Ann. Rept., 1901.
King, F. H.
Principles and conditions of the movements of ground water. U. S. G. S. 19th Ann. Rept., 1899, pt. 2, p. 59–294.
Meunier, Stanislas.
La géologie expériméntale. Paris, 1899, p. 71, fig. 9.
Nathorst.
The Nathorst-Saporta controversy. Neues Jahrb. f. Geol. Min. u. Pal. 1887, vol. 2, p. 204–209.
Salisbury, R. D.
See Chamberlin, T. C.
Smith, W. S. Tangier.
Some aspects of erosion in relation to the theory of the peneplain. Univ. Cal. Bull. Dept. Geol., 1899, vol. 2, p. 155–177.

Tarr, R. S.
The peneplain. Amer. Geol., 1898, vol. 21, p. 366.
Williamson, W. C.
On some undescribed tracks of invertebrate animals from the Yoredale rocks, and on some inorganic phenomena, produced on tidal shores, simulating plant-remains. Memoirs Manchester Lit. and Phil. Soc., 1887, Ser. 3, vol. 10, p. 19–29, pls. 1–3.

EXPLANATION OF PLATES.

PLATE 1.   Fig. 1.   Seepage streams in unstratified sand. Original 18 × 24 inches (.457 × .609 m).

Fig. 2.   Seepage stream in stratified sand. Original 18 inches (.457 m) delta front to head of stream.

PLATE 2.   Arborescent drainage formed by sudden flood run-off.

PLATE 3.   Stone's "Grand Canyon" model.   Original 48 inches (1.22 m) long.
Note.   Bend in lower bands is apparent, not real, and is due to the perspective of a curved surface in section.

PLATE 4.   Arborescent trickle-pattern formed by tilting a surface covered with liquid clay.   Tilt 45°.   Actual size.

PLATE 5.   Fig. 1.   Slime model, first stage. Sprayed with atomizers.   Original $6\frac{1}{2}$ × $8\frac{1}{2}$ inches (.165 × .216 m).

Fig. 2.   Slime model, second stage.

PLATE 6.   Slime model illustrating stream capture.

PLATE 1

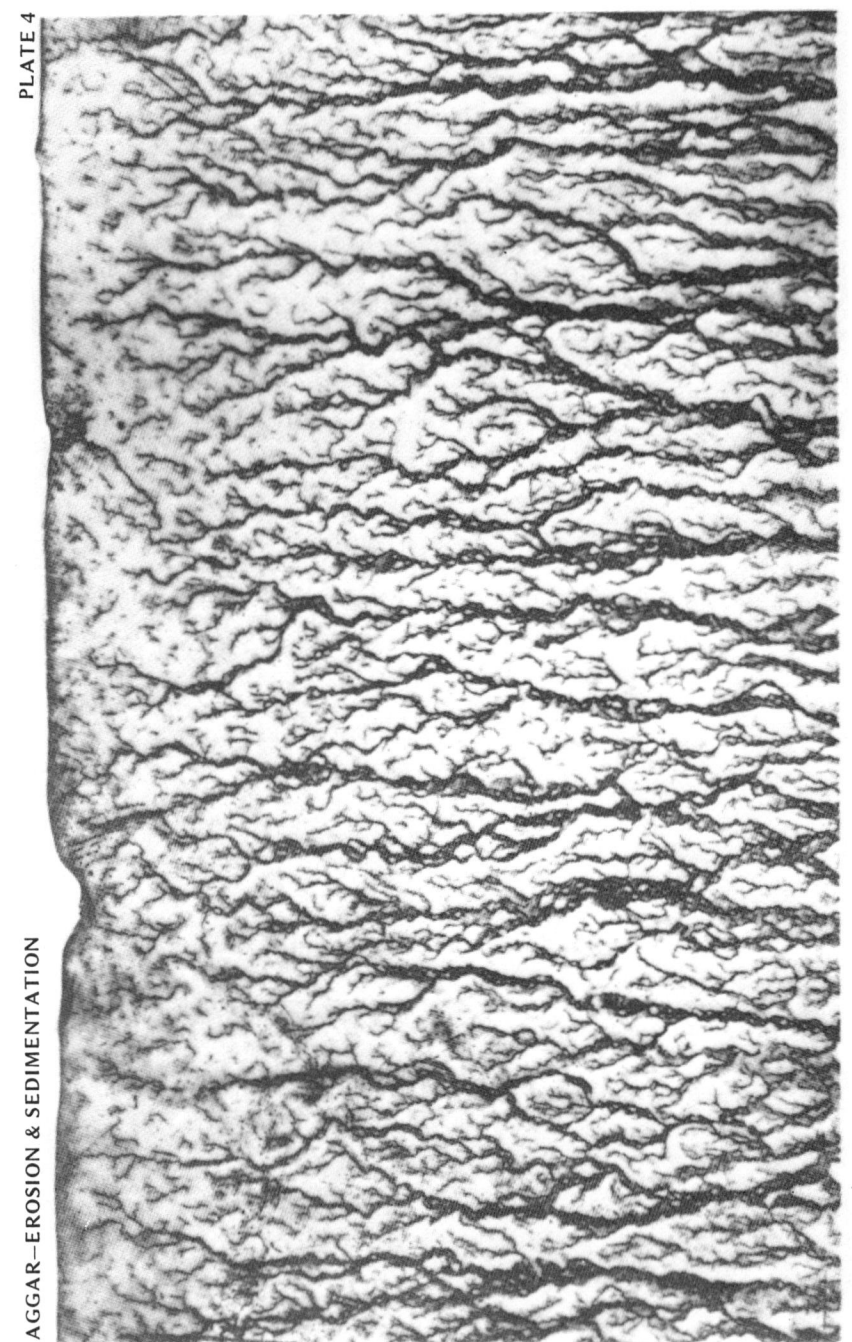

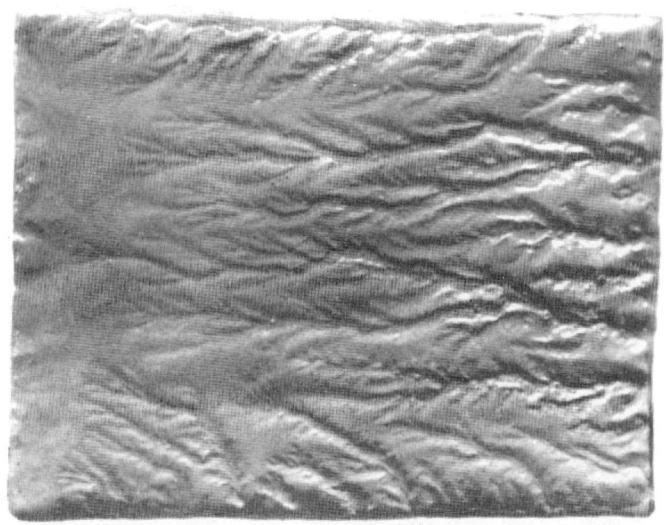

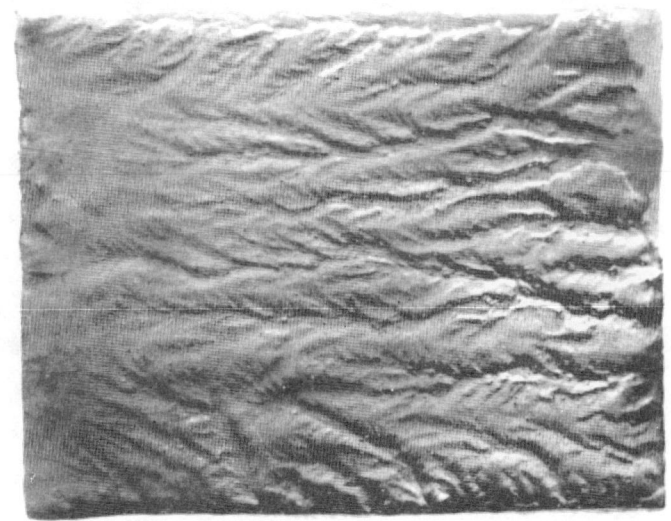

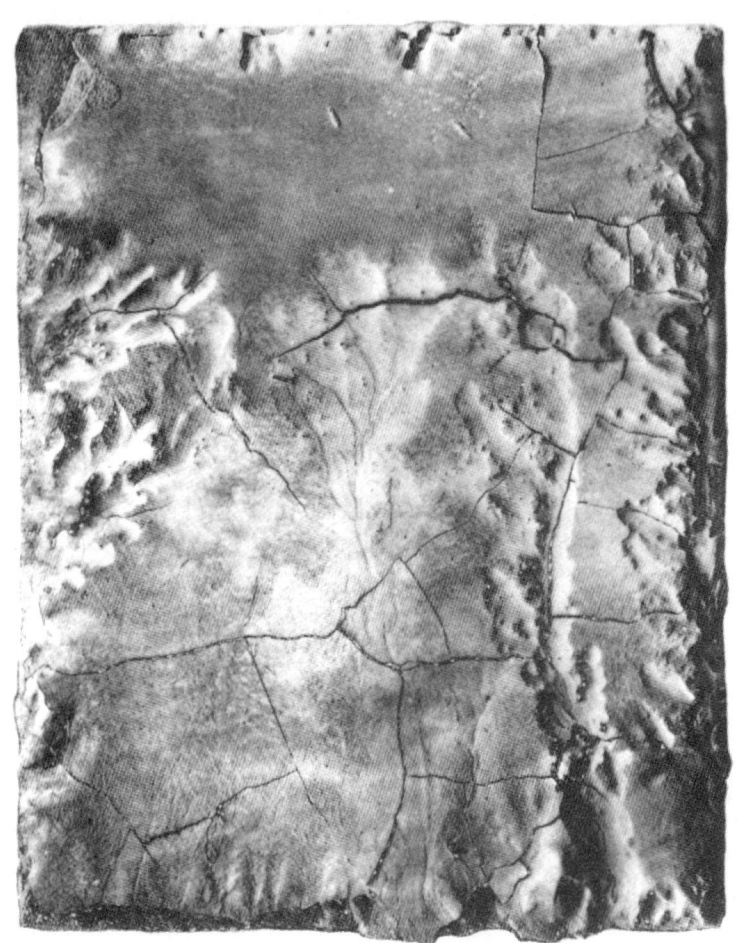

THE DEVELOPMENT OF DRAINAGE SYSTEMS: A SYNOPTIC VIEW*

Waldo S. Glock
Carnegie Institution of Washington

PHYSICAL geography has a dual nature.[1] It includes a static or passive phase involving the detailed form of the land surfaces and a dynamic or active phase involving the agents and processes modifying those surfaces. The first of these phases is known commonly as geomorphology; the second might well be called surficial geodynamics.

The present work desires to confine its attention to streams. It is the purpose of the paper to trace in a summary manner the sequence of pattern during the development of a drainage system after a fashion as nearly ideal as possible for a region of simple rock structure and of humid climate. The following treatment has no intention or wish to ignore the influence of rock and structure nor has it the desire to minimize the fascination in the study of land forms. It does, however, intend to adopt the dynamic viewpoint for the time being.

The sequential stages recognized in the evolution of a drainage system are "extension" and "integration": the first, a stage of increasing complexity; the second, of simplification. The assumption that extension begins on a "new" land surface will be entertained for the present.

THE STAGE OF EXTENSION

Extension may be held to include the genesis (initiation) of streams and thereby the birth of a drainage system, the headward growth (elongation) of the new streams into virgin territory theirs by right of inheritance, and the constant addition (elaboration) of tributaries of decreasing rank up to the time when the system is fully developed from the standpoint of drainage. The form of the drainage system assumed at initiation, which constitutes the first step in the establishment of the system, largely determines the general pattern of the drainage during the early phases of growth. The initial phase of

---

*A summary of parts of a paper presented before the Geological Society of America, Washington, D. C., December 28, 1929. The writer wishes to acknowledge his obligation to the Department of Geological Sciences at Yale University where the investigation into drainage systems was initiated, to the Research Committee of the Graduate School of Ohio State University for a grant of funds which made possible the continuation of the work, to the University's Department of Geology for many valuable favors, and to all three organizations for constant encouragement and material assistance.

[1] Waldo S. Glock: Dual Nature of Physiography, *Science*, No. 1853, Vol. 72, 1930, July 4, pp. 3-5.

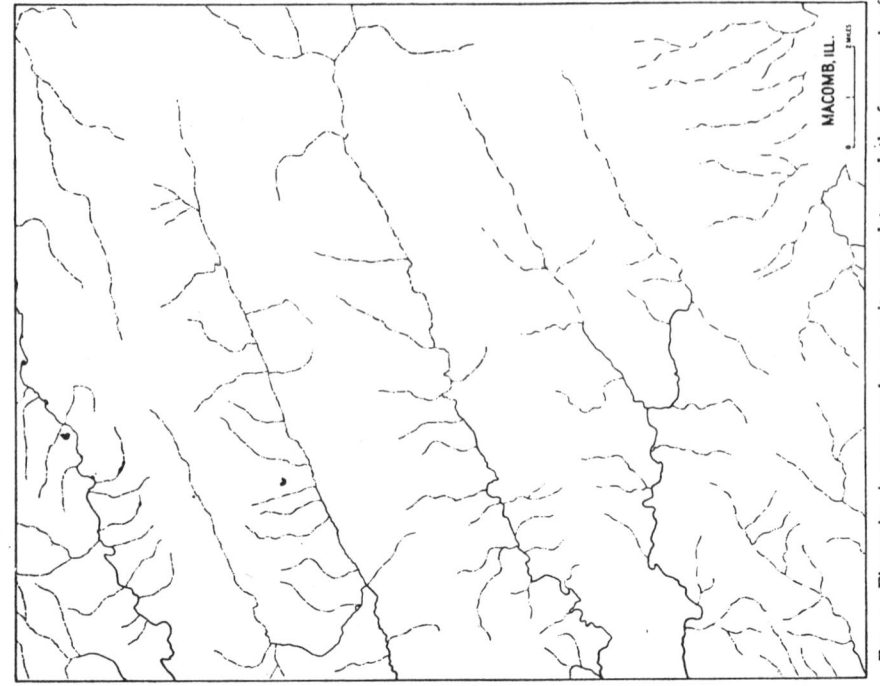

Fig. 2.—Elongation is very nearly, or quite, complete, and the framework of the system has been established. In contrast, elaboration has just begun.

Fig. 1.—The initiation of a drainage system. The first attempts at the formation of definite drainage courses, hence the very beginning of extension.

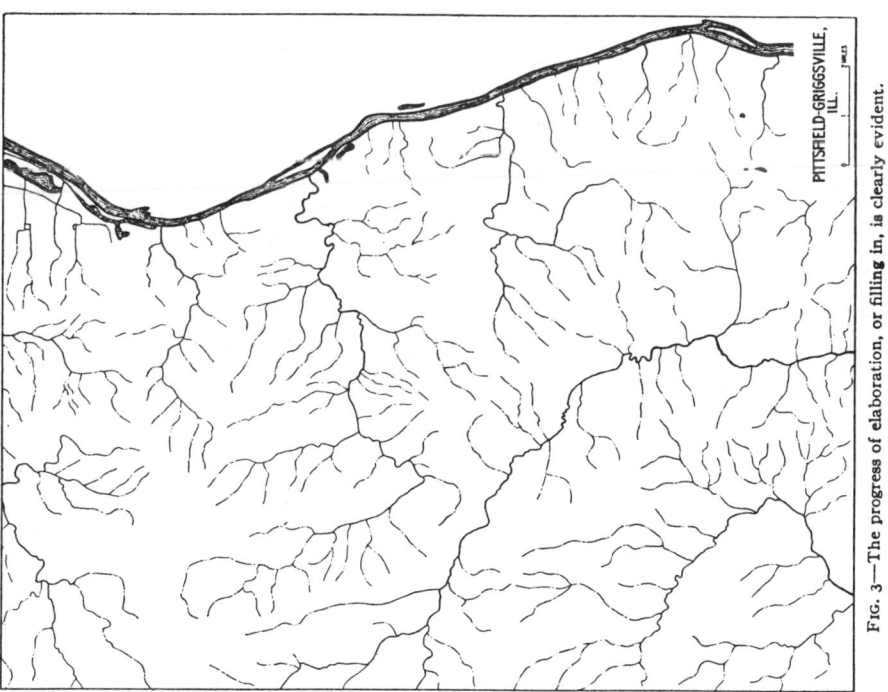

Fig. 3.—The progress of elaboration, or filling in, is clearly evident. PITTSFIELD-GRIGGSVILLE, ILL.

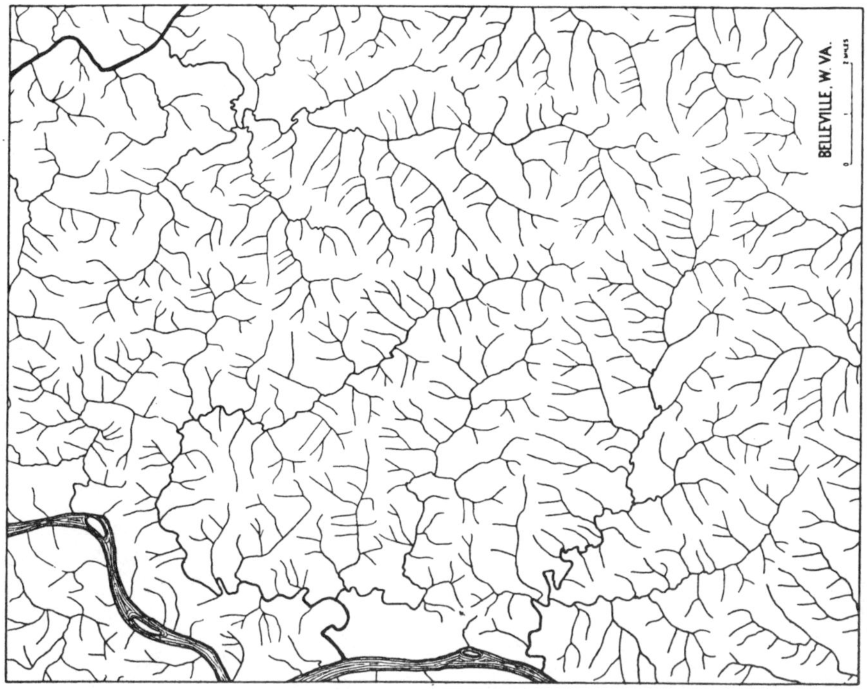

Fig. 4.—Elaboration completed and extension at a maximum. The territory has been wholly occupied by streams. BELLEVILLE, W. VA.

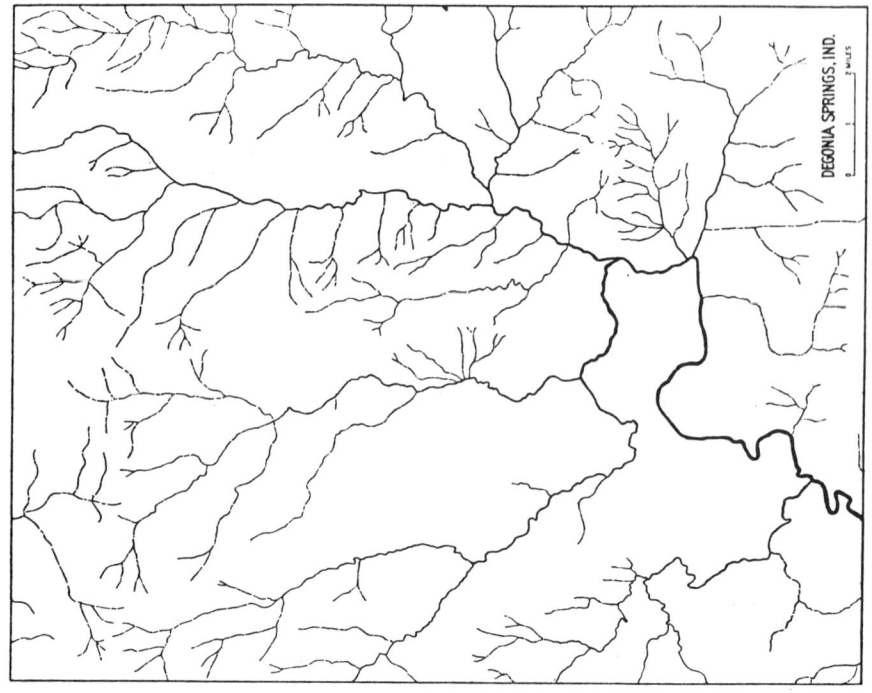

Fig. 6.—The reappearance of the skeletonized form, and integration rather well begun.

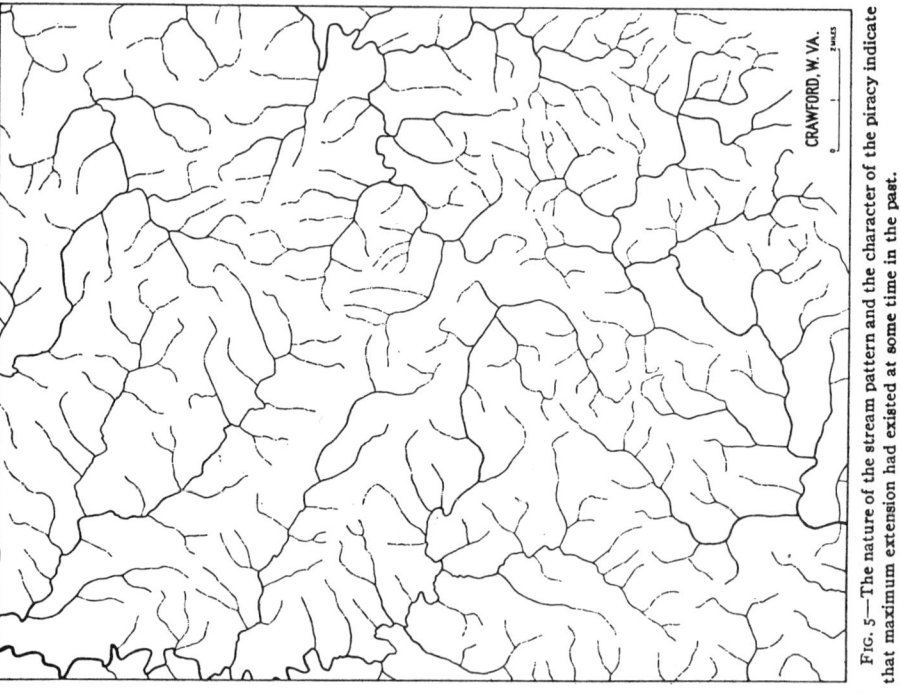

Fig. 5.—The nature of the stream pattern and the character of the piracy indicate that maximum extension had existed at some time in the past.

extension as illustrated by Figure 1 is characterized by a notable lack of streams over a large percentage of surface, by the indefinite termination of many streams without combining into a main stream, and by the failure on the part of the existing streams to have started that active conquest of territory so typical of their future histories. These features are highly typical of the phase initiation but are not necessarily habitual. There is in truth a temptation to say that the drainage system is quite amorphous as yet. However that may be, a time of increasing activity follows the genetic period of hesitation and indecision. The streams begin the conquest of territory, some in a highly aggressive fashion, others in a more or less passive way.

Abbreviated streams and skeletonized systems characterize the stage of extension in a broad way, and these two characteristics must be eliminated before the stage is ended. The growing streams reach out into the territory bequeathed them and take possession of their inheritance by means of headward elongation which begins shortly after the efformation of the system. They sketch in, as it were, the lineaments of the future drainage pattern (Fig. 2). After the area has been thus blocked out the process of elaboration gradually eliminates the skeletal form by means of the addition and growth of minor streams (Fig. 3). A plexus of these accessory tributaries spreads outward from the mains until a veritable network of drainage covers the area first possessed by the elongated streams. With the progress of extension the abbreviated form commonly disappears first while the elimination of the skeletonized pattern, although beginning more slowly, endures throughout the entire stage.

The attainment of complete elaboration may be simply called maximum extension. It is the time of completed territorial conquest and minute invasion—the time of the fully developed drainage system (Fig. 4). The land surface, in other words, has been entirely occupied. Theoretically, the exact position and nature of maximum extension may appear to present unexpected difficulties in view of the almost infinite variation to be found among individual streams. The question may arise whether the elimination of the abbreviated streams on the one hand and the skeletonized system on the other may not lead to different extensional maxima, especially if opposing systems enter into active competition with each other, an activity tending to elongate certain streams at the expense of others. Should lengthening accomplished by piracy after elaboration has been completed fall within the scope of extension? It seems not. Figure 5 shows that a system may grow at the expense of a neighbor after maximum extension has been attained, since the pattern of the captured streams themselves indicates that the territory had been completely occupied when the piracy occurred.

From a practical standpoint, also, the essential termination of the processes of elongation and elaboration may be more readily recognized and circumscribed if held to the simple concept inherent in the attempted description of extension. The scheme of classification

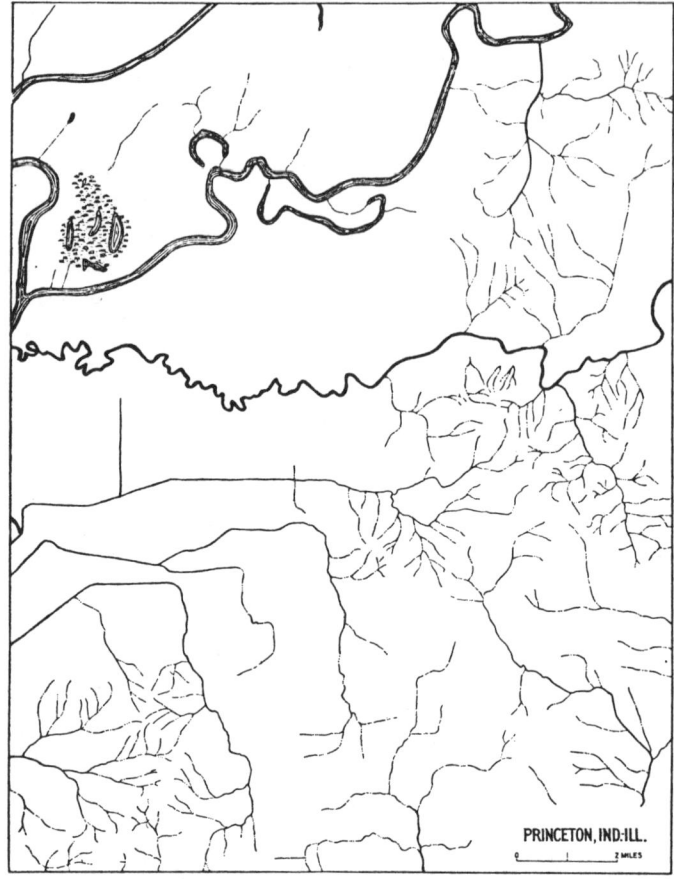

FIG. 7—Master streams and master tributaries late in the stage of integration.

to be desired is the one that is as simple and, at the same time, as adequate as possible. True maximum extension may be considered to refer to the completed invasion of the unoccupied surface within the jurisdiction of a particular drainage system after active elongation into rightfully inherited territory has added all possible area to the system. Hence, the intensity of intersystem strife may wax to a maximum either with, or after, the full development of the drainage systems, and extension may be at a maximum or may have definitely passed away when systems "come to an understanding about their drainage areas." Piracy between systems does not serve as a limiting criterion for extension. Treated, therefore, in the preferred fashion, the scheme is simple enough to conform to the standards which, it is believed, an ideal scheme should follow. Little doubt can exist that such an ideal scheme acts as an axis of variation about which natural phenomena appear to group themselves.

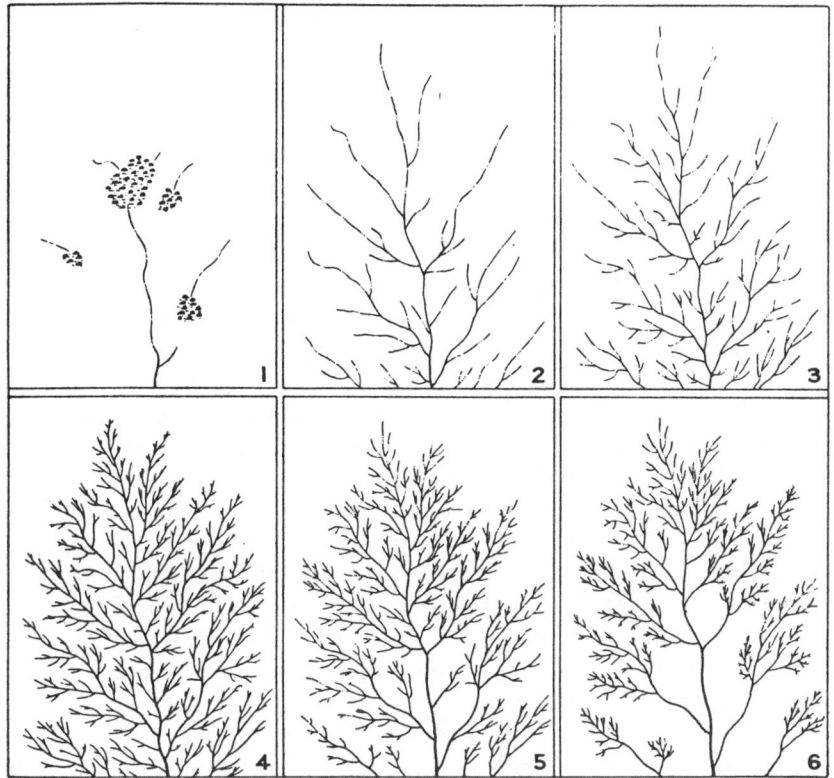

FIG. 8—An ideal diagrammatic summary of the development of a drainage system given for purposes of comparison only. The first four parts show extension, thus: 1, initiation; 2, elongation; 3, elaboration; and 4, maximum extension. Parts 5 and 6 represent steps during integration.

THE STAGE OF INTEGRATION

The processes responsible for integration may be designated as follows: (1) abstraction, the loss of identity suffered by a secondary stream at the hands of its primary; (2) absorption, the disappearance of a stream save immediately after rainfall; and (3) a sort of adjustment or aggression, the attempt made by the main stream to reach the sea by the shortest route consistent with regional slope. The reappearance of the skeletonized form out of the intricate plexus of streams some time after maximum extension definitely marks the existence of integration (Fig. 6). It constitutes the second and final stage in the developmental history of a drainage system.

Abstraction refers to the elimination of a secondary stream by its primary. As the stream swings from side to side it constantly increases its own drainage area and thus abstracts territory from its tributaries. The lower portions of a tributary system may be dismembered in this way, all of the shorter branches may be completely destroyed, and the lower part of the chief tributary may lose its identity and independence on the meander belt of the master.

With integration well begun many of the minor tributaries seem to be absorbed and thereafter come into existence only for the purpose of discharging the immediate rainfall. Their disappearance quite likely is involved in the flattening of the local ground-water surface during the progress of integration and the consequent sinking of that surface below stream level.

The third process is less easy, if not impossible, to observe to any great extent at the present day. As integration advances into the final phase, the master stream becomes more and more independent of its environment except for the general slope of its drainage area and its relation to the sea or point of discharge. Aggression is said to take place when the stream attempts to gain the shortest (valley) route to the sea governed solely by the general slope of its drainage area, and then only if that direct route did not exist at initiation or was abandoned for any reason later. A stream having worked out such a course for itself may conveniently be called an integrational consequent and undoubtedly finds abundant illustration in geologic history.

The three processes accomplish the elimination of a host of streams and result in the appearance of the simplified form of integration very similar to the skeletal outline of early extension. Master streams joined by master tributaries (Fig. 7) at last emerge from the plethora of previously existing streams and, if necessary, seek the shortest practicable route to the sea.

In some respects integration appears to reverse the procedure of extension, as for instance the disappearance of minor tributaries, the continuation of which appears constantly to accentuate the original framework of the system. The stages in reality do resemble each other in that the sequence of changes (with one exception) sweeps headward along the trunk streams and thence to the chief branches, arriving finally at the outermost confines of the system—a migration quite naturally anticipated. Here, however, resemblance ceases.

There is no intention to infer that the scheme of development including extension and integration, because of its simplicity, possesses definiteness and perfection to such a degree that each is a distinct stage by itself. In fact, the two may always exist at the same time from a regional standpoint; or, further, extension may be occurring locally among the headwaters while integration is taking place farther down stream. Minor cases of actual overlap should be expected in any detailed study of a drainage system.

12

Reprinted from pp. 597–598, 599–622, 636–641, and 645–646 of *Geol. Soc. America Bull.* 67:597–646 (May 1956)

EVOLUTION OF DRAINAGE SYSTEMS AND SLOPES IN BADLANDS AT PERTH AMBOY, NEW JERSEY

By Stanley A. Schumm

Abstract

To analyze the development of erosional topography the writer studied geomorphic processes and landforms in a small badlands area at Perth Amboy, New Jersey. The badlands developed on a clay-sand fill and were morphologically similar to badlands and areas of high relief in semiarid and arid regions. A fifth-order drainage system was selected for detailed study.

Composition of this drainage network conforms to Horton's laws. Within an area of homogeneous lithology and simple structure the drainage network develops in direct relation to a fixed value for the minimum area required for channel maintenance. Observed relationships between channel length, drainage-basin area, and stream-order number are dependent on this *constant of channel maintenance* which is in turn dependent on relative relief, lithology, and climate of any area.

Other characteristics of the drainage network and topography such as texture, maximum slope angles, stream gradients, drainage-basin shape, annual sediment loss per unit area, infiltration rate, drainage pattern, and even the morphologic evolution of the area appear related to relative relief expressed as a *relief ratio*, the height of the drainage basin divided by the length. Within one topographic unit or between areas of dissimilar but homogeneous lithology the relief ratio is a valuable means of comparing geomorphic characteristics.

Hypsometric curves are available for a series of 11 second-order drainage basins ranging in stage of development from initial to mature. Relief ratio and stream gradients attain a constant value when approximately 25 per cent of the mass of the basin has been eroded. Basin shape becomes essentially constant at 40 per cent of mass removed in accord with Strahler's hypothesis of time-independent forms of the steady state.

Comparison of the drainage pattern as mapped in 1948 with that of 1952 reveals a systematic change in angles of junction and a shift of the entire drainage pattern accompanying changes in the ratio between ground and channel slope.

Field observations and experimental studies suggest that badland slopes may retreat in parallel planes and that the rate of erosion on a slope is a function of the slope angle. The retreat of slopes may not conform to accepted concepts of runoff action as a function of depth and distance downslope. Runoff occurs as surge and subdivided flow which may be closely analogous to surficial creep.

Rills follow a definite cycle of destruction and reappearance throughout the year under the action of runoff and frost heaving.

At Perth Amboy, slopes are initiated by channel degradation and maintained by runoff and by creep induced through frost heaving. Runoff or creep may form convex divides, and both parallel and declining slope retreat are important in the evolution of stream-carved topography.

Hypsometric curves reveal that the point of maximum erosion within a drainage basin migrates upchannel and that the mass-distribution curve of any basin has a similar evolution to that of the longitudinal stream profile.

Comparative studies in badland areas of South Dakota and Arizona confirm conclusions drawn at Perth Amboy and show the importance of infiltration of runoff on topographic development and of subsurface flow in slope retreat and miniature pediment formation.

CONTENTS

TEXT

	Page
Introduction	599
Acknowledgments	600
General description of the Perth Amboy locality	600
Characteristics of the drainage network	602
Components of the drainage network	602
Limiting values of drainage components	607
Form of the drainage basins	612
Basin form related to geomorphic stage of development	614
Evolution of the drainage network	617
Effect of stage on angles of junction	617
Evolution of the Perth Amboy drainage pattern	620
Field observations and experimental studies on the development of badland topography	622
Field-erosion measurements	622
Experimental erosion measurements and study of runoff	627
Seasonal effects on erosion; the rill cycle	632
Cycle of development of erosional topography	634
Relation of stream profiles to slopes	634
Available relief and the development of landforms	636
Hypsometric study of geomorphic stages of development	638
Comparative studies in badland regions of the West	641
Topographic forms and erosion processes	641
Influence of regional upland slope on topography	644
Summary and conclusions	645
References cited	646

[*Editor's Note:* The list of illustrations has been deleted.]

INTRODUCTION

The factor of time has been a major difficulty in investigations of the evolution of landforms. Laboratory scale models which operate rapidly are unsatisfactory because scale ratios cannot be suitably maintained. An alternative exists, however, in studies of badland regions of the western United States, where detailed investigations (Smith, In press) within small areas undergoing rapid erosion have led to inferences concerning processes operating on morphologically similar but larger erosional landforms.

Small badland areas developing in clay pits on the Coastal Plain of New Jersey were also suitable for study and could be observed durng an annual cycle of climatic changes. A. N. Strahler and D. R. Coates of Columbia University had selected one area at Perth Amboy, New Jersey for detailed study and mapping in 1948. In 1951 the writer began his investigation of the area using Strahler's map as a basis for morphometric studies.

Several methods were used to evaluate the effectiveness of the erosive processes and to study the evolutionary development of badland topography: (1) Topographic and drainage maps, made with plane table and telescopic alidade, were sufficiently detailed for measurement of all components of the drainage net with planimeter and chartometer. A statistical analysis of the data followed. (2) Stakes placed along stream and slope profiles permitted repeated measurements of erosion within selected basins. (3) Photography from selected stations, combined with remapping, gave information on changes within the system. (4) Experiments made upon samples of fill from the Perth Amboy area showed the effect of slope angle on sediment loss and upon size of entrained particles of the eroded material. (5) Comparative field studies were made in selected badland areas of the western United States.

Landforms are functions of structure, process, and stage; all differences in landforms can be explained through combinations of these factors. Geomorphologists have emphasized striking form differences but have neglected the importance of persistent similarities among forms of geologically and climatically diverse areas. The morphology of the small badland areas is remarkably similar to youthfully dissected areas of high relief such as the recently uplifted mountains of the West, or any region of recent uplift that has not been severely glaciated. Scale of topography may have less influence than other factors in the creation of distinctive features of the landforms.

The project's primary objective was to study the development and modification of badlands, but it may also be considered a type of large-scale model study, with the hope that some of the conclusions reached at Perth Amboy may be extended in a tentative way to larger areas.

Fenneman (1922, p. 126) suggested the value of such studies in developing an understanding of the development of erosional landforms:

"The physiographer's conception of the progressive dissection of peneplains, as of other plains, is based mainly on the growth of gullies. Rapid downcutting, steep sides, V-shaped cross section, and branching enter most abundantly into his mental pictures of an upland of homogeneous material or horizontal rocks that is undergoing dissection, especially when erosion is vigorous."

ACKNOWLEDGMENTS

This investigation formed part of a quantitative study of erosional landforms sponsored by the Geography Branch of the Office of Naval Research as Project Number NR 389-042 under Contract N6 ONR 271, Task Order 30, with Columbia University.

The writer wishes to thank Prof. A. N. Strahler, who sponsored the project, for the use of data collected at Perth Amboy before 1951, and for the topographic map of the area (Pl. 1). Professor Strahler and members of the Seminar in Geomorphology at Columbia University during the years 1952 and 1953 gave much valuable criticism and advice. Messrs. J. T. Hack, L. B. Leopold, H. V. Peterson, and M. G. Wolman of the U. S. Geological Survey, Prof. John Miller of Harvard University, and Professor Strahler read and offered valuable suggestions for improvement of the manuscript.

During the field season of 1952 Mr. Iven Bennett of Rutgers University acted as field assistant. M. Rossics and A. H. Schumm also helped in the field. Mr. A. Broscoe kindly furnished the drainage map of the Chileno Canyon drainage basin.

GENERAL DESCRIPTION OF THE PERTH AMBOY LOCALITY

The Perth Amboy badlands, where most of the investigation was made, were located on the western boundary of Perth Amboy, New Jersey, on the north bank of the Raritan River between the two highway bridges over the river. The area was conspicuous because it was anomalous in the humid climate of New Jersey and resembled larger badlands of the more arid West (Pl. 2, figs. 1, 2, 3). The Perth Amboy badlands are not unique, however; badlands have developed elsewhere in humid climates. Rapid badland erosion is occurring in the Ducktown Copper Basin of Tennessee where $10\frac{1}{2}$ square miles are devoid of vegetation because of destruction by smelter fumes. It is considered the largest bare area in any humid region of the United States (Hursh, 1948, p. 2). Similar erosion occurs where volcanic ash covers the surface sufficiently to destroy vegetation (Segerstrom, 1950).

All the erosion forms in this badlands area were developed after 1929 when waste and overburden from other pits backfilled the abandoned clay pit at this site, producing a bench or terracelike deposit 40 feet high whose steep front and gently sloping upper surface rapidly gullied. The terrace as a whole was still in a youthful stage at the beginning of this study. Alluvial fans had been built along its base (Pl. 2, fig. 1, 2). The easily eroded terrace might be considered the initial stage of landform development of the theoretical Davisian cycle, with the elevated flat-terrace surface representing a rapidly elevated peneplain. Similar small badlands have developed elsewhere in the area where removal of vegetation and Pleistocene deposits have exposed the soft Cretaceous Raritan clays.

Plans for continued field studies to observe the cycle of development were unfortunately concluded when the area was leveled for construction purposes in the summer of 1953.

Reports of the nearest weather station at New Brunswick, New Jersey, $8\frac{1}{2}$ miles northwest, reveal that erosion up to 1948 was accomplished by a total of 844 inches of precipitation and that mean yearly precipitation was 43 inches (U. S. Weather Bureau, 1929–1948). These figures contribute to a statement of general climatic conditions but not to a quantitative evaluation of the effect of rainfall, because there are no data on the local intensities of precipitation during this period. Intensities here are not low, however, compared with storms in the badland regions of the West. The maximum intensities of precipitation of any storm over periods of time ranging from 2 to 100 years may be expected in the eastern portion of the humid regions of the United States (Yarnell, 1935). Undoubtedly the erroneous impression of higher intensities in the arid West is due to the destructive nature of runoff on sparsely vegetated slopes.

Temperatures during this period ranged from 0° to 95°F., indicating that frost heaving was probably important during the $5\frac{1}{2}$ months between the first and last severe frosts. The east-

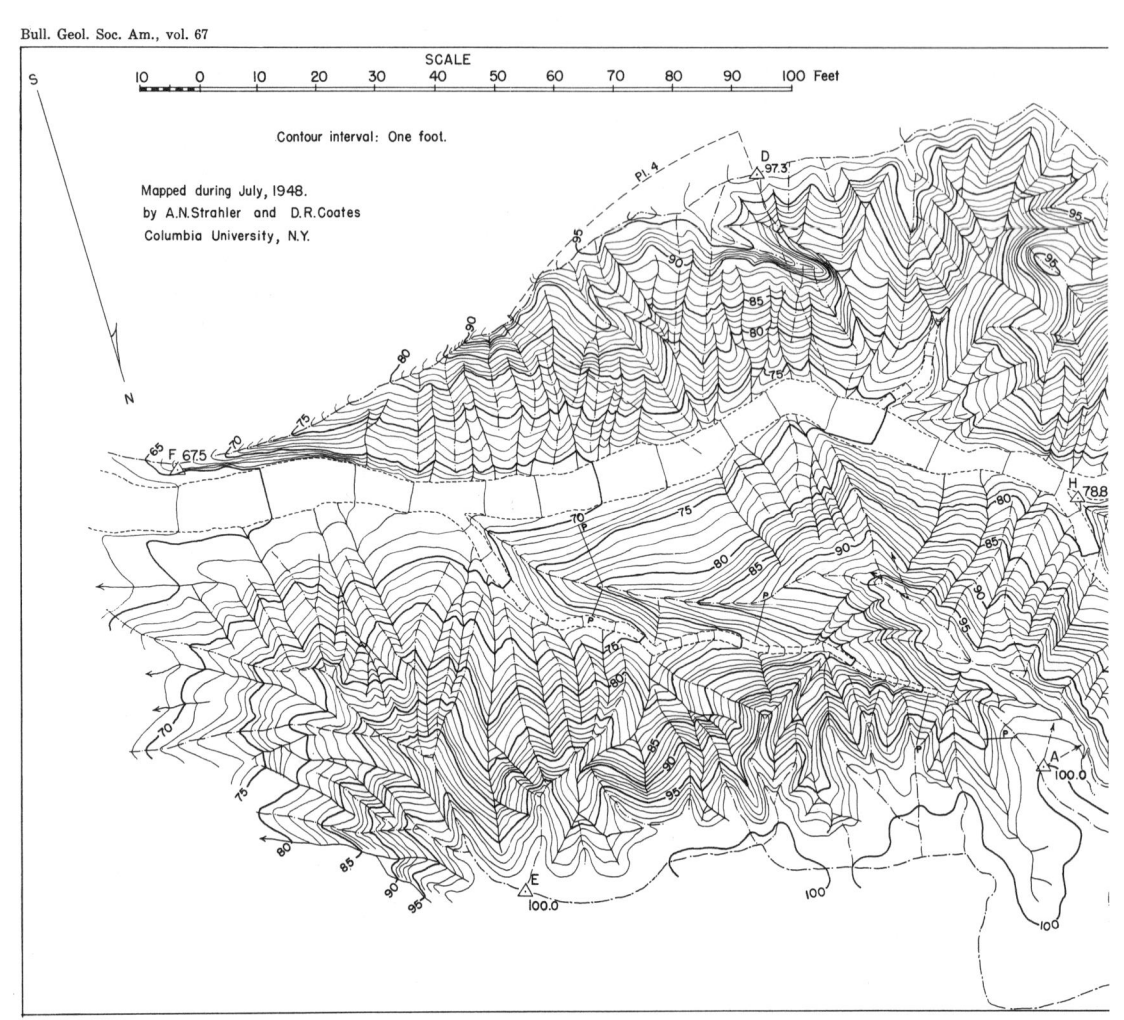

TOPOGRAPHIC MAP OF THE PERTH AMBOY BADLANDS

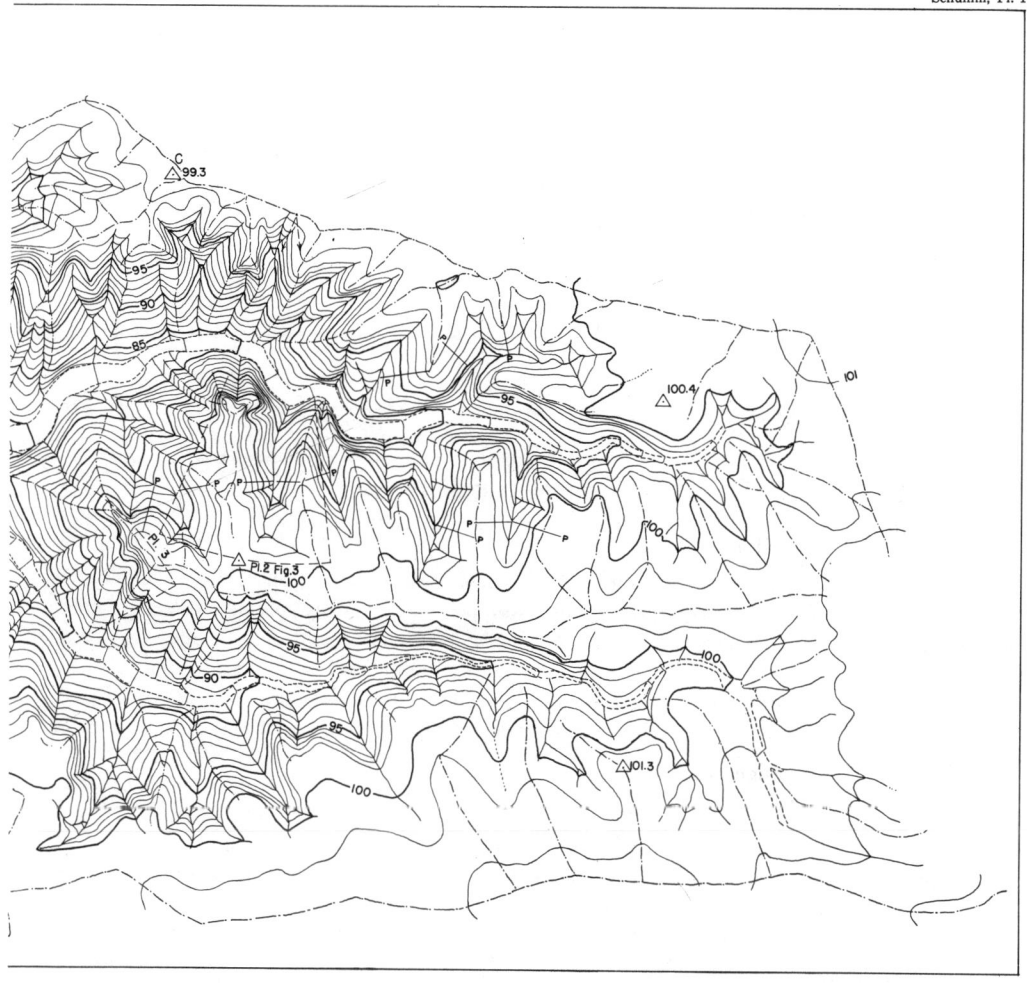

west orientation of the major drainage channels affects the microclimatic environment, subjecting the north- and south-facing slopes to different frequencies of freeze and thaw.

In spite of high precipitation the sterility of the fill and the rapidity of erosion prevented

It may be suspected that frost heaving probably modifies the topography somewhat because the soil has moderate to objectionable frost-heaving characteristics and the low temperatures at Perth Amboy cause frequent freezing during the winter months.

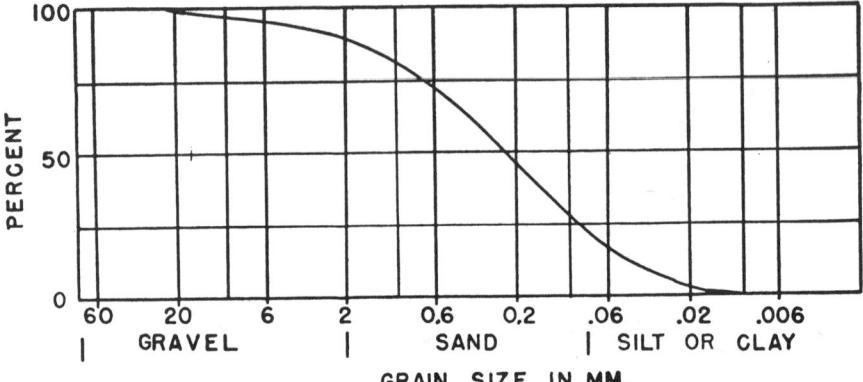

FIGURE 1.—GRAIN-SIZE DISTRIBUTION OF PERTH AMBOY FILL

the growth of vegetation within the basin studied, except for one wild cherry tree and a small patch of poison ivy on the western drainage divide.

The fill forming the terrace at Perth Amboy is essentially homogeneous and the topography shows no persistent control by structure or lithology. Mr. Richard Chorley of Columbia University made a grain-size analysis of the fill, showing that it is 67 per cent sand, 22 per cent silt and clay, and 11 per cent gravel. The almost equal amounts of silt and clay were determined by the low plasticity of the finer fraction of the sediment (Burmister, 1952, p. 102). A cumulative percentage curve was plotted for the sample (Fig. 1) and from this Hazen's "effective size", designated D_{10} by Burmister, was obtained. The quantity D_{10} is the grain size in millimeters for which 10 per cent of the material is finer. For the Perth Amboy fill D_{10} is 0.04 mm. Using this value the fill characteristics important to this study can easily be determined using Burmister's charts: (1) the drainage of the fill is fair with a coefficient of permeability approximately 0.004 cm/sec.; (2) the capillarity is moderate with an approximate rise of 5.0 feet; (3) the frost-heaving characteristics are moderate to objectionable.

The importance of the values for capillarity and permeability are apparent from Burmister's (1952, p. 83) statement that, during excavation, slopes seldom need to be cut flatter than 45° "in soils containing more than about 5% silt except where surface erosion of the slopes by the rapid runoff of rainwater is excessive." The 5 per cent silt causes enough capillary cohesion in the soil to maintain steep slope angles, and the permeability is low enough to aid runoff. Thus, steep slopes are maintained, but erosion is rapid.

The effect of capillary cohesion within the Perth Amboy fill makes the mean maximum slope angle for the area 48.8°, and thus the slopes are classified by Strahler (1950, p. 693) as high-cohesion slopes.

The drainage basin selected for intensive study was mapped in 1948 by Strahler and Coates of Columbia University on a scale of 1 inch equals 10 feet (Pl. 1). The contour interval is 1 foot. The drainage pattern was mapped with particular care so that all drainage-basin characteristics could be measured; all channels possessing recognizable drainage areas were considered permanent drainage features and mapped as such.

The drainage basin mapped was that of a fifth-order stream network. Streams are designated on the basis of orders; all unit or fingertip stream channels without tributaries are first-order streams (Horton, 1945, p. 281); the junction of two streams of the same order forms a stream of the next higher order.

In all the drainage basins studied the streams are assigned order numbers following the method outlined by Strahler (1952b, p. 1120) whereby the higher stream-order numbers are not extended headward to include smaller tributaries, but refer to segments of the main channel (Fig. 9). With the Horton classification, the higher stream-order numbers include the smallest headward extension of the main stream. Using Strahler's method, the two major channels joining at point H (Pl. 1) are third-order; using Horton's method, the south tributary would be the extension of the fifth-order channel and would be eliminated from studies involving third-order channels. This method will be referred to again in a discussion of channel lengths.

The fifth-order basin mapped includes 3531 feet of drainage channels within an area of 31,027 square feet. The drainage density (Horton, 1945, p. 283), equal to the sum of the channel lengths in miles divided by the area of the drainage basin in square miles, is 602, indicating that within an area of this type 602 miles of drainage channels occur for every square mile of drainage basin. This value is indicative of the fine texture of the area. Although the density is high compared to a typical value of 5 to 20 for humid regions, it is not high for badland topography.

Within the mapped area the first-, second-, and third-order stream basins show a transition from maturely developed topography near the mouth of the main stream, where the main channel has widened the valley until small segments of flood plain have developed, to progressively more youthful basins toward its head, where the tributaries are eroding into portions of the undissected surface of the fill.

The mean length of the first-order channels is 10.1 feet, and the mean drainage area is 85.0 square feet, indicating the small scale of the topography. The hypsometric integral (Strahler, 1952b) for the entire fifth-order network is 70 per cent, indicating that erosion has removed a minimum of 30 per cent of the total mass of the basin. This figure is reasonably accurate because the upper surface of the terrace into which the system developed is still preserved in the headwater areas. Although the terrace is not a natural deposit and the drainage is developing on a small scale, investigation of principles of drainage-network development is aided by knowledge of several factors not available in the study of the geomorphic evolution of other areas.

The homogeneity of the fill aided development of an insequent drainage pattern on the terrace. The rapid erosion developed youthful V-valleys with steep straight slopes descending from convex or sharp-crested divides. The longitudinal profile of the main channel, typical of streams growing headward into an upland surface, had a concave lower segment and an upper convexity where degradation was most rapid. Tributary profiles varied with stage from convex-up, where the streams were unable to maintain themselves against the rapid degradation of the main channel in the headwater areas, to concave-up where the main channel appeared to be at grade.

Stream-channel erosion with sheet and rill erosion on the slopes were the dominant geomorphic processes observed. Wind erosion was negligible, but frost action became important during the winter months.

CHARACTERISTICS OF THE DRAINAGE NETWORK

Components of the Drainage Network

Morphometric studies of drainage-network components at Perth Amboy included measurements of all stream-channel lengths and drainage-basin areas for all stream orders, so that each component could be studied independently. Stream-order analysis permits comparison of the drainage network developed on the Perth Amboy terrace with patterns originating under natural conditions. Horton (1945) proposed certain laws of drainage composition which assume an orderly development of the geometrical qualities of an insequent drainage system. These laws were applied to data obtained from morphometric measurements on the Perth Amboy map (Pl. 1) to determine whether they conformed; if they did, conclusions from the Perth Amboy study might apply to other larger areas. Geometry of two other fifth-order basins was measured for comparison with Perth Amboy basin (Table 2): Chileno Canyon basin (Chileno Canyon, California,

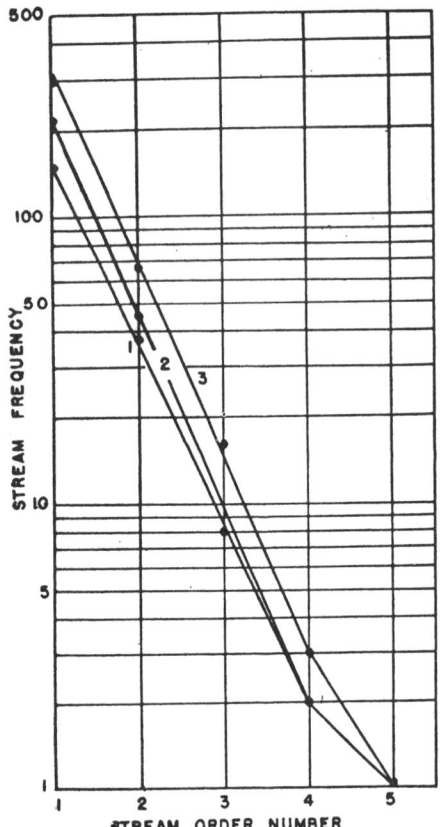

FIGURE 2.—RELATION OF NUMBER OF STREAMS OF EACH ORDER TO ORDER NUMBER
(1) Perth Amboy, (2) Chileno Canyon, (3) Hughesville area

quadrangle) and Mill Dam Run basin (Hughesville, Maryland, quadrangle).

The law of stream numbers, first of Horton's laws of drainage composition, is stated as follows (Horton, 1945, p. 291): "The numbers of streams of different orders in a given drainage basin tend closely to approximate an inverse geometric series in which the first term is unity and the ratio is the bifurcation ratio." If a geometric series exists, a straight-line series of points results where the numbers of streams of each order are plotted on a logarithmic scale on the ordinate against order numbers on an arithmetic scale on the abscissa. This has been done in Figure 2, in the manner of Horton's graphs. All three sets of points show a marked up-concavity at the lower end, suggesting that the geometric progression is not closely observed in the higher orders, but the Perth Amboy data show general similarity with the rest

TABLE 1.—METHOD OF DERIVING WEIGHTED MEAN BIFURCATION RATIO

1 Stream order	2 Number of streams	3 Bifurcation ratio	4 No. of streams involved in ratio	5 Products of columns 3 and 4
1	214			
		4.78	259	1238.0
2	45			
		5.63	53	298.4
3	8			
		4.00	10	40.0
4	2			
		2.00	3	6.0
5	1			

Total number of streams used in Col. 4 = 325.
Sum of products of Col. 5 = 1582.4.
Weighted mean bifurcation ratio = $\frac{1582.4}{325}$ = 4.87.

and there is no reason to believe that any fundamental dissimilarity exists.

The weighted mean of the Perth Amboy bifurcation ratio is 4.87. Bifurcation ratio is the ratio of the total number of streams of one order to that of the next higher order (Horton, 1945, p. 280), e.g., a basin with 20 second-order channels and 60 first-order channels would have a bifurcation ratio between these two orders of 3. Because of chance irregularities, bifurcation ratio between successive pairs of orders differs within the same basin even if a general observance of a geometric series exists. To arrive at a more representative bifurcation number Strahler (1953) used a weighted-mean bifurcation ratio obtained by multiplying the bifurcation ratio for each successive pair of orders by the total number of streams involved in the ratio and taking the mean of the sum of these values (Table 1).

The second law stated by Horton (1945, p. 291) concerns stream lengths: "The average lengths of streams of each of the different orders in a drainage basin tend closely to approximate a direct geometric series in which the first term is the average length of streams of the first order."

lengths although the value of the length of the fifth-order stream is low in two cases Because integer values only are used for order numbers continued channel development might be ex-

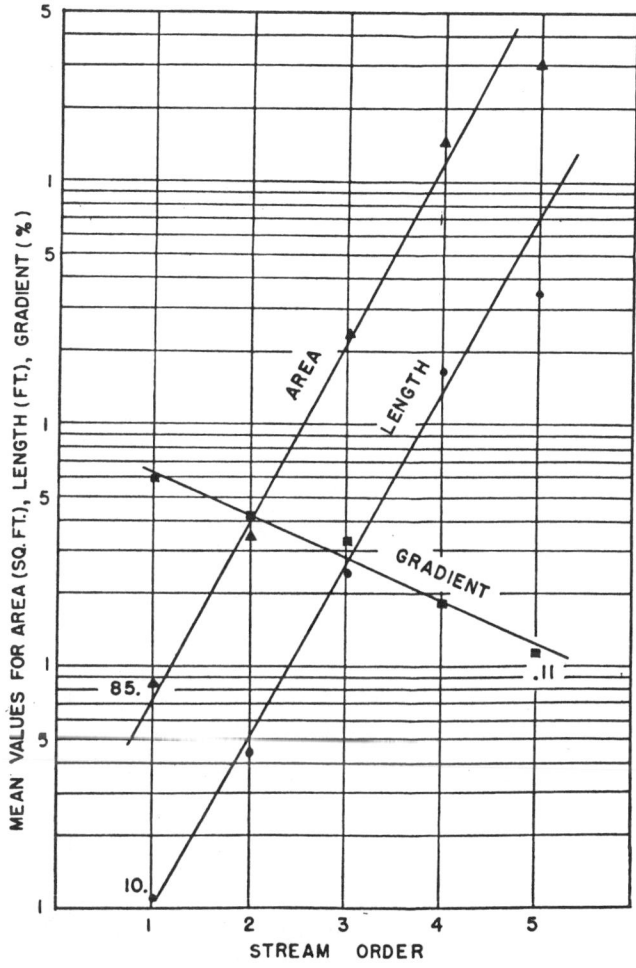

FIGURE 3.—RELATION OF MEAN BASIN AREA, MEAN STREAM LENGTH, AND MEAN STREAM GRADIENT TO STREAM ORDER

The length of streams of each order was obtained by measuring all the drainage channels within a basin of a given order; the length of the fifth-order stream at Perth Amboy is the total length of all the channels within the basin. This method differs from Horton's, but the total channel lengths may be more meaningful when considered within the area of each drainage basin. Using this method, however, the mean stream-length plots (Figs. 3, 5) for the Perth Amboy, Chileno Canyon, and Mill Dam Run systems adhere to Horton's law of stream

pected to raise the value of the fifth-order length, bringing it closer to the fitted regression line. The shortness of the fifth-order segment at Perth Amboy may be due to the truncation of the drainage pattern at the front of the terrace. In Figure 4 both the Hughesville and Perth Amboy drainage patterns have short fifth-order segments in comparison to that of the Chileno Canyon area. This may explain the low fifth-order channel lengths and areas for those two basins (Figs. 3, 5).

Horton's third law (1945, p. 295) states:

"There is a fairly definite relationship between slope of the streams and stream order, which can be expressed by an inverse geometric series law." The Perth Amboy stream slopes appear to conform (Fig. 3). In this case the gradient is obtained by dividing stream length measured from mouth to headwaters by the elevation difference.

Horton's laws may require revision because he obtained his data from old maps of small scale on which he measured as stream channels only the blue drainage symbols, thus omitting a large part of the first- and second-order channet network. His statements are sound, however, in the light of investigations made on modern topographic maps, either mapped for the purpose (Perth Amboy basin) or selected because of their large scale and detailed representation of topography (Hughesville and Chileno Canyons quadrangles). These undoubtedly afford data more precisely representative of the natural development of drainage systems than the old maps.

The writer compared the Hughesville and

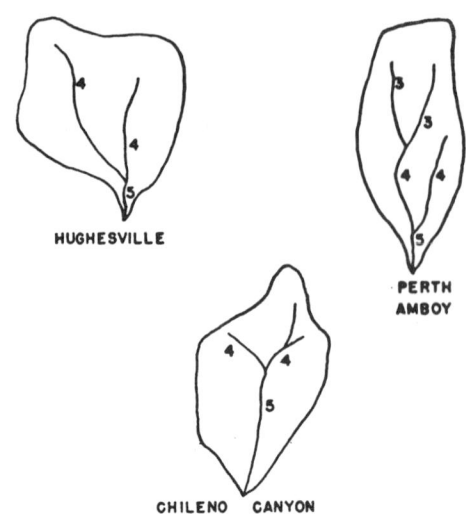

FIGURE 4.—COMPARISON OF SHAPE OF BASIN AND MAIN DRAINAGE ELEMENTS OF THREE AREAS

Numbers indicate order of main drainage channels

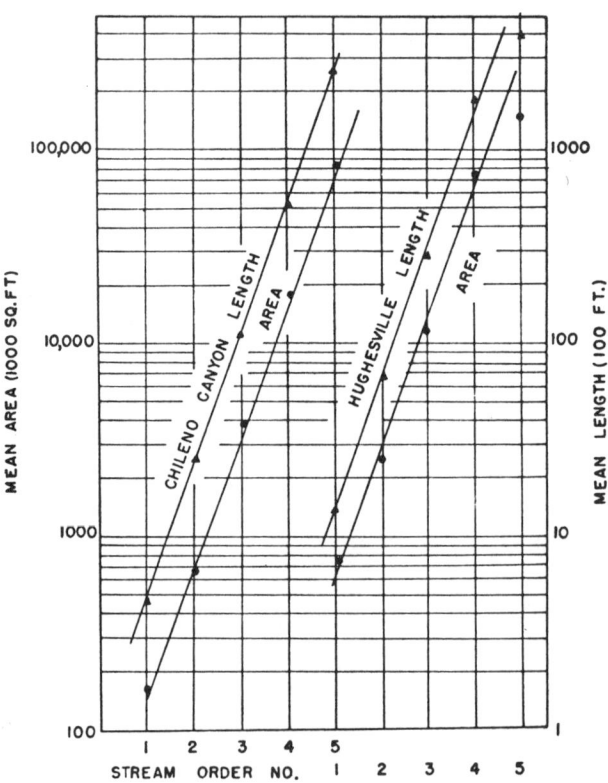

FIGURE 5.—RELATION OF MEAN BASIN AREA AND MEAN STREAM LENGTH TO STREAM ORDER

TABLE 2.—DRAINAGE-NETWORK CHARACTERISTICS

Basin	Order number	Number of streams	Mean length (ft.)	Mean area (sq. ft.)	Mean gradient (%)
Perth Amboy	1	214	10.1	85.0	59.9
	2	45	40.4	343	40.6
	3	8	242	2360	33.7
	4	2	1660	14600	18.2
	5	1	3530	31000	11.1
Chileno Canyon	1	296	482	167000	....
	2	66	2560	872000	....
	3	16	11400	3890000	....
	4	3	51100	18100000	....
	5	1	254000	86100000	....
Hughes-ville	1	150	1420	781000	....
	2	37	6860	2540000	....
	3	8	28400	12000000	....
	4	2	180000	78800000	....
	5	1	397000	154000000	....

length of channels close to the maximum value, perhaps obtainable only by detailed remapping in the field.

Further investigations included map measurement by polar planimeter of all drainage-basin areas. Horton (1945, p. 294) inferred that mean drainage-basin areas of each order should form a geometric series. A plot of the mean areas of stream basins of each order for the three basins compared above (Figs. 3, 5) reveals this relationship. A fourth law of drainage composition may therefore be formulated in the style set by Horton: the mean drainage-basin areas of streams of each order tend to approximate closely a direct geometric series in which the first term is the mean area of the first-order basins. It could be assumed that such a relationship would exist if there were any connection between the length of a stream and the size of its drainage basin.

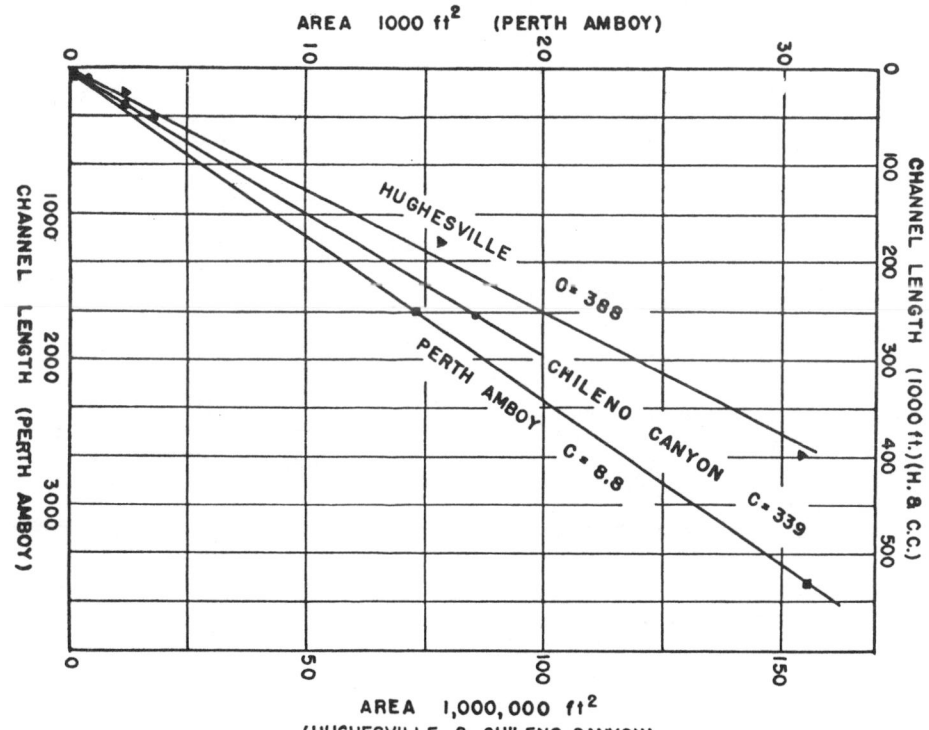

FIGURE 6.—RELATION OF MEAN STREAM LENGTH OF EACH ORDER TO MEAN BASIN AREA OF EACH ORDER

Chileno Canyon maps with aerial photographs so that the blue drainage lines could be extended to what appeared to be the correct length and small tributaries were also added to the drainage pattern. This method brought the

In Figures 3 and 5 the parallelism of the plots of mean stream length and mean drainage-basin area is striking and suggests a directly proportional relationship between the two. Figure 6 shows a plot of the mean drainage-basin areas

279

and mean stream-channel lengths for the three areas. The scatter of the Perth Amboy data is slight around a regression line fitted by the method of least squares and is described by the regression equation $Yc = 56.8 + 8.77X$. The ratio between mean area and length values is thus approximately 9. The calculated ratio for the Chileno Canyon basin is 339 and for the Mill Dam Run basin 388.

The significance of the ratio is that it represents in square feet the area required to maintain 1 foot of drainage channel. It is the quantitative expression of one of the most important numerical values characteristic of a drainage system: the minimum limiting area required for the development of a drainage channel. This value, *the constant of channel maintenance*, is a measure of texture similar to drainage density; it is in fact, equal to the reciprocal of drainage density multiplied by 5280 (because the channel-maintenance ratio is expressed in square feet while drainage density is expressed in miles). Along with drainage density this constant is of value as a means of comparing the surface erodibility or other factors affecting surface erosion and drainage-network development. A related texture measure is Horton's (1945) length of overland flow, the distance over which runoff will flow before concentrating into permanent drainage channels. The length of overland flow equals the reciprocal of twice the drainage density.

The discovery of the above relationship permits statement of a fifth law of drainage composition: the relationship between mean drainage–basin areas of each order and mean channel lengths of each order of any drainage network is a linear function whose slope (regression coefficient) is equivalent to the area in square feet necessary on the average for the maintenance of 1 foot of drainage channel. This law requires an orderly development of any drainage network, for the extension of any drainage system can occur only if an area equal to the constant of channel maintenance is available for each foot of lengthening drainage channel.

Limiting Values of Drainage Components

In addition to a lower limiting area necessary for channel maintenance there may be expected upper limits to basin areas and stream lengths of each order beyond which new tributaries or bifurcation occurs, forming new basins. These

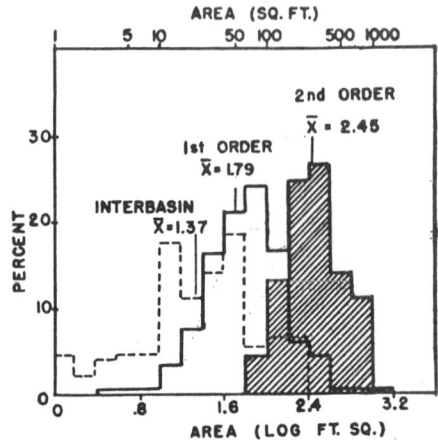

FIGURE 7.—FREQUENCY-DISTRIBUTION HISTOGRAMS OF THE LOGS OF DRAINAGE-BASIN AREA

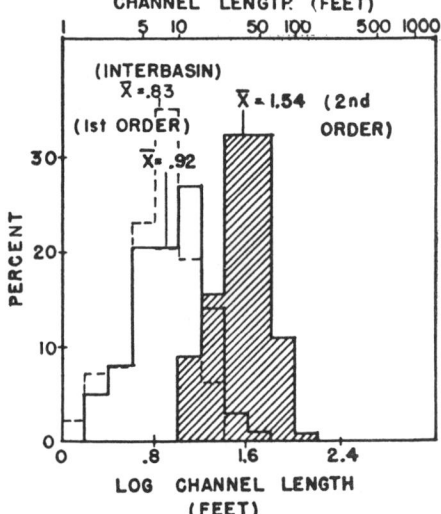

FIGURE 8.—FREQUENCY-DISTRIBUTION HISTOGRAMS OF THE LOGS OF STREAM-CHANNEL LENGTHS

relationships would appear in frequency-distribution histograms of the basin areas and stream lengths of each order and further confirm the principle of a channel-maintenance constant.

Frequency-distribution histograms of the stream lengths and basin areas show a marked right skewness, which appears to be corrected by plotting log values on the abscissa (Figs. 7, 8; Tables 3, 4). All measurements are made on a topographic map and are therefore taken from the horizontal projection of the drainage-basin elements rather than from true lengths and surface areas. Frequency-distribution study is limited to the first two orders by the small number of streams in the third and higher orders. A study of the first- and second-order basin areas and interbasin areas may be adequate, however, to determine if a transition phase exists between orders.

Between adjacent drainage basins are *interbasin areas*, those roughly triangular areas which have not developed a drainage channel (Fig. 9), but which drain directly into a higher-order channel. The histograms of first- and second-order basin areas and interbasin triangular areas are superimposed in Figure 10 (Table 5); the histograms of first- and second-order stream lengths are compared with maximum interbasin-slope lengths in Figure 11 (Table 6). Figures 7 and 8 compare the histograms of the logarithms of basin area, channel length, interbasin areas, and interbasin maximum lengths. The discussion of limiting values of drainage components may be followed on either set of figures.

An overlap between the areas of each histogram suggests that transformation from first to second order takes place within a wide range of values. In Figure 10 interbasin areas show a sharp decrease in frequency for areas above 50 square feet, which is well below the mean of the first-order areas. Of the 27 interbasin areas over 50 square feet, 12 seemed capable of developing a channel at any time; the remaining 15 were irregular, wider than long, or were on rounded spurs where the divergence of orthogonals downslope prevents the concentration of runoff. From this investigation alone it is difficult to set limiting area above which channel development may be expected on the interbasin areas, especially since the comparison of triangular interbasin areas with elliptical first-order basins is questionable.

Areas of first-order basins rise sharply at the 10-square-foot class limit. Two first-order basins of less than 10 square feet were mapped by Strahler and Coates, but a field check revealed that these did not contain permanent drainage channels. The fact that areas of less than 10 square feet are remarkably free of drainage channels coincides with the concept of a constant of channel maintenance of about 9 (8.77). Thus, no permanent channel will develop without a drainage area of about 10 square feet, while the channel can lengthen only with the average increment of 9 square feet of area for each additional foot of length.

Most of the overlap between the first- and second-order areas falls between 50 and 150 square feet, although some first-order areas range up to 650 square feet. Again an inspection

TABLE 3.—FREQUENCY DISTRIBUTIONS OF LOGS OF CHANNEL LENGTHS*

Sample	Class mid-values in logs of lengths in feet										$\bar{X}$	s	N
	0.3	0.5	0.7	0.9	1.1	1.3	1.5	1.7	1.9	2.1			
First order channels	11	17	45	45	58	30	6	2	...	...	.92	.302	214
Second-order channels	...	...	...	...	4	7	14	14	5	1	1.54	.21	45

* In this and all following tables and figures, $\bar{X}$ is the arithmetic mean, s is the standard deviation, and N is the number of items in each sample.

TABLE 4.—FREQUENCY DISTRIBUTIONS OF LOGS OF DRAINAGE-BASIN AREAS

Sample	Class mid-values in logs of area in square feet															$\bar{X}$	s	N
	0.3	0.5	0.7	0.9	1.1	1.3	1.5	1.7	1.9	2.1	2.3	2.5	2.7	2.9	3.1			
First-order areas	...	2	2	6	17	34	45	50	34	12	8	1	1	1	...	1.79	.19	214
Second-order areas	...	...	...	...	...	...	...	...	2	7	11	12	7	5	1	2.45	.284	45

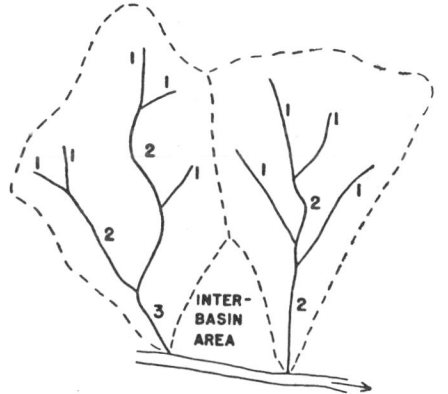

FIGURE 9.—POSITION OF INTERBASIN AREA AND METHOD OF CLASSIFYING STREAMS BY ORDER NUMBER

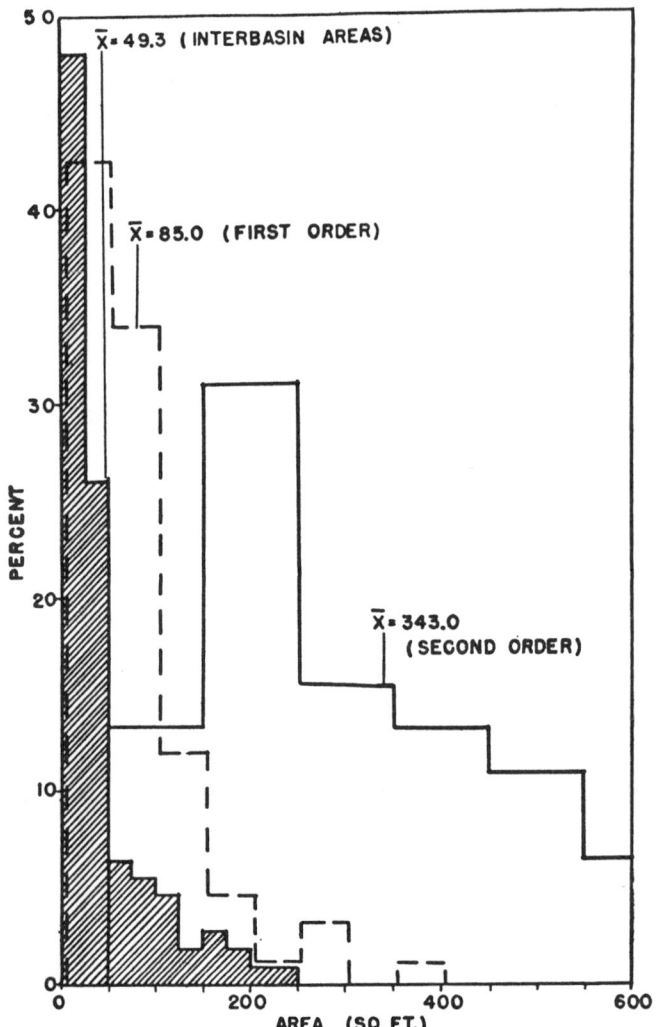

FIGURE 10.—FREQUENCY-DISTRIBUTION HISTOGRAMS OF FIRST- AND SECOND-ORDER BASIN AREAS AND INTERBASIN AREAS

TABLE 5.—FREQUENCY DISTRIBUTIONS OF FIRST- AND SECOND-ORDER DRAINAGE-BASIN AREAS AND INTERBASIN AREAS

Sample	Class mid-values in square feet													$\overline{X}$	s	N
Mid-values	30	80	130	180	230	280	330	380	430	480	530	580	630			
First-order areas	91	73	26	10	3	7	0	2	0	0	0	1	1	85	80	214
Mid-values	100	200	300	400	500	600	700	800	900	1000	1100	...	...			
Second-order areas	6	14	7	6	5	3	2	0	1	0	1	...	...	343	218	45
Mid-values	12.5	37.5	62.5	87.5	112.5	137.5	162.5	187.5	212.5	237.5	..	...	...			
Interbasin areas	51	28	7	5	5	3	4	3	1	1	..	...	...	49.3	49	108

of individual basin characterstics within the zone of histogram overlap is profitable. Of the 46 first-order areas greater than 110 square feet, 29 are of very youthful basins including basins it seems a fair generalization that first-order channels with areas greater than 100 square feet are unstable and ready for subdivision.

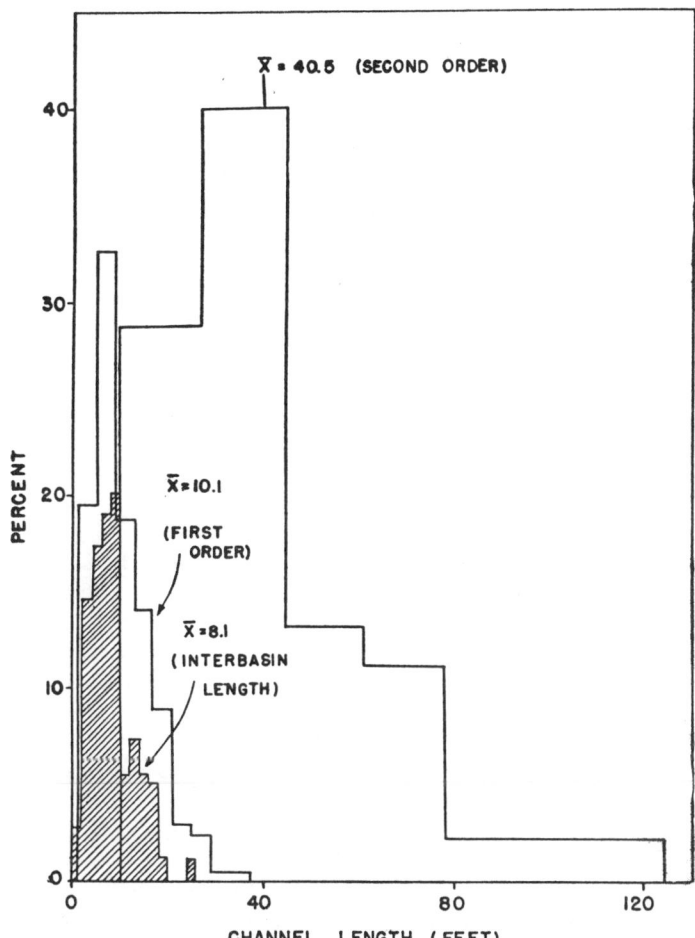

FIGURE 11.—FREQUENCY-DISTRIBUTION HISTOGRAMS OF FIRST- AND SECOND-ORDER CHANNEL LENGTHS AND MAXIMUM INTERBASIN LENGTHS

broad, gently sloping surfaces with only traces of channels on the flat undissected divide areas. With further development these would evolve to a higher order, for their longitudinal profiles are still essentially convex-up, retarding rapid tributary extension into their headwater areas. Thirteen of the 46 are narrow almost rill-like basins unable to broaden because of adjacent more aggressive basins. The remaining 4 of the 46 first-order basins larger than 110 square feet have no obvious reason for not developing into second-order channels. Although it is difficult to explain peculiarities of individual drainage

The smallest second-order channel area is 65 square feet. The class limits of the second-order areas within the overlap are 50 and 150 square feet (Fig. 10). Within this size range are six second-order basins which have developed tributaries recently and are capable of enlarging by headward extension so that the lowest-frequency class of the second-order area histogram would disappear unless replaced by new units created by bifurcation of first-order channels. Youthfulness of the entire system at Perth Amboy prevents the recognition of narrow transition zones between orders. A similar study in a fully extended mature drainage system might show sharper distinctions.

In accordance with the fifth law of drainage composition, stream-length frequency distributions are similar to the area distributions. Maximum interbasin-slope lengths cannot be directly compared to actual stream lengths because a channel developing on the interbasin surface will not extend the entire length of the slope. Nevertheless, a sharp drop in frequency at 10 feet suggests that at lengths above this runoff surfaces are unstable in form and will tend to develop channels (Fig. 11). Twenty-six interbasin areas with lengths greater than 10 feet had no channels. Seven were very narrow with little drainage area. The remaining 19, as previously noted under the discussion of interbasin areas, are irregular or on rounded spurs, while 4 seem capable of developing channels.

The lower values for first-order stream lengths are not significant because all channels must originate from a point and then lengthen. The region of transition between first- and second-order stream lengths lies between 9 and 17 feet, but 17 feet is not the upper limit of first-order lengths. Of 27 streams longer than 17 feet, 20, within basins considered previously under the discussion of areas, were in very youthful or narrow basins; the remaining 7 seemed capable of change. All but 1 of the 12 second-order channels between 10 and 17 feet will continue to develop, eliminating these streams from the frequency class. Youthfulness of the area probably masks a more distinct transition zone.

Within the Perth Amboy drainage network there are recognizable limits to the areas and lengths of streams of each order. First-order streams have an upper limiting length between 9 and 17 feet and an upper limiting area between 65 and 110 square feet. The limited number of stream orders considered and the subjective evaluation of parts of the data make it more appropriate to set a lower limit below which higher orders cannot exist. The first-order streams require more than 10 square feet for development; second-order streams will not normally evolve from first orders until the drainage area is equal to 65 square feet and the first-order channel is longer than 10 feet.

The writer remapped the drainage pattern in 1952 and compared it with the pattern mapped in 1948, aiding the study of channel alterations within the zones of transition. In all cases the addition of channels occurred only in basins above the size limits set from the frequency-distribution analysis. No channels developed on areas less than 10 square feet. Four new channels developed on interbasin areas, all but one (46.5 sq. ft.) greater than 50 square feet. Twelve new tributaries developed on first-order channels, forming several new second-order basins. Each new basin was youthful (developing headward into the as yet undissected fills), and almost all exceeded 110 square feet. Four were within the transition zone between first- and second-order areas. The newer field study, therefore, seems to confirm the existence of the zones of transition and upper limiting values of development related to the constant of channel maintenance. The constant of channel maintenance, therefore, may be applied to the as yet undissected portions of a drainage system to aid in the prediction of areas of future sediment loss.

TABLE 6.—FREQUENCY DISTRIBUTIONS OF FIRST- AND SECOND-ORDER CHANNEL LENGTHS AND INTERBASIN LENGTHS

Sample	Class mid-values in feet									$\bar{X}$	s	N
Mid-values	3	7	11	15	19	23	27	31	35			
First-order lengths	42	70	40	30	19	6	5	1	1	10.1	6.08	214
Mid-values	18.5	35.5	52.5	68.5	86.5	103.5	120.5					
Second-order lengths	13	18	6	5	1	1	1			40.4	23.4	45
Mid-values	1	3	5	7	9	11	13	15	17	19	21	23	25
Interbasin lengths	3	16	18	21	23	5	8	5	5	2	0	0	2	8.06	4.17	108

Form of the Drainage Basins

In addition to indices of drainage-network composition based on stream orders other important geomorphic characteristics are shape of the basins, relief, surface slope, drainage density, and stage of geomorphic development. Geomorphic development can be evaluated by means of the hypsometric integral (Strahler, 1952b). If each characteristic had a numerical value, comparisons could be made between topographic units. It may be appropriate to set up standards of comparison from the available information which can be modified later or rejected if unacceptable.

Relief is analyzed by a *relief ratio*, defined as the ratio between the total relief of a basin (elevation difference of lowest and highest points of a basin) and the longest dimension of the basin parallel to the principal drainage line. This relief ratio is a dimensionless height-length ratio equal to the tangent of the angle formed by two planes intersecting at the mouth of the basin, one representing the horizontal, the other passing through the highest point of the basin. Relief ratio allows comparison of the relative relief of any basins regardless of differences in scale of topography. Recent field studies, however, reveal that residuals or abnormally high points on the divide should be ignored when obtaining the total relief of a basin (Hadley and Schumm, In preparation).

The shape of any drainage basin is expressed by an *elongation ratio*, the ratio between the diameter of a circle with the same area as the basin and the maximum length of the basin as measured for the relief ratio. This ratio is the same as the Wadell sphericity ratio used in petrology (Krumbein and Pettijohn, 1938, p. 284), where the ratio approaches 1 as the sediment grain, or in this case the shape of the drainage basin, approaches a circle. Miller (1953, Ph.D. dissertation, Columbia University) used a similar measure, the *circularity ratio*, which is the ratio of circumference of a circle of same area as the basin to the basin perimeter.

Table 7 compares Strahler's data (1952b, p. 1134) on five mature drainage basins with the writer's data obtained from the more youthful Perth Amboy and Hughesville areas and the Chileno Canyon area.

The writer compared relief ratio and drainage density for fourth- and fifth-order channels (Fig. 12, Table 7) and found a definite positive trend in the mature basins. Points for the youthful Hughesville and Perth Amboy basins displace upward to positions well above the

TABLE 7.—DRAINAGE-BASIN CHARACTERISTICS

Area	Drainage density (miles/sq. mi.)	Relief ratio	Elongation ratio	Gradient (%)	Mean maximum slope angles (%)
1. Gulf Coastal Plain	4.6	.008	.975	0.33	5.9
2. Piedmont	6.9	.025	.935	1.13	17.5
3. Ozark Plateau	13.8	.062	.692	3.52	53.7
4. Verdugo Hills	26.2	.245	.594	22.46	99.0
5. Great Smoky Mts.	14.2	.267	.760	12.33	86.7
6. San Gabriel Mts.	15.6	.220	.675	17.2	73.4
7. Hughesville	13.6	.006	.730	0.22	7.0
8. Perth Amboy	602.0	.117	.602	11.1	110.8

trend line. When the values for the individual third-order basins of each of the two youthful areas are plotted (Fig. 13), the points show a positive trend similar to the plot of the mature basins. Thus, within homogeneous areas of similar development the drainage density is a power function of the relief ratio.

In Figure 14 the relief ratio shows a close correlation with stream gradient. The gradient values are means for the entire stream length and thus would approach the value of the relief ratio if the stream length was measured to the drainage divide. In general, the gradient so measured will be less than the relief ratio, for meandering or the usual lack of straightness of a channel will increase the stream length beyond the drainage-basin length.

Valley-side slope angles are also clearly related to the relief ratio. In Figure 15 three values would lie well to the right of a line fitted to the other points. This may be the result of obtaining the mean slope values from topographic maps in these cases; all the data were not measured in the field, and slopes measured on maps usually are lower than field measurements (Strahler, 1950, p. 692).

In Figure 16 the shape of the drainage basin is plotted against relief. The trend is negative and may indicate that as the relief ratio increases the drainage basin becomes more elongate. The data are not conclusive, but the steeper the slope on which small basins develop

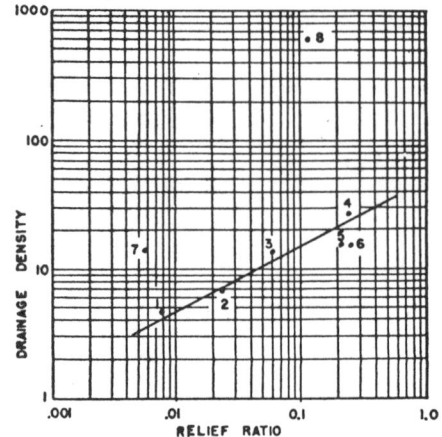

FIGURE 12.—RELATION OF DRAINAGE DENSITY TO RELIEF RATIO
Numbers refer to basins described in Table 7

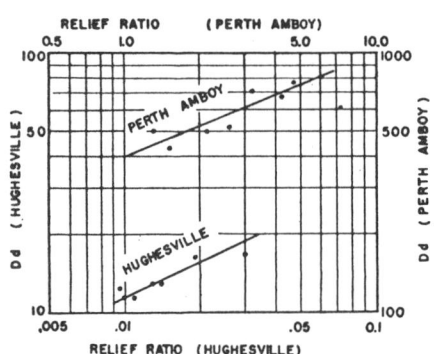

FIGURE 13.—RELATION OF DRAINAGE DENSITY TO RELIEF RATIO OF THIRD-ORDER BASINS

the more closely spaced are the drainage channels, resulting in more elongate basin shapes.

One practical application of the relief ratio is in estimation of sediment loss. Figure 17 shows the direct relation between mean relief ratio for several areas in Utah, New Mexico, and Arizona and mean annual sediment loss as estimated from sedimentation in small stock reservoirs. Once the characteristic regression trend has been established for a region the investigator may select areas of high potential sediment production from topographic maps (Schumm, 1955).

Various interrelationships among drainage-basin characteristics have been previously determined. Langbein (1947, p. 125) states that steep land slopes are generally associated with steep channel slopes and fine texture, and that

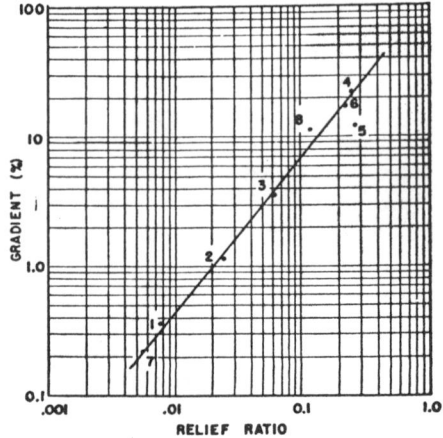

FIGURE 14.—RELATION OF MEAN STREAM GRADIENTS TO RELIEF RATIO
Numbers refer to drainage basins described in Table 7

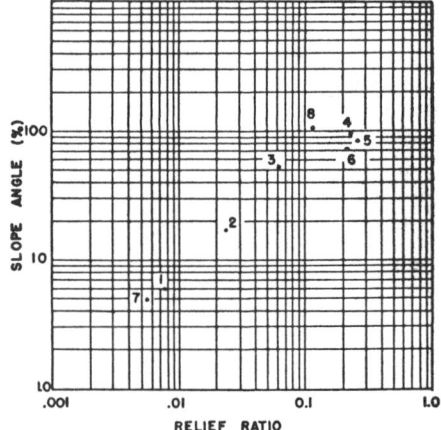

FIGURE 15.—RELATION OF MEAN MAXIMUM-SLOPE ANGLES TO RELIEF RATIO
Numbers refer to drainage basins described in Table 7

altitude of a basin above its outlet increases with steepening land and channel slopes. Paulsen (1940, p. 440) found that infiltration increases with decrease in mean land slope, explaining in part the increase of sediment loss with the relief ratio. Strahler (1952b, p. 1136) observed that the hypsometric integral decreased as basin height, slope steepness, gradient, and drainage density increased.

Although more data are desirable the relationships observed suggest that the geomorphic character and even rates of erosion may be predicted from the relief ratio, although it is only a geometrical element which is probably related to lithology, structure, stage, vegetation, and climate.

The above relationships when considered in the light of recent discussions of the quasi equilibrium between the hydraulic and geomorphic characteristics of stream channels (Leopold and Maddock, 1953; Wolman, 1955; Leopold and Miller, in press) suggest that when more information becomes available this concept of quasi equilibrium in graded and ungraded stream channels may extend to the landforms adjacent to the stream channels, and close interrelationships may be found among the geomorphic, hydrologic, and hydraulic characteristics of a topographic type.

Basin Form Related to Geomorphic Stage of Development

During geomorphic development basin forms change with time. According to the classic Davisian analysis, relief, slope of valley walls, stream gradients, and drainage density increase rapidly during youth to a maximum in early maturity, then decline slowly throughout later maturity and old age. A unique opportunity to study stage changes was afforded by the developmental sequence of drainage basins tributary to the main channel at Perth Amboy. Eleven second-order drainage basins forming a

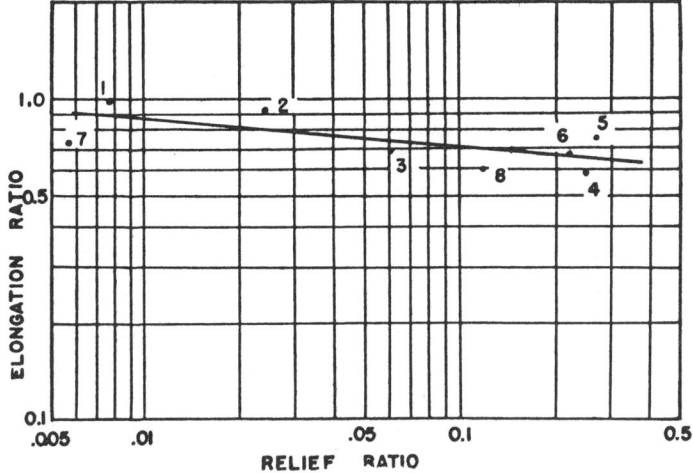

FIGURE 16.—RELATION OF ELONGATION RATIO TO RELIEF RATIO
Numbers refer to drainage basins described in Table 7.

sequence from earliest youth to late maturity were selected for map study and a comparison of hypsometric integrals. The hypsometric integral is a measure of stage (Strahler, 1952b) because it expresses as a percentage the mass of the drainage basin remaining above a basal plane of reference.

Figure 18 shows hypsometric curves plotted for each basin in the sequence. Data are obtained from the topographic map by measuring the total area of each basin with a planimeter, then measuring the area between each contour and the basin perimeter above it. Each area is converted into a percentage of total basin area, so that a cumulative percentage curve can be plotted, each area value corresponding to a percentage of the total height of the basin. Using this hypsometric curve it is possible to read the percentage of total basin area above any percentage of total height. The area-altitude relations of the basin are thus revealed by a curve illustrating in dimensionless co-ordinates the distribution of mass within the drainage basin (Langbein, 1947, p. 140; Strahler, 1952b). Area under the hypsometric curve is the hypsometric integral, expressed as per cent.

An integral of 60 per cent indicates that erosion has removed 40 per cent of the mass of the basin between reference planes passing through summit and base. Strahler (1952b) discussed in more detail the hypsometric curve and its use in geomorphic research.

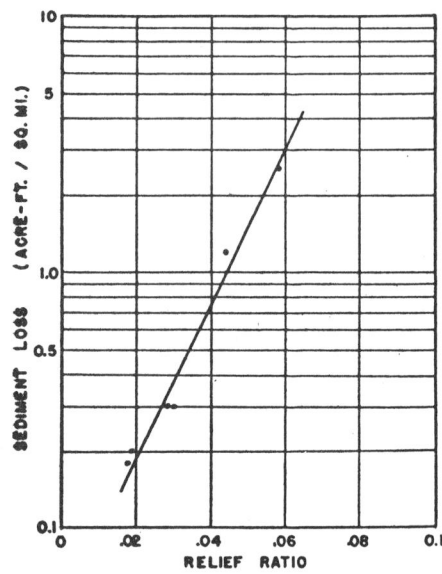

FIGURE 17.—RELATION OF MEAN ESTIMATED ANNUAL SEDIMENT LOSS TO RELIEF RATIO
From the U. S. Geological Survey studies of reservoir sedimentation in New Mexico, Arizona, and Utah

The 11 second-order basins selected for this study were of approximately the same area so that the series of hypsometric curves illustrate basin development with time accompanying lowering of the main channel through 40 feet of total relief. The convex curves with high integrals (Fig. 18) reveal youthful inequilibrium; the curves of more mature basins show the beginning of the typically mature sigmoid curve. The percentage curves cannot show continued down-wasting of the basin, because when maturity is reached curves tend to stabilize between integrals of 40 and 60 per cent. Strahler uses this stable integral as the point of onset of the equilibrium stage of drainage-basin development. Only basins containing monadnocks of resistant rock develop integrals markedly less than 40 per cent.

A better picture of the sequence of natural basin changes in terms of total erosional reduction may be obtained by plotting percentage of area against percentage of total elevation of the terrace at Perth Amboy rather than against

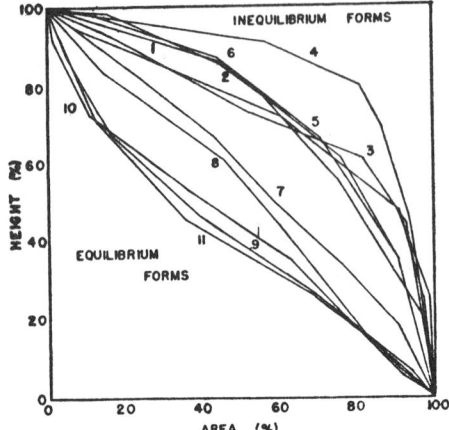

FIGURE 18.—SEQUENCE OF SECOND-ORDER HYPSOMETRIC CURVES
From Perth Amboy.
Numbers increase from youthful to mature basins

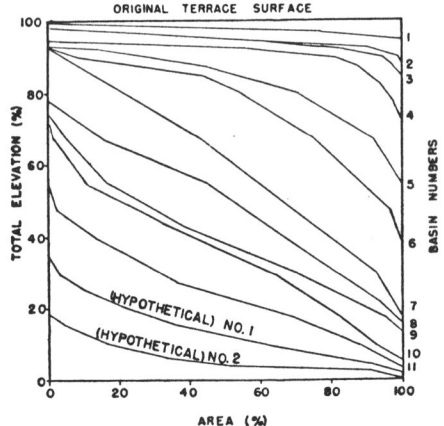

FIGURE 19.—SEQUENCE OF SECOND-ORDER HYPSOMETRIC CURVES
Per cent area is plotted against per cent of total relief at Perth Amboy. Numbers increase from youthful to mature basins and are the same areas as shown in Figure 18.

total relief within each basin. This method (Fig. 19) is more satisfactory because the square in which the curves are plotted may be visualized as a vertical section through the entire terrace at Perth Amboy. Each curve occupies its true relative vertical position within that mass and reveals the degradational history of the basin. The 100 per cent elevation line should be visualized as the upper surface of the terrace, the base as the level of the main stream's mouth. The right edge of the chart is the locus of points of junction of second-order tributaries with some higher-order stream. Each

TABLE 8.—DRAINAGE-BASIN CHARACTERISTICS OF THE SECOND-ORDER SEQUENCE

Basin number	Per cent mass removed	Relief ratio	Elongation ratio	Gradient (%)	Drainage density (mi./sq. mi.)
1	2.4	.049	.993	4.0	553
2	4.9	.121	.648	14.4	504
3	5.6	.158	.595	18.6	270
4	9.5	.156	.645	21.8	241
5	17.2	.330	.783	42.0	672
6	23.5	.575	.725	52.0	560
7	39.8	.590	.507	58.3	610
8	50.8	.660	.473	67.5	895
9	60.8	.710	.474	65.5	1230
10	64.6	.620	.478	51.5	1150
11	77.0	.690	.530	50.7	1320

line represents the distribution of mass within a second-order basin at a different stage of development, the position of its mouth controlled by the degrading stream to which it is tributary.

To determine the nature of basin-form changes with time, or stage of evolution, an important index is the percentage of mass removed at each position in the sequence of basins. This value, obtained by measuring the area above each curve and comparing it to the total area of the diagram, is a measure of the mass removed in relation to the total available for removal.

Percentage of mass removed, relief ratio, stream gradient, basin shape, and drainage density were determined for each of the 11 basins whose curves are drawn in Figure 19. The data for each basin in the sequence (Table 8) are plotted against corresponding per cent of mass removed (Fig. 20). The plot of relief ratio with per cent removed (A) reveals that with initial dissection the relative relief rapidly increases. A sharp break in the continuity of the plot occurs when approximately 25 per cent of the mass of the basin is removed, after which the relief ratio remains almost constant to 80 per cent of mass removed; beyond 80 per cent, data are lacking.

The stream-gradient plot (B) shows a similar form. The rapid increase in gradient is checked at approximately 25 per cent of mass removed; a decrease in gradient sets in at the upper part of the plot after a maximum value reached at 50 per cent removal. This agrees with the typical descriptive concept of stream development, but, in comparison to the rapid early increase in gradient, the portion of the plot above 25 per cent is essentially constant.

Maximum slope angles were not obtained for each basin, but, because these values cluster closely about a mean value for any homogeneous area (Strahler, 1950, p. 685), and because the close relationship between stream gradients and maximum slope angles has been established (Strahler, 1950, p. 689), any plot of slope angles and mass removed would be expected to approximate the gradient curve.

The relationships of basin shape and drainage density to stage (Fig. 20C, D) are less clear, but after early variations in which the basin is close to a circular shape the influence of increased relief is felt and the elongation ratio decreases to a constant of about 0.5 at 40 per cent mass removal, indicating that the basin maximum length is twice the diameter of a circle of the same area. The drainage-density plot is not regular, probably because of a high degree of length variability in the low order of the streams used. Nevertheless, the plot suggests rapid early increase in drainage density, followed by a decreasing increment. Probably continued headward development of the drainage channels continues until late in the erosion cycle, lagging behind the early stability of other basin characteristics. If other series of basins could be studied similarly, the additional data might lead to the establishment of a general system of basin evolution.

In summary, the form of the typical basin at Perth Amboy changes most rapidly in the earliest stage of development. Relief and stream gradient increase rapidly to the point at which about 25 per cent of the mass of the basin has been removed, then remains essentially constant. Because relief ratio elsehwere has shown a close positive correlation with stream gradient, drainage density, and ground-slope angles, stage of development might be expected to have little effect on any of these values once the relief ratio has become constant.

The Perth Amboy data thus support the concept of a steady state of drainage-basin development as outlined by Strahler (1950, p. 676) who compares a graded drainage system with an open dynamic system in a steady state:

"In a graded drainage system the steady state manifests itself in the development of certain topographic form characteristics which achieve a time-independent condition. (The forms may be described as "equilibrium forms"). Erosional and transportational processes meanwhile produce a steady flow (averaged over periods of years or tens of years) of water and waste from and through the landform system.... Over the long span of the erosion cycle continual readjustment of the components in the steady state is required as relief lowers and available energy diminishes. The forms will likewise show a slow evolution."

This hypothesis of time-independent forms and basin characteristics is supported by the constancy of the values of the basin parameters in the Perth Amboy sequence of second-order basins, once the amount of mass removed has exceeded 25 per cent.

Evolution of the Drainage Network

Effect of Stage on Angles of Junction

In the early stage of basin development stream channels grow headward until they establish major drainage divides; the relief ratio then reaches a fixed value, but changes in channel network continue until a large portion of the basin mass is removed. Thus, the relief ratio becomes fixed before other network characteristics become constant. This is especially true of drainage density in the Perth Amboy area.

Angles of junction of tributaries, resurveyed in 1952, showed some marked differences from corresponding angles in the 1948 map. The writer reasoned that systematic changes were occurring in the drainage pattern as a normal part of the erosional development of the basins. Horton (1945, p. 349) recognized that the course followed by a new tributary is governed by both the slope of the ground over which it flows and the gradient of the channel to which it is tributary. Where the ground slope is great in relation to the gradient of the master stream a tributary joins at almost right angles; where master-stream gradient and valley-side slope are almost the same the tributary almost parallels the main channel, joining it at a small angle. Horton expresses this as follows:

$$\cos Zc = \tan Sc/\tan Sg$$

where the cosine of the entrance angle or angle of junction, measured between the tributary and main channel above the point of junction,

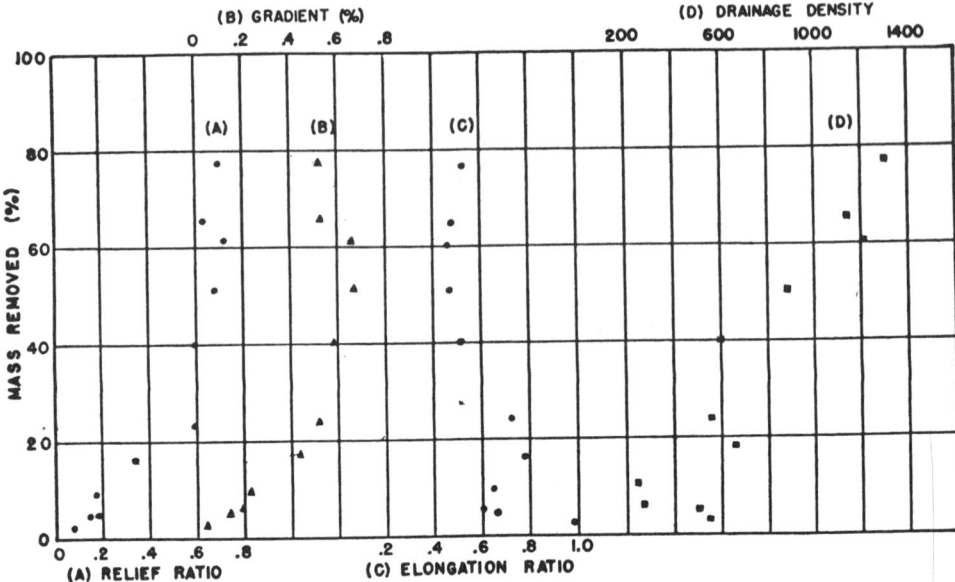

Figure 20.—Relation of Mass Removed within a Basin to Relief Ratio, Gradient, Elongation Ratio, and Drainage Density

equals the ratio of the tangent of the main channel gradient to the tangent of the gradient of the tributary stream or of the ground slope over which the tributary flows. It follows that during the early part of basin development, stream-entrance angles change with stream gradients.

Thus, a tributary will develop with an initially large angle of junction; then as the ratio between the two gradients increases the angle of junction decreases. Horton notes that as the ratio increases from 0.3 to 0.9 angles decrease from 72.3° to 25.5°. The decrease is accomplished by lateral migration of the tributary toward the main channel and down-valley shift of the junction.

The writer measured 61 entrance angles on the 1948 Perth Amboy map. The frequency-distribution histogram is broad and flat-topped with angles ranging from 24° to 90° (Fig. 21C; Table 9).

If the assumed changes occur, then by classifying all entrance angles according to stage of development of their tributary drainage basins a significant difference should occur between

TABLE 9.—FREQUENCY DISTRIBUTIONS OF ANGLES OF JUNCTION

Sample	Class mid-values in degrees							$\bar{x}$	s	N
	25	35	45	55	65	75	85			
Combined angles	4	8	8	14	8	10	7	56.8	17.6	59
Mature angles	3	7	5	7	2	1	1	65.2	15.3	26
Young angles	1	1	3	7	6	9	6	46.2	14.9	33

the means of youthful and more mature basins. The basins were separated into two groups on the basis of the existence of flat, undissected areas within the drainage areas, classifying as young basins capable of headward extension or having undissected areas within their drainage areas. The frequency-distribution histograms of each group (Fig. 21) show an expected overlap, but the means of the two groups are significantly different as judged by a t-test. The mean of the youthful class is 65°, that of the older group 46°. The probability that such a difference or greater would occur by chance alone is about 1 in 10,000. A reasonable explanation of the observed difference in angles is the shifting of tributary channels in response to changes in the gradient ratio.

A similar test was applied to angles of bifurcation, defined as the angles between two approximately equal first-order branches. In this case, the stream has bifurcated at its upper end, whereas in the tributary junction referred to above a branch has grown from the trunk of an existing major drainage line. Twenty angles of bifurcation were measured from youthful drainage basins having undissected areas. The mean is 62.1°, compared with the mean of the youthful angles of tributary junction, 65.2°. The frequency distributions of both samples have such great dispersions that this observed difference in means is not significant.

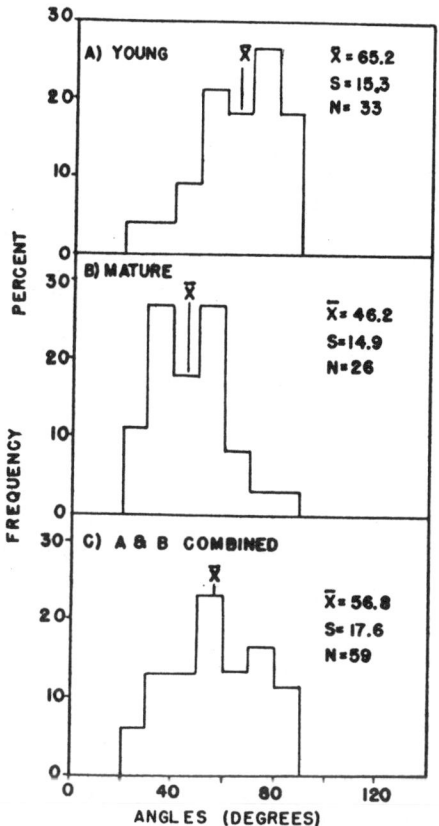

FIGURE 21.—FREQUENCY-DISTRIBUTION HISTOGRAMS OF YOUNG, MATURE, AND COMBINED ANGLES OF JUNCTION

Remapping of the drainage pattern revealed changes in the values of tributary entrance angles and angles of bifurcation. Table 10 shows data for mean entrance angles and angles of bifurcation measured from the 1948 and 1952 drainage maps. There is a decrease of 5.3° in the

TABLE 10.—ANGLES OF BIFURCATION AND ANGLES OF JUNCTION

	Mean angles (degrees)		Standard deviation (s)		Number in sample (N)	
	1948	1952	1948	1952	1948	1952
Angles of bifurcation	62.1	53.3	13.4	17.5	20	12
Angles of junction:						
Total	56.8	53.6	17.6	18.5	59	46
Young	65.2	59.9	15.3	17.1	33	29
Mature	46.2	43.0	14.9	14.9	26	17

mean of the youthful tributary-junction angles, but the standard deviation of each distribution is so large that a statistical test of the significance of difference between the means shows that such a difference would be expected through chance alone 20 per cent of the time and is not significant. This is true also of the difference between the mature angles, 3.2°.

The means of the young angles in both 1948 and 1952 are significantly different from those of the mature angles. It is interesting to note that the means for the total, youthful and mature angles decreased by several degrees during the 4-year period. The difference in each case is not statistically significant but suggests that with more time a significant change might occur.

Only the angles of bifurcation showed a significant reduction, 8.8°, between 1948 and 1952. Only 12 of the 20 original angles could be recognized and measured in 1952. The extreme youthfulness of the newly formed drainage basins, with rapid lowering of channel gradients in progress, is the cause of the great change in bifurcation angle.

A comparison of the drainage patterns showed marked drainage changes. Twelve new tributaries were added to the drainage system between mappings. Coincidentally, 12 others were eliminated, 6 by abstraction or lateral expansion of a more competent neighbor, 2 by angle reduction to the minimum with collapse of the divide and union of the streams, while the remaining 4 were in small, shrinking basins surrounded by headward-growing channels. Two of these channels were originally near the lower limiting area of channel formation, 11.8 and 15.3 square feet. It is interesting to note that both stages of Glock's (1931) drainage-development series are represented here: extension and integration, with abstraction as the major process of integration. Capture occurred in two other instances. Examples of the straightening of the stream channels were numerous.

One other change of pattern noted is the lateral shift of the major tributaries toward the center of the basin. This migration toward a common axis within the system is gradual, but the asymmetry of all high-order transverse-valley profiles testifies to its presence.

A series of drainage patterns traced from the 1948 and 1952 maps (Fig. 22) illustrates some of the changes during that period. The basins illustrated have steep channel gradients, and erosion would be rapid. In addition, the fill is easily eroded and presents few structural obstacles to drainage-channel modifications.

The following generalizations summarize changes in the drainage network at Perth Amboy: A tributary to a channel of higher order develops with an entrance angle dependent on the ratio between channel and ground slope. Because of relatively slower degradation of the main channel, a downstream migration of the point of junction occurs with lessening of the entrance angle. If the ratio between main-channel gradient and tributary gradient remains constant (steady state), no changes in junction will occur except those caused by chance structural irregularities in the fill. As channel gradation spreads throughout the entire system the main-channel gradient will first reach an essentially constant value, but the tributary gradient will continue to lower, with a lessening of the junction angle. When the junction angle becomes very small, lateral planation removes the intervening divide, and the junction migrates upstream. Comparable evolution of stream-entrance angles and drainage patterns in other regions may occur only in youthful areas with a high relief ratio, but similarities between Perth Amboy and other areas in other aspects of drainage-basin morphology suggest that similar

In the initial stage the steep front of the terrace was probably strongly rilled. Because the upper surface of the terrace drained toward the front the rills quickly advanced across the lip of the terrace onto the essentially flat upper

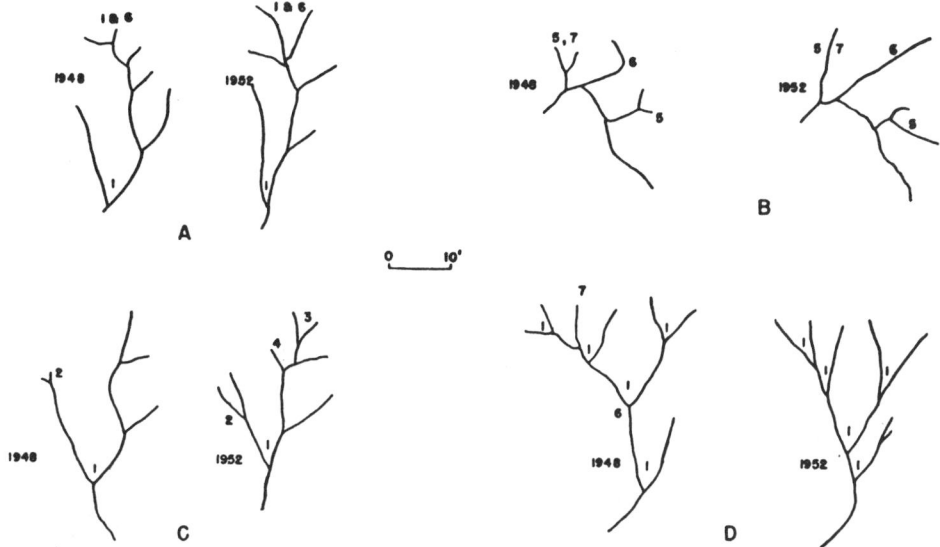

FIGURE 22.—DRAINAGE-PATTERN CHANGES IN SELECTED BASINS BETWEEN 1948 AND 1952 AT PERTH AMBOY

Basins A, C, and D are steep gradient streams. Basin B is a youthful basin on the upper surface of the terrace. Drainage changes are indicated by numbers on the figures:

1 Angle of junction change
2 Migration of junction
3 Bifurcation
4 Addition of tributary

5 Angle of bifurcation change
6 Channel straightening
7 Elimination of tributary

changes although perhaps less obvious, are nevertheless slowly occurring in all expanding drainage systems.

Because the observed drainage-pattern changes were occurring mainly as the stream channels were rapidly downcutting, any uplift of a land surface might initiate the same changes. Studies of drainage patterns on the Pleistocene terraces of the Atlantic Coast, for example, might indicate that height above base level and stage are correlatable with angles of junction.

Evolution of the Perth Amboy Drainage Pattern

From the observed systematic drainage changes at Perth Amboy and the known development of a network within the limiting values of basin area, it may be possible to deduce from the existing pattern the initial and future patterns.

surface. The channels most favored by chance encounter with weak patches of fill deepened and grew rapidly toward the divides of the individual small watersheds. These deeply cut permanent channels followed the path of initial drainage concentration manifested as faint channel traces on the upper surface. Channel traces of this type were observed at Perth Amboy in areas of headward channel development. The permanent channels follow these faint swales on the original surface because there the discharge of runoff is concentrated from the entire watershed. A headward developing incised channel is hydrophilic, advancing always toward maximum water supply.

The most vigorously developing initial rill channel thus dominated its less effective neighbors and established itself as the axis of a broadening ovate drainage basin. Its permanence was decided initially by a favored position in line with the axis of a shallow watershed on the terrace surface from which it was supplied with more runoff than its competitors. It may also have struck zones of weaker material in its bed. The added runoff allowed

stages of the Perth Amboy development, thus forming the first major tributary (Fig. 24, 1, 2). The tributary end grew normal to the main channel until it came under the influence of the forward slope of the terrace when its growth

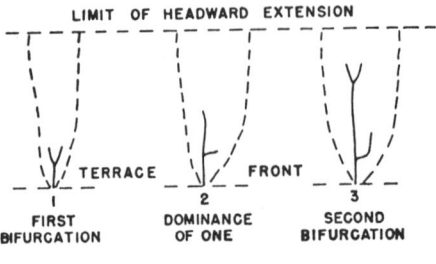

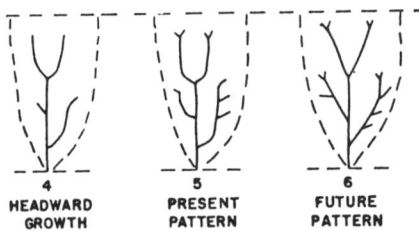

FIGURE 23.—POSSIBLE DEVELOPMENT OF ANGLES OF BIFURCATION

(1) Angle remains unchanged; (2) One channel becomes dominant; (3) On steepest slopes angles decrease and channels unite.

FIGURE 24.—SUGGESTED EVOLUTION OF THE PERTH AMBOY DRAINAGE PATTERN

deepening of the drainage channel with corresponding oversteepening and collapse of its valley-side slopes. As soon as lateral expansion of the drainage basin produced sufficiently long slopes, tributary development set in on these slopes.

As the channel outstripped its neighbors the expanding drainage area permitted its bifurcation. The comparison of angles of bifurcation on the 1948 and 1952 maps suggests three predictions of possible future development of a bifurcated channel: (1) both segments of the bifurcated channel continue to grow headward unchanged in angle (Fig. 23, 1); (2) one segment becomes dominant and straightens its channel, while the other segment becomes tributary (Fig. 23, 2); (3) on steep slopes the angle of bifurcation reduces in accordance with the Sc/Sg ratio, and the two segments become one (Fig. 23, 3).

It is postulated that (2) occurred in the early

direction altered and its upper segment developed parallel to the main channel (Fig. 24, 3). Perhaps the next permanent bifurcation was as indicated in Figure 24, 3, followed by headward growth (Fig. 24, 4) and other branches and bifurcations to yield the major elements of the present pattern (Fig. 24, 5; Pl. 1).

As these streams incised their channels secondary tributaries formed on the slopes of the valley walls. These tributaries were under the influence of the rapidly degrading main channels; many still are in the youthful headwater areas. Figure 25 is a frequency-distribution histogram of the angles measured between tributaries and the segments of the main channel (Table 11). The modal class lies between limits of 90° and 100°, indicating a right-angle pattern in accordance with a low Sc/Sg ratio. The earliest-formed of these tributaries became the most important and hindered the development of younger neighbors on adjoining slopes. This is borne out by the fact that the mean distance separating first-order streams along the main channel (4.3 feet) is smaller than the mean distance separating first- and second-order stream channels (7.6 feet). Order number thus provides a rough means of classifying channels

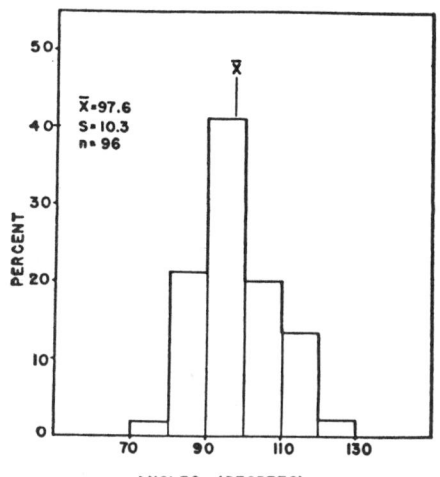

FIGURE 25.—FREQUENCY-DISTRIBUTION HISTOGRAMS OF ANGLES BETWEEN TRIBUTARIES AND SEGMENTS OF THE MAIN CHANNEL

TABLE 11.—FREQUENCY DISTRIBUTION OF ANGLES BETWEEN TRIBUTARIES AND MAIN CHANNEL

Sample	Mid-values in degrees						$\bar{x}$	s	N
	75	85	95	105	115	125			
Angles of junction	2	20	40	19	13	2	97.6	10.3	96

according to age; the oldest tributary channels have the higher order number.

The present faintly trellised pattern of the principal large-order channels (Fig. 24, 5; Pl. 1) is not considered permanent. It is supposed that the angles of junction become smaller with the increased Sc/Sg ratio, and the lateral shifting of the larger tributaries toward the main channel would result in the acute-angled dendritic drainage pattern that is typical of mature areas of simple structure. This change would involve considerable lateral planation and channel straightening, with a modified final pattern perhaps like that in Figure 24, 6.

If, as previously noted, a positive relationship exists between stream gradients, maximum slopes, and relative relief (expressed as the relief ratio), it follows that the Sc/Sg ratio should vary as the relief ratio. Because entrance angles, and therefore the total drainage pattern, are dependent on this Sc/Sg ratio, similar areas differing only in relative relief probably have recognizable differences in drainage pattern, at least in the early stages of development.

[*Editor's Note:* A discussion of hillslope erosion has been deleted.]

Available Relief and the Development of Landforms

Because a correlation exists between relative relief and other drainage basin parameters at Perth Amboy and other areas it may be possible to predict some effects of dissection in areas of widely different degrees of available relief but similar in lithology, structure, and climate. Glock (1932, p. 75) defines available relief as the vertical distance from the initial upland surface to the lowest level of channel degradation; it is the total relief available to the stream channels for dissection. It is reasonable to suppose that within one topographic type basins of high relief ratio would also have high available relief so that whatever relationships exist between relief ratio and other morphological elements (maximum slope angles, stream gradients, drainage density, and basin shape) would also exist for available relief.

Glock (1932) compared profiles of areas of high, moderate, and low relief; his main conclusion was that an "upper flat", the undissected upland surface, disappears with increased relief. His discussion of the evolution of an area of high relief agrees with W. Penck's (1953) that in the condition of "uniform development" constant relief may be maintained dur-

PLATE 2.—PERTH AMBOY BADLANDS

FIGURE 1.—GENERAL VIEW OF WEST FRONT OF TERRACE AT PERTH AMBOY
Total relief here is about 40 feet. Intricately dissected terrace front is bordered by alluvial fans

FIGURE 2.—EAST FRONT OF TERRACE AT PERTH AMBOY FROM POINT NEAR MOUTH OF A FIFTH-ORDER STREAM

FIGURE 3.—TYPICAL SMALL SECOND-ORDER DRAINAGE BASIN WITH STEEP STRAIGHT SLOPES
Flat undissected remnants of the upland surface are visible

FIGURE 1

FIGURE 2

FIGURE 3

PERTH AMBOY BADLANDS

ing dissection. Glock states that downcutting of a thick mass maintains a region of high available relief until the lower limit of channel degradation is reached, as illustrated in Figure 38a where the retreating steep slopes intersect in a sharp-crested divide which is lowered at the same rate as the channel.

The writer chose the following as illustrative examples based on observed quantitative relationships between relief ratio and drainage density in mature areas of the fifth order: an area of high available relief with relief ratio greater than 0.15 (Fig. 38a); an area of moderate relief with relief ratio of 0.04 (Fig. 38b); an area of low relief with relief ratio of 0.009 (Fig. 38c). The range in Figure 12 supplied the relief-ratio values, and corresponding drainage-density values were read. This is a reasonable procedure, for in areas of similar lithology drainage density is related to relief ratio (Fig. 13). The values for the length of overland flow (distance water would flow from the divide to a stream channel) are half the reciprocal of drainage density (Horton, 1945, p. 284). Twice the length of overland flow is the average distance between adjacent stream channels, and is a mean value for the spacing of stream channels in any area. The three cross sections of areas of high, low, and moderate relief (Fig. 38) were constructed so that the distance between divide and channel is proportional to the drainage density which is in turn proportional to the relief ratio for each section. It is now possible to compare the three hypothetical areas, remembering that the fundamental difference is available relief; lithology is assumed to be the same in each.

The area of high relief has a relief ratio of 0.15, a drainage density of 20, and a length of overland flow of 0.025 miles. Therefore, the following characterize the area (Fig. 38a):

(1) Degradation of stream channel will be rapid.

(2) Straight slopes will develop early at a steep angle characteristic of the area (Stage 1).

(3) Parallel retreat of the slope will maintain the steep maximum angle during most of the geomorphic evolution of the area (Stages 1–7).

(4) Slopes will quickly intersect, eliminating the "upper flat" (Stage 4).

(5) Local relief will remain constant during much of the evolution; the mature stage dominates during the topographic evolution (Stages 4–6).

(6) With cessation of channel downcutting lateral planation may maintain the characteris-

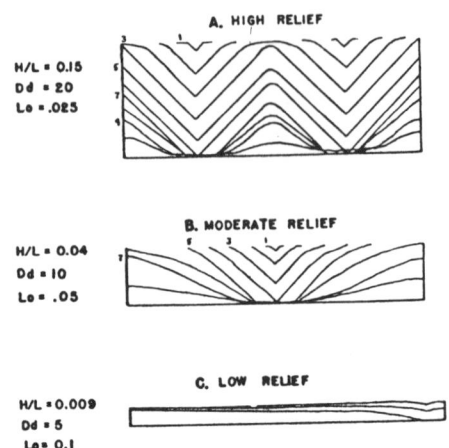

FIGURE 38.—DEVELOPMENT OF TOPOGRAPHY IN AREAS OF HIGH, MODERATE, AND LOW RELIEF
Horizontal and vertical scales: about 1 inch equals 0.025 mile or 132 feet

tic steep angle, but relief will be reduced (Stage 7).

(7) Slope angles will lower as basal deposition occurs. Strahler (1950, p. 813) compared mean maximum-slope angles of slopes protected at their base with slopes with actively degrading streams at their base and found a significant decrease in angle (Stage 8).

(8) Continued deposition at slope base will result in a concave-convex profile (Stages 9, 10). Erosion at this late stage may approach Lawson's (1932) postulation that runoff would remove lune-shaped segments from the crest of hills. Development of drainage basin form at Perth Amboy is an example of a region of high available relief.

The hypothetical area of moderate available relief (Fig. 38b) has a relief ratio of 0.04, a drainage density of 10, and a length of overland flow of 0.05 miles, differing as follows from areas of high available relief:

(1) Degradation will be less rapid.

(2) The intersection of straight parallel retreating slopes will not quickly eliminate the "upper flat".

(3) The stage of maturity (maximum relief) will be briefer (Stages 4, 5).

(4) Stream channels will be twice as far apart.

Evolution is not markedly different from the region of high relief in early stages, but the

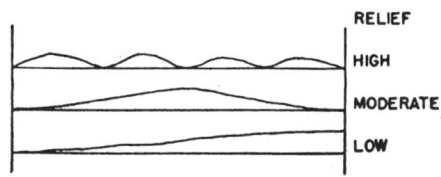

FIGURE 39.—POSSIBLE TOPOGRAPHIC DIFFERENCES, AT A LATE STAGE OF DEVELOPMENT, BETWEEN AREAS OF ORIGINALLY HIGH, MODERATE, AND LOW RELATIVE RELIEF

area is soon one of rounded slopes with decreasing slope angles (Stages 6-9).

The hypothetical area of low available relief has a relief ratio of 0.009, a drainage density of 5.0, and a length of overland flow of 0.1 mile, differing from the others as follows:

(1) Distance between adjacent channels will be four times that of the area of high relief.

(2) Channel degradation is assumed to be slow, and incision will be slight.

(3) Straight slopes never develop, and broad convexities characterize the area. (*Cf.* Fig. 36, profile 2.)

This hypothetical discussion takes into account the effects of increased relief on topographic development within areas of the same lithology. In late stages closer spacing of channels may distinguish areas of originally high relief from areas of originally lower relief. Hypothetical sections of the three areas (Fig. 39) illustrate the expected topography. It may be, however, that integration of the systems would reduce drainage density, as relief declined, and interstream spacing in the later stages, so that no basis of comparison would remain.

This discussion is not intended to set forth any cycle of slope development but rather to show that diverse slope-profile forms are rational parts of sequences in which both parallel and declining slope retreat play parts, depending on the relative relief at the start of denudation and the stage to which reduction of relief has progressed.

Hypsometric Study of Geomorphic Stages of Development

The classic Davisian geomorphic cycles list criteria for delimiting each stage of development of landscape. Many geomorphologists have studied the relative time required or the proportionate mass removed from any basin in each stage. Johnson (1932, p. 488) guessed that the youthful stage takes up 15 per cent of the total time required for the complete cycle. Johnson (1933) also used per cent of total erosion as a basis of stage classification, stating that youth persists until 30-40 per cent of the available mass is removed, on the supposition that the initial surface would at this point be largely destroyed. Strahler (1952b, p. 1130) proposed that the youthful stage persists until the hypsometric integral reduces to about 60 per cent, implying about 40 per cent removal of mass.

Strahler (1952b, p. 1130) states:

"From the standpoint of hypsometric analysis, the development of the drainage basin in a normal fluvial cycle seems to consist of two major stages only; (1) an inequilibrium stage of early development, in which slope transformations are taking place rapidly as the drainage system is expanded and ramified. (2) An equilibrium stage in which a stable hypsometric curve is developed and maintained in a steady state as relief slowly diminishes. The monadnock phase with abnormally low hypsometric integral, when it does occur, can be regarded as transitory, because removal of the monadnock will result in restoration of the curve to the equilibrium form.... The hypsometric curve of the equilibrium stage is an expression of the attainment of a steady state in the processes of erosion and transportation within the fluvial system and its contributing slopes."

The writer studied these concepts in a series of second-order basins whose hypsometric curves form a series of increasing mass removal (Figs. 18, 19). Detailed description of these curves has already been presented. To the 11 basin curves plotted in Figure 19 two hypothetical curves were added in accordance with the supposed further reduction of the basins.

The curves illustrating the distribution of mass within each basin (Fig. 19) demonstrate that as erosion continues the following changes occur: (1) The point of maximum erosion migrates toward the head of the basin. The vertical distance between curves is greatest near the mouth of the basins (right side of illustration) in early stages but is greatest near the basin head in later stages (left side of illustration), beginning with curve 10. (2) The rate of lowering of the main channel decreases as the point of maximum erosion advances upchannel. Vertical distance between curves at right side of illustration, or mouth of basin, decreases downward indicating reduced rate of lowering of mouth of basin or reduced degradation of main channel. (3) The profile of the hypsometric curves changes with mass removal from convex-up to essentially straight to concave-up.

Although these hypsometric curves are not stream profiles they seem to reflect the activity of the stream in each case. Channel degradation has not greatly affected Curve 1, for the channels are mere traces on the upland surface. Curve 2 gives the first evidence of channel degradation in the abrupt down bend at the right. Curves 2, 3, 4, 5, and 6 reveal successively that the deeply incised channel is growing headward. At the same time the channel to which these basins are tributary is lowering rapidly and there is no stable base level at the basin mouth. Challinor (1930) stated that a stream not controlled by a fixed base level will have such a convex profile.

In the stage of Curves 7 and 8 the main channel has ceased rapid degradation and the tributaries immediately respond by removing convexities in their channels. Continued adjustment of basin forms brings a concavity to the hypsometric curves. Figure 40 shows that stream-channel profiles follow a similar evolution.

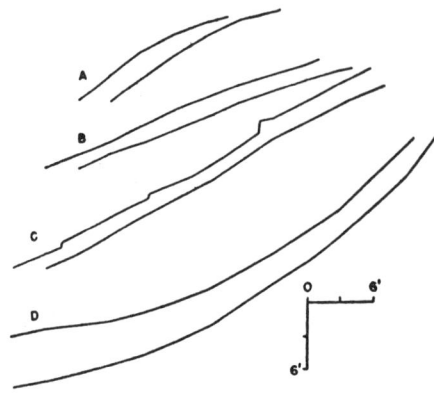

FIGURE 40.—CHANGES IN LONGITUDINAL STREAM PROFILES AT PERTH AMBOY

The profiles were measured on four stream channels ranging from very youthful to fully mature. The lower profile of each pair is a resurvey 1 year after the upper profile was surveyed. Relative vertical spacing of each pair of profiles has no significance in terms of actual vertical lowering.

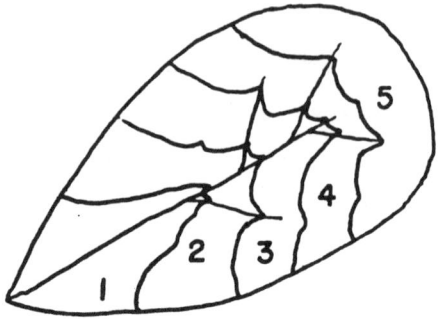

FIGURE 41.—METHOD OF DIVIDING A DRAINAGE BASIN INTO FIVE EQUAL AREAS AND THE SYSTEM OF NUMBERING EACH 20 PER CENT AREA

Comparing erosion in different areas of a basin at successive times further illustrates the progress of erosion. The mass removed from any portion of a basin can be determined by measuring the area above the hypsometric curve within limits of that percentage area class of the basin. The basins were divided into five equal-area classes bounded by the vertical lines for each 20 per cent increase of area. For example, the mass removed in the lower 20 per cent of the basin area was determined by measuring with a planimeter the area remaining above each curve limited by the ordinates for 80 and 100 per cent of basin area. The basins were thus divided into five equal areas (Fig. 41) numbered from lowest to highest. To illustrate the progress of erosion within each 20 per cent unit of basin area, each drainage basin represented by a curve in Figure 19 was replotted as a vertical series of points (Fig. 42). The percentage of total mass removed from each 20 per cent area unit of one basin is a point on the new curve. Five new curves are thus drawn, corresponding to each 20 per cent area, numbered in accordance with Figure 41 where the lowest area, nearest the mouth, is number 1. These curves then show the changes in the relative magnitude of mass removed from each 20 per cent area during the progress of the erosion cycle.

Comparison of Figure 42 with Figure 19 shows that the shape of each hypsometric curve in Figure 19 depends upon which area begins, shown by the steepening of area unit curve 5. Beyond 40 per cent total mass removed the hypsometric curves (Fig. 19) become increasingly concave from increased erosion at the basin heads.

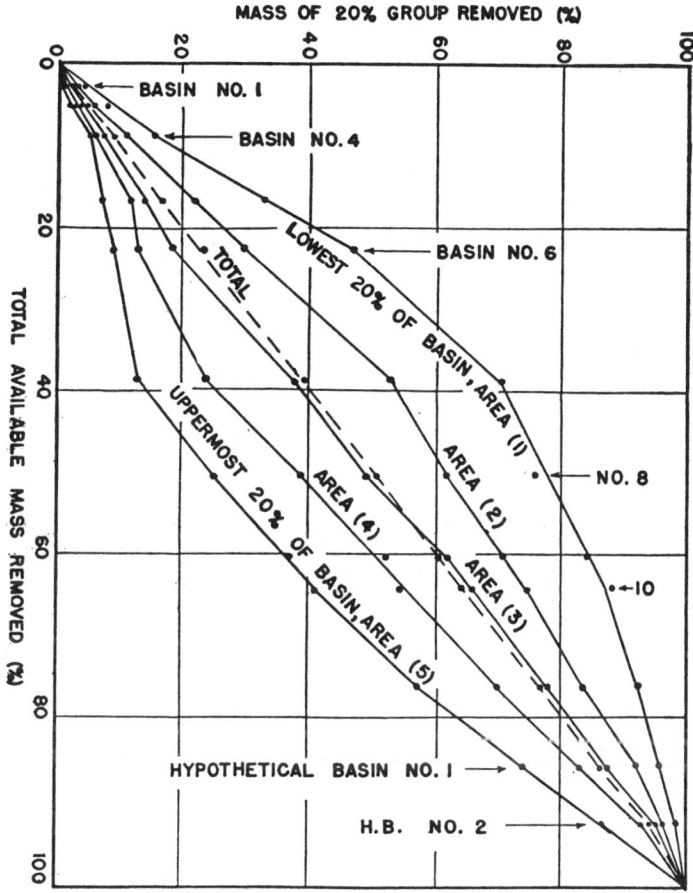

FIGURE 42.—RELATION OF EROSION WITHIN EACH 20 PER CENT AREA OF A BASIN TO TOTAL EROSION DURING A COMPLETE GEOMORPHIC CYCLE

segment is experiencing maximum erosion. Erosion in area unit 1 (Fig. 42) is greatest during the earliest stages of basin development but begins to decrease at about 25 per cent total mass removed (Basin 6) and is markedly decreased at 40 per cent (Basin 7). In Figure 19 hypsometric curve 6 is the last of the series showing a pronounced up-convexity. In Figure 42, at 40 per cent total mass removed, area unit 1 has lost 70 per cent of its mass, whereas area unit 5 has lost only 13 per cent. Erosion in area unit 5 is minor until 40 per cent total mass removed (Basin 7), when an increase in erosion

A comparison of curves in Figure 42 shows that erosion is greatest at first near the basin mouth (area unit 1) but dominant later in area unit 5. Erosion in area unit 1 decreases after removal of 25 per cent of the total mass. This is the same percentage at which the relief ratio becomes constant, for the relief ratio may reach a fixed value when the mouth of the stream has lowered to local base level. Erosion progresses upchannel into the higher area units causing modification of the other basin characteristics until later in the cycle.

Beyond 40 per cent total available mass removed, all curves of Figure 42 converge toward 100 per cent mass removal or peneplanation. This suggests that in this sequence there is no sudden change to an equilibrium stage in distribution of erosion but rather a zone of transition between 25 and 40 per cent of mass removal.

Strahler's selection of hypsometric integral of 40 per cent as the beginning of the equilibrium stage agrees roughly with the curve inflections in Figure 42, although the fortuitous occurrence of a basin at 40 per cent total available mass removed may give the misleading impression that abrupt changes in erosion occur at this point. It is safer to say that major changes occurred between 30 and 40 per cent.

Strahler's equilibrium stage includes both maturity and old age stages of the Davisian cycle. Observations of basin-form parameters suggest subdividing the equilibrium stage into two parts. Therefore the equilibrium stage will have two phases, one in which the mean values for relief ratio, gradient, and basin shape are constant or essentially constant, and a later phase when these values decrease. The phase of constant values would vary in duration; it would be shorter in areas of moderate available relief and absent in areas of low relief. Maturity and old age are thus recognizable in the constant and decreasing phase respectively, within the equilibrium stage of hypsometric development.

[*Editor's Note:* A discussion of badland hillslope erosion has been deleted.]

Summary and Conclusions

The drainage system at Perth Amboy conforms to Horton's laws of drainage composition; further, the mean drainage areas of streams of each order also form a geometric progression. Within an area of homogeneous lithology and simple structure the scale of the drainage-network elements is determined by a characteristic value for the minimum area required for channel maintenance. Relationships between channel lengths, drainage-basin area, and stream-order number are dependent on this constant of channel maintenance, which is in turn dependent on relative relief, lithology, and climate. The constant of channel maintenance is valuable as a measure of texture and decreases with increasing erodibility. Other characteristics of the drainage network which vary with relative relief are: drainage density, maximum valley-side slope angles, stream-channel gradient, drainage-basin shape, rate of sediment loss, and drainage pattern. Hadley and Schumm (In preparation) found relief ratio related to the infiltration rate of the soil on a lithologic unit and infiltration rates to texture of the topography. This suggests that infiltration rates are one of the most important factors influencing topographic development.

The mass-removal sequence of drainage basins strongly supports Strahler's concept of an open system in a steady state as related to drainage-basin development. At Perth Amboy the equilibrium stage begins at 25 per cent of mass removal as indicated by the constancy of drainage-basin parameters beyond this stage.

A comparison of young and mature angles of junction shows a significant difference suggesting that the angles change during the geomorphic cycle. During the reduction of a landmass of high relief much erosion may occur by lateral shifting of tributaries toward main channels.

Experimental studies and field observations reveal that erosion and the development of slope profiles may not conform to certain widely assumed concepts of the action of runoff as a function of depth and distance from a divide. Because runoff occurs as subdivided and surge flow, runoff erosion is more analogous to creep than to transport by fluid flow in channels.

Parallel and declining slope retreat may both be important in the evolution of erosional topography, depending on the available relief and perhaps on infiltration rates of the soil.

At Perth Amboy slopes are initiated by channel degradation and maintained by runoff and creep induced by frost heaving. Convex divides may be formed either by runoff or by rainbeat and creep. Rills follow a definite cycle of destruction and reappearance under the action of frost heaving and runoff.

Two types of topography appearing in the Badlands National Monument resemble arid and humid landscapes, the first with sharp-crested, steep, straight slopes bordered by pediments, the second with more gentle slopes and convex summits. The topographic differences are explained by the differences of infiltration rates of the two formations on which they form. This distinction may apply generally to differences between arid and humid topography. Vegetation and thick-soil horizons aid infiltration which promotes creep and the development of the rounded humid-cycle landforms characterized by declining slope retreat in later stages. In arid regions sparse vegetation and meager soils aid rapid runoff and the formation of steep, parallel retreating slopes and pediments.

Much work must be done before a workable theory of erosional landform development can be projected from detailed field examination and measurements of topographic characteristics and geomorphic processes, but there is no alternative to an empirical quantitative approach if these relationships are to be clarified in realistic terms.

REFERENCES

Challinor, J. 1930. The curve of stream erosion. *Geol. Mag.* **67**:61-76.

Davis, W. M. 1909. *Geographical essays:* N.Y.: Ginn and Co.

Fenneman, N. M. 1922. Physiographic provinces and sections in western Oklahoma and adjacent parts of Texas. *U.S. Geol. Survey Bull.* **730**:115-134.

Glock, W. S. 1931. The development of drainage systems: A synoptic view. *Geog. Rev.* **21**:475-482.

———. 1932. Available relief as a factor of control in the profile of a landform. *Jour. Geol.* **40**:74-83.

Hadley, R. F., and Schumm, S. A. 1961. Studies of erosion and drainage basin characteristics in the Cheyenne River Basin. *U.S. Geol. Survey Water-Supply Paper 1531-B.*

Horton, R. E. 1945. Erosional development of streams and their drainage basins; Hydrophysical approach to quantitative morphology. *Geol. Soc. America Bull.* **56**:275-370.

Hursh, C. R. 1948. Local climate in the Copper Basin of Tennessee as modified by the removal of vegetation. *U.S. Dept. Agriculture Circ. 774.*

Johnson, D. 1932. Streams and their significance. *Jour. Geol.* **40**:481-497.

———. 1933. Available relief and texture of topography: A discussion. *Jour. Geol.* **41**:293-305.

Krumbein, W. C. and Pettijohn, F. J. 1938. *Manual of sedimentary petrography.* N.Y.: Appleton-Century Co.

Langbein, W. B. 1947. Topographic characteristics of drainage basins. *U.S. Geol. Survey Water-Supply Paper 968-C*:99-114.

Lawson, A. C. 1932. Rain-wash in humid regions. *Geol. Soc. America Bull.* **43**:703-724.

Leopold, L. B. and Maddock, T. Jr. 1953. The hydraulic geometry of stream channels and some physiographic implications. *U.S. Geol. Survey Prof. Paper 252.*

———. and Miller, J. P. 1956. Ephemeral streams—hydraulic factors and their relation to the drainage net. *U.S. Geol. Survey Prof. Paper 282-A.*

Paulsen, C. G. 1940. Hurricane floods of September 1938. *U.S. Geol. Survey Water-Supply Paper 867.*

Penck, W. 1953. *Morphological analysis of landforms.* Translated by Hella Czeck and K. C. Boswell: London: Macmillan and Co., Ltd.

Schumm, S. A. 1955. The relation of drainage basin relief to sediment loss. *Internat. Union Geodesy and Geophysics*, 19th General Assembly (Rome), trans. **1**:216-219.

Segerstrom, K. 1950. Erosional studies at Paricutin, State of Michoacan, Mexico. *U.S. Geol. Survey Bull. 965-A.*

Smith, K. G. 1958. Erosional processes and landforms in Badlands National Monument, South Dakota. *Geol. Soc. America Bull.* **69**:975-1007.

Strahler, A. N. 1950. Equilibrium theory of erosional slopes approached by frequency distribution analysis. *Am. Jour. Sci.* **24S**:673-696, 800-814.

———. 1952a. Dynamic basis of geomorphology. *Geo. Soc. America Bull.* **63**:923-938.

———. 1952b. Hypsometric (area-altitude) analysis of erosional topography. *Geol. Soc. America Bull.* **63**:1117-1142.

———. 1943. *Revisions of Horton's quantitative factors in erosional terrain.* Paper read before Hydrology Section of Am. Geophys. Union, Washington, D.C., May 1953.

Wolman, M. G. 1955. The natural channel of Brandywine Creek Pennsylvania. *U.S. Geol. Survey Prof. Paper 271.*

THE CONCEPT OF ENTROPY IN LANDSCAPE EVOLUTION

Luna B. Leopold and Walter B. Langbein

[*Editor's Note:* Preceding discussions of entropy, open systems, longitudinal river profiles, and hydraulic geometry have been deleted.]

THE DRAINAGE NETWORK

As has been shown in the example of the longitudinal profile, the probability that a random walk will fall in certain positions within the given constraints can be ascertained. There is also a mean or most probable position for a random walk within those constraints. This statement suggests the possibility that a particular set of constraints might be specified that would describe the physical situation in which drainage channels would develop and meet, eventuating in the drainage network.

Let us postulate that precipitation falling on a uniformly sloping plain develops an incipient set of rills near the watershed divide and that they are oriented generally downhill. As the rills deepen with time, crossgrading begins owing to overflow of the shallow incipient rills in the manner postulated by Horton (1945, p. 337). The direction that the crossgrading takes place and the micropiracy of incipient rills is, as Horton implies, a matter of chance until the rills deepen sufficiently to become master rills.

This randomness in the first stages of crossgrading might be approximated in the following conceptual model which is amenable to mathematical description. Consider a series of initial points on a line and equidistant from one another at spacing a. Assume random walks originating at each of these points. In each unit of time each random walk proceeds away from the initial line a unit distance. But let us specify that each walk may move forward, left or right at any angle, but may not move backward. The accumulation of moves will produce in time sufficient cumulative departures from the orthogonal to the direction of the original line of points that some pairs of paths might meet. After such a junction only one walk proceeds forward in like manner and would behave similarly, just as when two stream tributaries join the single stream proceeds onward.

The physical situation described is analogous to the statistical model called the "gambler's ruin" (Feller, 1950, chap. 14). This model treats the probability that in a certain number of consecutive plays, a gambler playing against the "house" will lose all of his capital to the house. In the model, the capital assets of the gambler and the house are specified. If the gambler has a lucky run he accumulates capital at the expense of the house. If the reverse takes place, he gradually may lose his capital completely.

The statistical statement of the probable duration of the game (number of plays) is given by $D=Z(A-Z)$ provided the probabilities of a win and loss are equal, and where Z is the capital of the gambler, and A is the total capital in the game (house plus gambler). The size of Z and A are measured in units of the size of equal wagers.

The analogy to the physical situation of the postulated developing drainage network may now be specified quantitatively. It is desired to compute the distance two random walks will proceed before joining. In this model each walk has the same probability that it will accumulate a deviation of some given distance from the mean or orthogonal path.

This would be analogous to the condition in the game where the house and gambler begin with the same capital, that is, $Z=\frac{1}{2}A$. If the size of each wager is equal, then $D=\frac{1}{4}A^2$.

The duration of the game under the conditions stated is proportional to the square of the size of the initial capital. However, this describes the duration of a game with fixed boundaries. In the stream case, contiguous streams may join either on the right or left, so that the distance travelled before joining would not necessarily increase as the square of the separation distance at the last junction. On the other hand, the distance travelled would increase at least as the first power of the separation distance. Hence, there are two limits

$$D \propto A^2$$

or

$$D \propto A.$$

In the stream case, the quantity A is equivalent to the average distance between streams, and we may assume that on the average the streams are evenly spaced (equivalent to the statement $Z=\frac{1}{2}A$).

Consider the geometry of a simple and regular network of joining streams. Let us follow the definition introduced by Horton (1945) for stream order. The smallest unbranched tributary is called first order, the stream receiving as tributaries only streams of first order is called second order. A stream receiving as a tributary a second-order stream is called third order, and so on.

The geometry of joining streams may be described by the relation $A=a2^{R-1}$ where A is the mean distance between streams of order R, and a is the mean distance between the first-order streams. Hence, we can state that the average stream length from one junction to the next varies with the mean interfluve distance, in the form

$$L \propto 4^R$$
$$L \propto 2^R$$

depending on whether the length varies as the square or the first power of the distance between streams of a given order. We deduce therefore that the logarithms of stream length vary linearly as the order number, and that the mean ratio between the lengths of streams differing in order by 1 will be between 2 and 4. This result derived from considerations of entropy (most probable state) is in agreement with the findings of Horton's (1945) analysis of field data on river systems.

The mathematical relationship derived may perhaps best be visualized by actual trial, constructing random walks on cross-section paper using a table of random numbers.

We constructed a stream network from the following elemental structure: Beginning with a series of points 1 inch apart on a line, a random walk was constructed, limited, however, only to motion right, left, or ahead. Backward or uphill steps were excluded. The several random walks when extended far enough made junctions. The resulting walk proceeding onward again joined and so on. A portion of the work sheet on which the random walks were plotted is shown as figure 6.

To test how this constructed network compared with nature, we have made what is called a Horton analysis. This consists of plotting the logarithms of the average length of streams of various orders against the order number. Horton (1945, p. 298) found the length of streams of a given order increased in constant ratio with increasing order number. Figure 7 presents the results obtained from the random walks plotted, and shows that stream lengths increased by a ratio of about 2.8 between consecutive order numbers. This result lies within the range of 2 and 4 as suggested previously. Horton obtained, from field data he studied, ratios between 2.5 and 3.7. Ephemeral channels near Santa Fe, N. Mex., were found to exhibit a ratio of about 2.1 (Leopold and Miller, 1956, p. 18, fig. 13). Thus the random-walk results give results in general quantitative agreement with field data.

The constraints used in construction of the random walks in figure 6 may seem too limited. Another type of construction is as follows: Using a sheet of rectangular cross-section graph paper, each square is presumed to represent a unit area. Each square is to be drained, but the drainage channel from each square has equal chance of leading off in any of the four cardinal directions, subject only to the condition that, having made a choice, flow in the reverse direction is not possible. Under these conditions, it is possible for one or more streams to flow into a unit area, but only one can flow out. By the random selection of these directions under the conditions specified, a stream network was generated as shown on figure 8 which has striking similarities to natural drainage nets. Divides were developed, and the streams joined so as to create rivers of increasing size. One can say that the random pattern represents a most probable network in a structurally and lithologically homogeneous region. The Horton analysis of the network pictured on figure 8 is presented as figure 9.

Figure 10 shows the relation between stream length and drainage area. The slope of the line indicates that for this random-walk model length increases as the 0.64 power of the drainage area. Simple geometric relations would suggest that stream lengths vary as the square root of the drainage area. This model accounts for the higher values that Langbein (1947, p. 135) and Hack (1957, p. 66–67) found for natural streams.

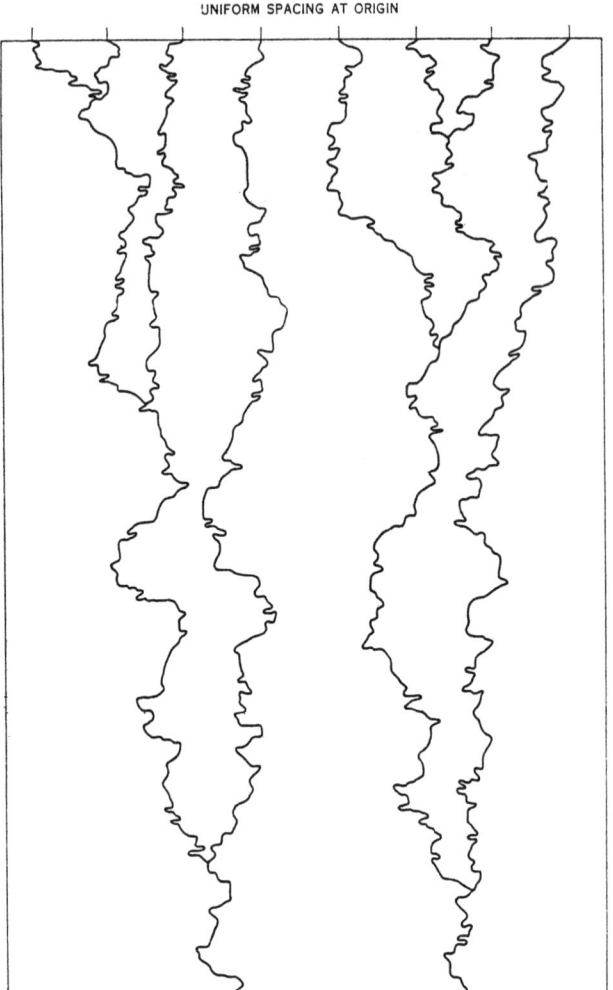

FIGURE 6.—Portion of random-walk model of stream network.

Although in this example each elemental area is a square, we note that the diagram might have been drawn on an elastic sheet. If this sheet were stretched in various ways to alter the geometric figures, none of the relations between numbers and lengths with orders, shown on figure 9, would have been changed.

Horton also showed that in natural stream networks, the logarithms of the number of streams of various orders is proportional to the order number. More specifically, the number of streams of a given order decreased in constant ratio with increasing order number. This ratio he called the "bifurcation ratio." Figure 9 shows the relation between numbers of streams of the orders 1 to 5 that were developed by the streams on figure 8. When compared with similar diagrams from data on actual rivers, given by Horton in his original treatment of the subject, one may note that in each case there is a logarithmic relationship between

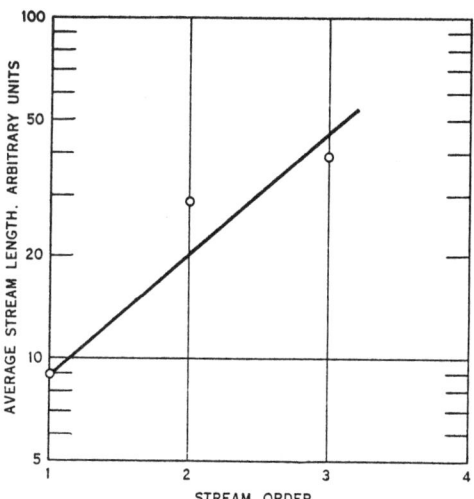

FIGURE 7.—Relation of length of streams to stream order.

orders and numbers of streams, and moreover, the bifurication ratios are quite comparable.

The essential point of these demonstrations is that the logarithmic increase, both of stream length and number of streams with order number as found in natural stream networks, accords with the geometrical properties and the probabilities involved. The logarithmic relationship is thus one of optimum probability. Optimum probability, in this sense, represents maximum entropy.

DISCUSSION AND GEOMORPHIC IMPLICATIONS

In statistical mechanics the probability aspect of entropy has been demonstrated. The word entropy has, therefore, been used in the development of systems other than thermodynamic ones, specifically, in information theory and in biology. Its use is continued here.

With this understanding of the term, the present paper shows that several geomorphic forms appear to be explained in a general way as conditions of most probable distribution of energy, the basic concept in the term "entropy."

It is perhaps understandable that features such as stream profiles which occur in nature in large numbers should display, on the average, conditions that might be expected from probability considerations because of the large population from which samples may be drawn. The difficulty in accepting this proposition is that there is not one but many populations owing to the variety of local geologic and lithologic combinations that occur. Further, the geomorphic forms seen in the field often are influenced by previous conditions. Stream channels, in particular, show in many ways the effects of previous climates, of orogeny, and of structural or stratigraphic relations that existed in the past. In some instances the present streams reflect the effects of sequences of beds which have been eradicated by erosion during the geologic past.

In a sense, however, these conditions that presently control or have controlled in the past the development of geomorphic features now observed need not be viewed as preventing the application of a concept of maximum probability. Rather, the importance of these controls strengthens the usefulness and generality of the entropy concept. In the example presented here we have attempted to show that the differences between patterns derived from averaging random walks result from the constraints or controls imposed upon the system. We have, in effect, outlined the mathematical nature of a few of the controls which exist in the field. The terms in the genetic classification of streams reflect the operation of constraints. Terms like "consequent" and "subsequent" are qualitative statements concerning constraints imposed on streams which, in the absence of such constraints, would have a different drainage net and longitudinal profile.

In a sense, then, much of geomorphology has been the study of the very same constraints that we have attempted to express in a mathematical model.

The present paper is put forward as a theoretical one. It is not the purpose of a theoretical paper to compare in detail the variety of field situations with the derived theoretical relations. Rather, it is hoped merely to provide the basis for some broad generalizations about the physical principles operating in the field situations.

On the other hand, the random-walk models used here are simple demonstrations of how probability considerations enter into the problem. They are intended to exemplify how the basic equations can be tested experimentally. Thus the present paper should not be considered to deal with random walks. We hope it is concerned with the distribution of energy in real landscape problems. The random-walk models exemplify the form of the equations, but the equations describe the distribution of energy in real landscapes, simplified though the described landscape may be.

One of the interesting characteristics of the experiments with random walks is the relatively small number of trials that need be carried out to obtain average results that closely approximate the final result from a large number of trials. This suggests, though it does not prove, that random processes operating in the landscape within certain constraints develop rather quickly

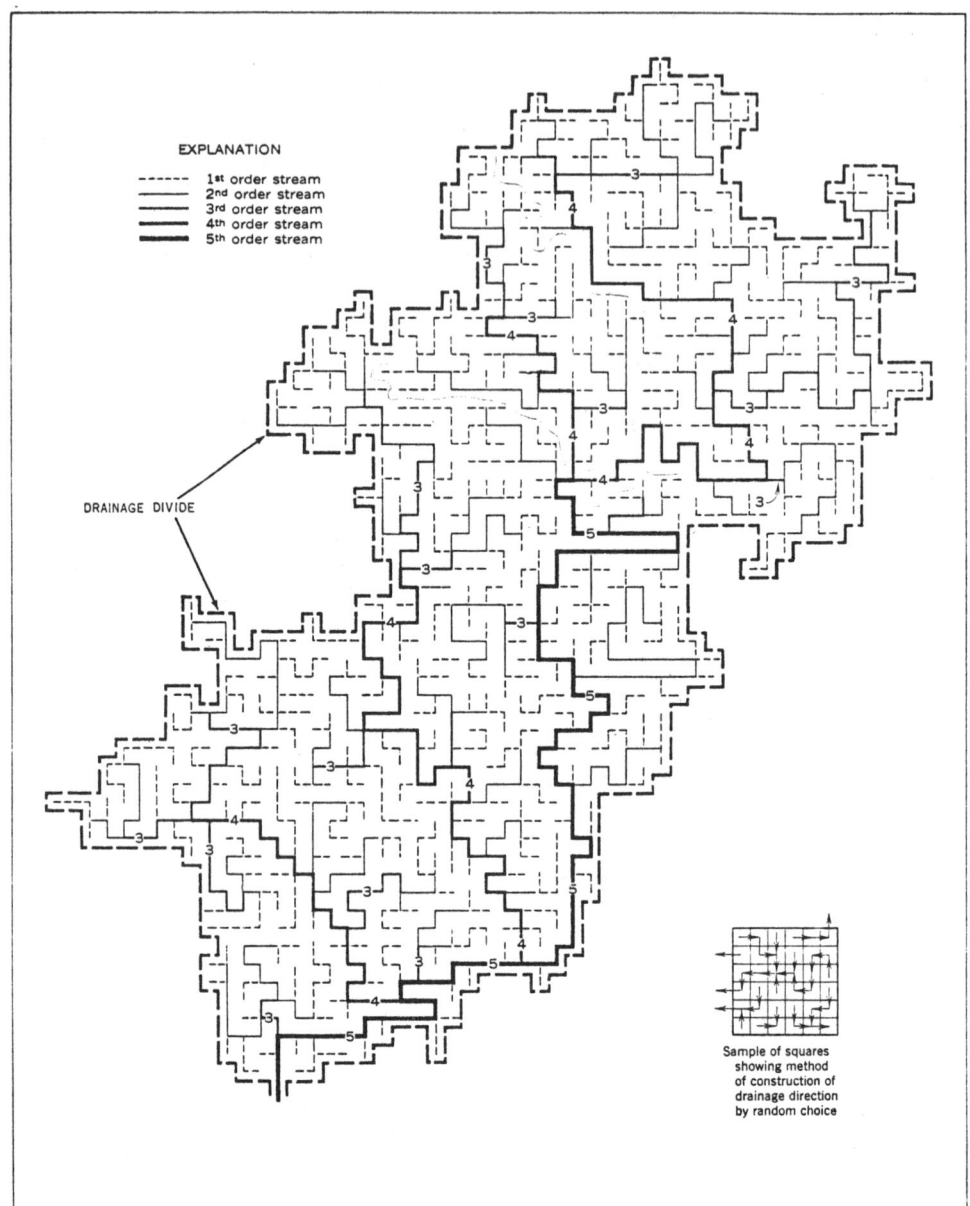

FIGURE 8.—Development of a random-walk drainage basin network.

the characteristics that obtain after a much longer period of time.

Further, the relative insensitivity of the results to the lack of exactly equal probabilities among alternatives suggests that the approach to the most probable condition is at first very rapid. In the mathematical models the final elimination of minor deviations from the condition of maximum entropy requires a very large number of samples from the population of possible alternatives. In terms of the field development of forms, it seems logical that this may be equated to a long time-period required to eliminate minor variations from the theoretic most probable state.

These observed characteristics of the averages of samples used in the mathematical models make it difficult to believe that in field conditions, time periods measured in geologic epochs could elapse before fluvial systems approach the condition of quasi-equilibrium. This line of reasoning quite fails to lend support to the Davisian concept that the stage of geomorphic youth is characterized by disequilibrium whereas the stage of maturity is characterized by the achievement through time of an equilibrium state. Rather, the reasoning seems to support the concept recently restated by Hack (1960) that no important time period is necessary to achieve a quasi-equilibrium state.

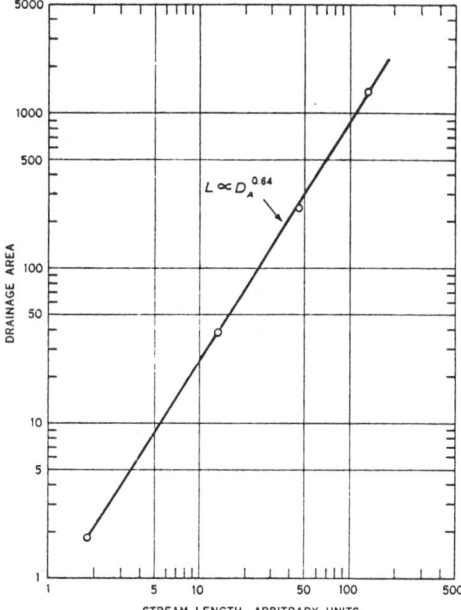

FIGURE 10.—Relation between stream length and drainage area for random-walk model.

The differing results obtained from varying the constraints imposed in the mathematical models lead us to following view as a working hypothesis: Landscape evolution is an evolution in the nature of constraints in time, maintaining meanwhile and through time essentially a dynamic equilibrium or quasi-equilibrium.

This conclusion also is in agreement with the view of Hack (1960), though arrived at by a quite different line of reasoning.

Whether or not the particular inferences stated in the present paper are sustained, we believe that the concept of entropy and the most probable state provides a basic mathematical conception which does deal with relations of time and space. Its elaboration may provide a tool by which the various philosophic premises still characterizing geomorphology may be subjected to critical test.

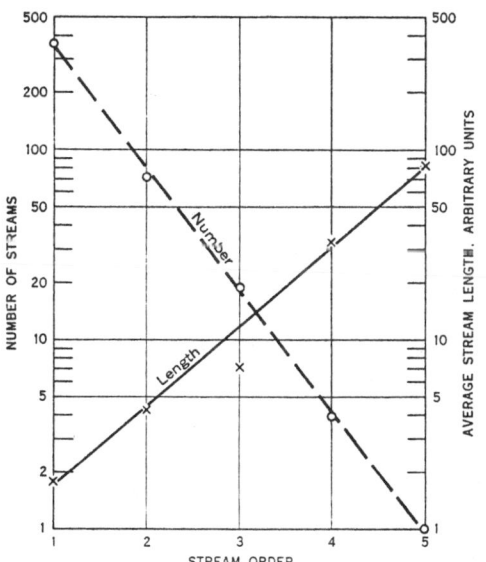

FIGURE 9.—Relation of number and average lengths of streams to stream order for model in figure 8.

311

REFERENCES

Feller, W. 1950. *An introduction to probability theory and its applications.* New York: John Wiley and Sons.

Hack, J. T. 1957. Studies of longitudinal stream profiles in Virginia and Maryland. *U.S. Geol. Survey Prof. Paper 294-B*:46-94.

———. 1960. Interpretation of erosional topography in humid temperate regions. *Am. Jour. Sci. Bradley Volume* **258**-A:80-97.

Horton, R. E. 1945. Erosional development of streams and their drainage basins; Hydrophysical approach to quantitative morphology. *Geol. Soc. America Bull.* **56**:275-370.

Langbein, W. B. 1947. Topographic characteristics of drainage basins. *U.S. Geol. Survey Water-Supply Paper 968-C*:125-155.

Leopold, L. B. and Miller, J. P. 1956. Ephemeral streams—Hydraulic factors and their relation to the drainage net. *U.S. Geol. Survey Prof. Paper 282-A*:1-36.

Part V

APPLICATIONS AND DIRECTIONS

Editor's Comments on Papers 14 Through 17

14   HADLEY and SCHUMM
Excerpts from *Sediment Sources and Drainage Basin Characteristics in Upper Cheyenne River Basin*

15   MORISAWA
Excerpts from *Quantitative Geomorphology of Some Watersheds in the Appalachian Plateau*

16   HICKOK, KEPPEL, and RAFFERTY
Hydrograph Synthesis for Small Arid-Land Watersheds

17   SCHUMM
Geomorphic Thresholds and Complex Response of Drainage Systems

    The applicability and significance of drainage basin geomorphology to the interpretation of geologic conditions is obvious (Paper 3). The manner in which a drainage basin functions is closely related to its morphology and therefore it should also be possible to predict, at least in part, the hydrologic characteristics of the basin from its morphology. This approach is particularly valuable in areas where hydrologic data is at a minimum. For example, existing information on sediment yields and runoff can be related to drainage basin morphology and these empirical relations can be used to estimate sediment yields and runoff for ungaged drainage basins. An example of this is the reconnaissance study in the Cheyenne River Basin of Wyoming by Hadley and Schumm (Paper 14). The lack of topographic maps and the poor quality of aerial photographs in this area hampered the investigation, and only very simple measurements could be made on photographs and in the field to compare basin morphology and hydrology. Similar but more sophisticated studies have been made to estimate peak discharge and hydrograph characteristics by Morisawa (Paper 15), and Hickok et al. (Paper 16). This subject has been thoroughly reviewed by Gray (1965), Gregory and Walling (1973), and Ward (1971), and further treatment here is not necessary.

Of equal importance is the need to estimate both the landform changes and hydrologic changes that result from man's modification of drainage basins. Removal of vegetation can have a significant effect, but even more significant is the morphologic modification of drainage basins during mining and urbanization (Gregroy and Walling, 1973). Further work is required to determine the response of drainage basins to both natural and man-induced change. Recent experimental and field studies indicate that during part of the erosional evolution of a drainage basin, erosion may be episodic rather than progressive (Schumm, 1977).

The response of the drainage basin to change is complex and this is of great significance to the agricultural engineer and land manager as well as to the Quaternary stratigrapher. Future geomorphic research can be concentrated on those topics with both practical and scientific benefit. Paper 17, therefore, is a summary of these ideas by Schumm and it provides a direction for future research endeavors. It also provides a basis for the renewal of the link between geomorphology and geology, as the significance of geomorphic thresholds and the complex response of drainage basins to stratigraphy and sedimentology is realized. The erosional evolution of a drainage basin through geologic time determines the quantity and nature of the sediment moving from that basin. Therefore, the mobilization and eventual concentration of heavy minerals in fluvial placers is also closely related to drainage basin morphology (Schumm, 1977).

REFERENCES

Gray, D. M. 1965. Physiographic characteristics and the runoff pattern. In *Research Watersheds*, pp. 147-164. Hydrology Symposium Proc. 4, National Res. Council Canada.

Gregory, K. J. and Walling, D. E. 1973. *Drainage basin form and process.* New York: John Wiley and Sons.

Schumm, S. A. 1977. *The fluvial system.* Wiley Interscience, in press.

Ward, R. C. 1971. *Small watershed experiments: An appraisal of concepts and research developments.* Univ. of Hull, Occasional Papers in Geography No. 18.

14

Reprinted from pp. 169–176, and 177 of *U.S. Geol. Survey Water-Supply Paper 1531-B*:137–197 (1961)

Sediment Sources and Drainage Basin Characteristics In Upper Cheyenne River Basin

By R. F. HADLEY *and* S. A. SCHUMM

[*Editor's Note:* Descriptive sections on the general features of the upper Cheyenne River Basin and on major sources of sediment have been deleted.]

RELATION BETWEEN GEOMORPHIC AND HYDROLOGIC CHARACTERISTICS OF SMALL DRAINAGE BASINS

The discussion of erosion and sedimentation in the Cheyenne River basin thus far has been limited mainly to measurements of sediment accumulation, as determined by deposition in stock-water reservoirs, and in tracing the deposition of sediment as it moves from the uplands to the stream channels and flood plains. In the course of obtaining this information some observations were made in an effort to determine the relations between the hydrologic and geomorphic characteristics of small drainage basins. Such a relationship, if it were dependable, would provide a ready means for developing an erosion classification of rangelands on the basis of reconnaissance surveys.

GEOMORPHIC CHARACTERISTICS OF SMALL DRAINAGE BASINS

In order to obtain some quantitative understanding of the topography of small drainage basins, some of their characteristics were measured as described below.

Drainage density in particular was considered to be an important index of drainage basin character. Drainage density, or what can be thought of as the texture of topography, was calculated by dividing the length of stream channels, in miles within each basin, by basin area in square miles (Horton, 1945). The stream length and basin area were measured on aerial photographs of an approximate scale 1:30,000. Undoubtedly, on photographs of this scale many of the small first and second order channels cannot be measured accurately, and so the values for drainage density may be low, that is, shorter drainage channels per unit area.

In figure 29, the total channel length in miles is plotted against drainage area in square miles for 81 drainage basins on the Fort Union formation. This plot illustrates the variation in texture to be expected among small drainage basins developed on one lithologic unit.

Although drainage density is an important characteristic of a drainage system, it gives only a two-dimensional indication of basin character. For a more complete picture of the basin, relief should also be considered. Absolute relief alone may not be significant but it was demonstrated (Schumm, 1956a) that several geometrical properties of a maturely developed drainage system (valley-side slope angle, stream gradients, basin shape) appear to be related to a topographic index expressed as a dimensionless relief ratio.

The relief ratio was obtained for small drainage basins by dividing the difference in elevation between the spillway of the reservoir and the headwater divide, by the length of the basin. The relief as measured does not include abnormally high points on the divide, and the length is measured essentially parallel to the main drainage channel within the basin and may not be the maximum basin length.

In the Cheyenne River basin only the reliefs of small drainage basins were obtained in the field. Basin length was measured from aerial photographs.

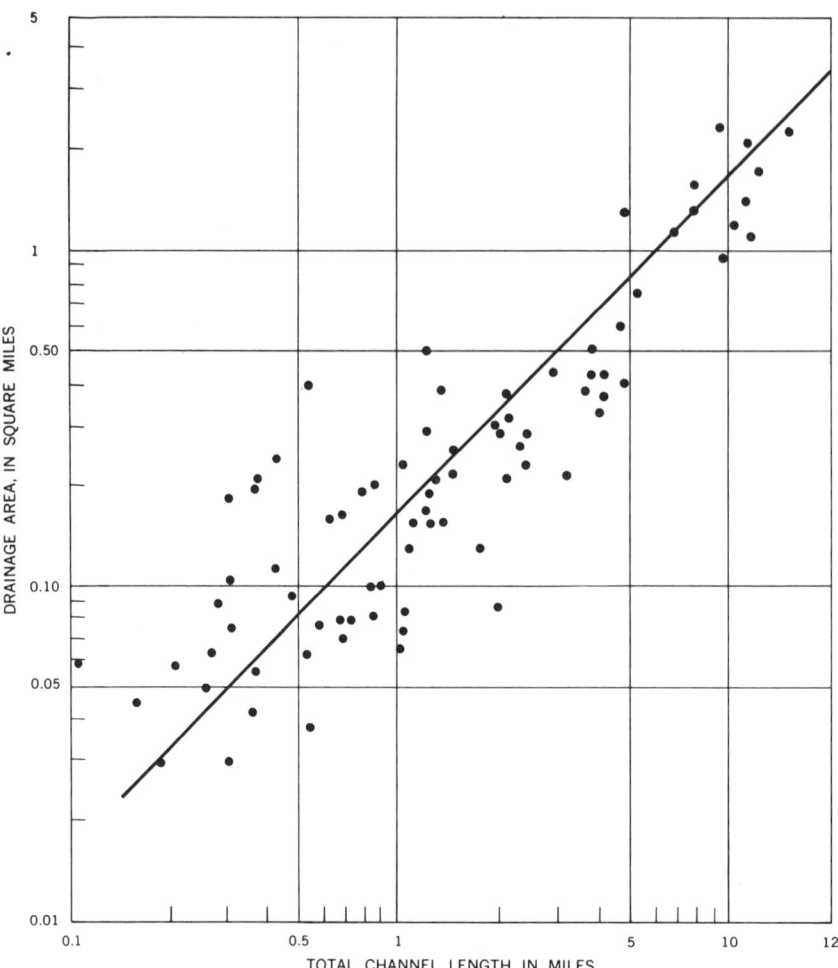

FIGURE 29.—Relation of total channel length to drainage area for basins on the Fort Union formation.

Another characteristic that would be important in the evaluation of topography, and which may not be compensated for by the relief ratio, is the condition of the drainage channels. For example, drainage channels may be differentiated into grassed and raw, or bare. A comparison of the types of channels within a drainage system shows that sediment yield increases rapidly with an increase in the density of raw channels for densities greater than 2.0, that is, more than 2 miles of raw channels per square mile of drainage area.

Also, many of the drainage basins contain numerous discontinuous gullies within tributary channels. Most of the discontinuous channels have a headcut at the upstream end; some are actively advancing while others are partly stabilized by vegetation. The density of headcuts of all character, or number per square mile of drainage area, was determined for each basin.

Field examination showed that many headcuts in a single channel are separated by long, grassed reaches having low gradients, where most of the sediment contributed by headcut advancement is redeposited in aggrading reaches, whereas others are joined directly to the reservoir by well defined, raw channels which provide a better opportunity for transporting eroded material through the basin. Headcuts of the latter type were listed separately on the premise that they would increase the rate of sediment accumulation. A graphical analysis of the field data collected shows that regardless of the location of the headcut or density value for a single basin, there is either no apparent relationship to the rate of sediment accumulation, or, if any, it is masked by other factors.

In summary, it appears that drainage density and relief ratio are easily obtained drainage basin characteristics, which may be of use in attempting to relate runoff and sediment accumulation rates to the geomorphic characteristics of small drainage basins.

HYDROLOGIC CHARACTERISTICS OF SMALL DRAINAGE BASINS

Within the Cheyenne River basin two groups of small reservoir drainage basins were selected, one for measurement of annual sediment accumulation and the other for measurement of runoff (p. 85–90). In the following sections the geomorphic character of both groups of small basins as expressed by relief ratio and drainage density are related to annual sediment accumulation and runoff.

In addition to the present investigation, mean annual sediment accumulation has been calculated from measurements of sediment trapped in the stock reservoirs on the Navajo Indian Reservation in Arizona and New Mexico (Hains, Van Sickle, and Peterson, 1952) and on the San Rafael Swell in Utah (King and Mace, 1953). These data were used to compare probable long-term sediment accumulation with rock type of the drainage basin, and a general relationship between rock type and erosion rates was found to exist.

Complete observations of basin dimensions for the small areas studied made the calculation of the relief ratio for each basin possible (Schumm, 1955). Mean values of the relief ratio were obtained for each rock type for comparison with mean annual sediment accumulation. Accessory data contained in the source reports reveal that for 4 years preceding the survey in the New Mexico–Arizona area both summer precipitation and runoff exceeded that of previous years. Runoff, assumed by Hains, Van Sickle, and Peterson (1952) to occur when precipitation exceeded 0.5 inch per day, was especially high during the 4 years preceding the study. In view of these

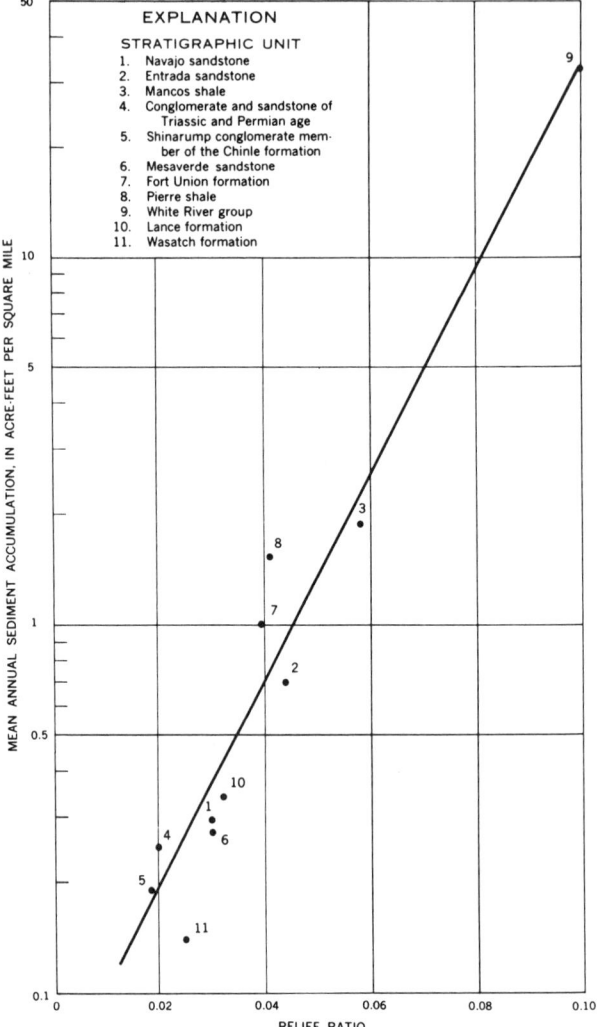

FIGURE 30.—Relation between mean annual sediment accumulation and relief ratio for basins on indicated rock units.

climatic data it was decided to eliminate from this analysis the data from any reservoirs in operation 5 years or less. Six mean values for reservoirs which were in operation for periods ranging from 10 to 15 years remain from the Arizona–New Mexico and Utah studies. These values of mean annual sediment yield, as well as those for each of the rock units in Cheyenne River basin, are plotted against the mean relief ratio in figure 30.

The good correlation of these mean values led to the plotting of individual basin values for 26 drainage basins located on the Fort Union formation in the Cheyenne River basin (fig. 31).

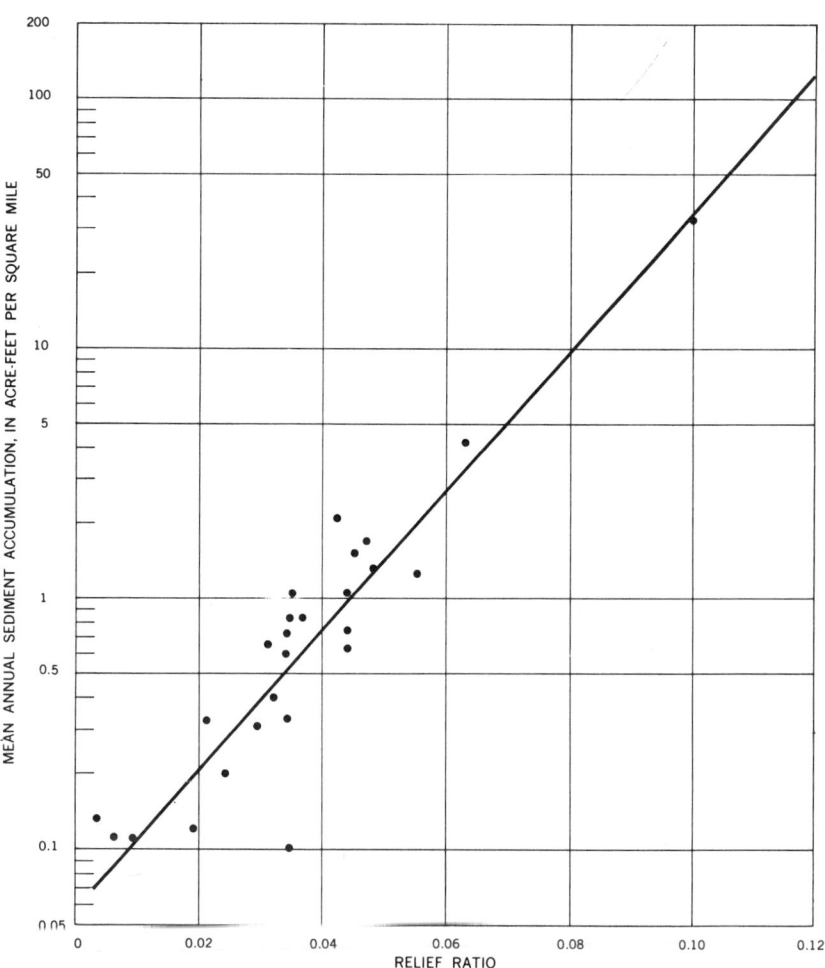

FIGURE 31.—Relation between mean annual sediment accumulation and relief ratio for basins on Fort Union formation.

In spite of the good correlations presented in figures 30 and 31 some exceptions were noted. It was found, for example, that in basins which contained two distinct types of topography, the relief ratio was not a satisfactory measure of geomorphic character or erosion rates. This was exemplified in some basins underlain by the White River group at the base of the Pine Ridge. In a typical example the upper part of the drainage basin was composed of badlands, and the lower part toward the reservoir was a smooth plain of aggradation. In this example, the sediment trapped in the reservoir was less than that indicated by the relationship in figure 30. However, the data for a reservoir in sec. 13, T. 33 N., R. 54 W., Sioux County, Nebr., at the edge of the badlands zone, shown by observation 9 (fig. 30), falls essentially on the average line for all the observations.

A mean of drainage density for the small basins also shows a general relationship with sediment yield from the different rock units, wherein the sediment accumulation increases with an increase in relief ratio, as shown in the following table.

Relation between drainage density and mean sediment yield classified by rock units

Rock unit	Drainage density (miles per sq. mi.)	Mean annual rate of sediment yield (acre-ft. per sq. mi.)
Wasatch formation	5.4	0.13
Lance formation	7.1	.5
Fort Union formation	11.4	1.3
Pierre shale	16.1	1.4
White River group	[1] 258.0	1.8

[1] Area of comparable dissection in Badlands National Monument, S. Dak. (Smith, 1958, p. 1001).

The correlation between mean annual sediment accumulation and the relief ratio suggest that a practical approach to an erosion classification of lands similar to those in the Cheyenne River basin area may be approximated by a quantitative analysis of the geomorphic characteristics of the region. Many other factors are, of course, important, but they may only modify what is essentially a geomorphic control.

Runoff measurements in 30 reservoir drainage basins within the Cheyenne River basin were obtained during the period 1951–54 (p. 85–90). The relief ratio and drainage density were measured for several of these basins and then were analyzed with respect to the runoff. Many of the basins, for which the relief ratio was obtained, subsequently had to be eliminated from this analysis because of diversion of runoff from its natural course by roads and dams or because the record was too short.

In figure 32 the texture expressed as drainage density is plotted against mean annual runoff. A relationship is apparent, suggesting that with additional information it may be possible to estimate quantitatively mean annual runoff for small drainage basins within one climatic type.

Plotting of relief ratio and runoff, however, shows no such correlation but only a general trend of increasing runoff with relief ratio that is too poorly defined to be of value.

The mean precipitation, during the 4 years in which the runoff records were collected, was 7.5 inches on the Wasatch formation, 10.0 inches on the Lance formation, 13.0 inches on the Fort Union formation, and 15.7 inches on the Pierre shale. Records of longer duration at the same stations show that the long-term mean precipi-

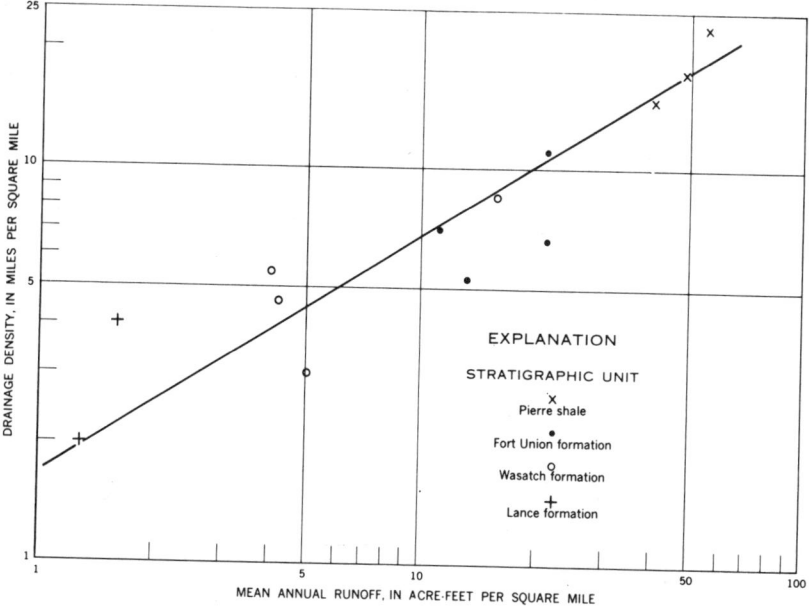

FIGURE 32.—Relation of mean annual runoff to drainage density for small basins.

tation ranges from about 13 to 15 inches from west to east across the Cheyenne River basin with a mean of 14.5 inches for the entire basin. Some adjustment of runoff should be made to compensate for the range in precipitation during the 4-year period. Runoff was adjusted by increasing runoff on each stratigraphic unit proportionally as the mean precipitation was above or below the 14.5-inch mean. The adjusted runoff rates are plotted against drainage density in figure 33.

Comparisons among the relations shown in figures 30, 31, and 32 suggest that the geomorphic character of the small basins has an important influence on the hydrologic character. Additional studies will be needed to clarify the existing relations.

[*Editor's Note:* The remainder of the discussion of the hydrologic character of small drainage basins and the significance of aggradational features has been deleted.]

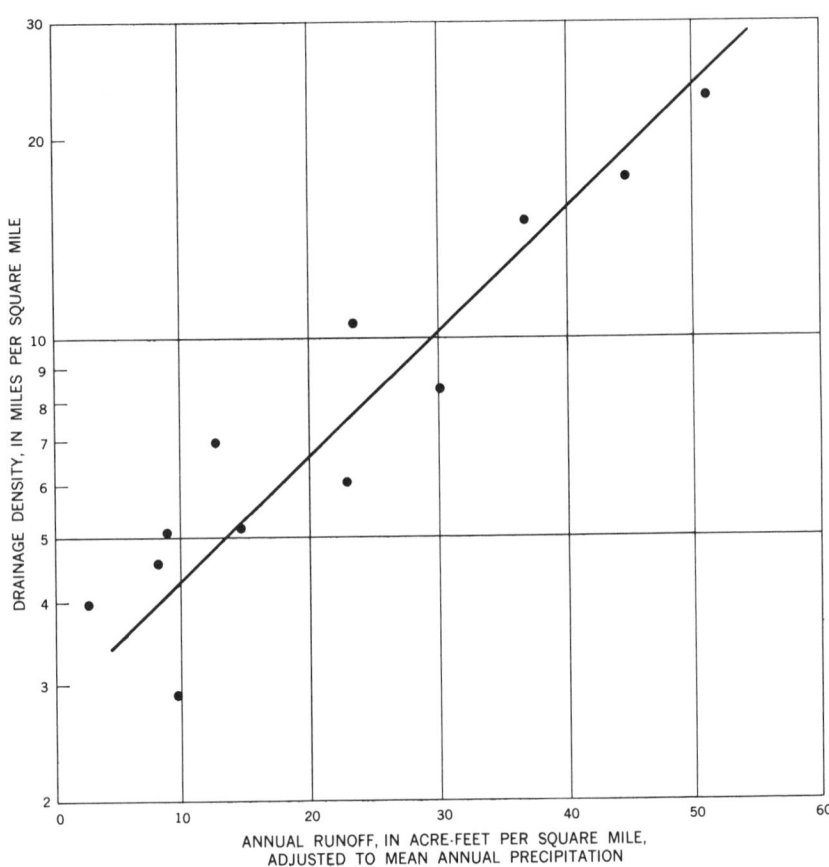

FIGURE 33.—Relation of adjusted mean annual runoff to drainage density for small basins.

REFERENCES

Hains, C. F.; VanSickle, D. M.; and Peterson, H. V. 1952. Sedimentation rates in small reservoirs in the Little Colorado River Basin. *U.S. Geol. Survey Water-Supply Paper 1110-D*:129-155.

Horton, R. E. 1945. Erosional development of streams and their drainage basins; Hydrophysical approach to quantitative morphology. *Geol. Soc. America Bull.* 56:275-370.

King, N. J. and Mace, M. M. 1953. Sedimentation in small reservoirs on the San Rafael Swell, Utah. *U.S. Geol. Survey Circ. 256.*

Schumm, S. A. 1955. The relation of drainage basin relief to sediment loss. *Assoc. Internat. Hydrology (Rome) Pub. 36.* 1:216-219.

———. 1956. Evolution of drainage systems and slopes in badlands at Perth Amboy, New Jersey *Geol. Soc. America Bull.* 67:597-646.

Smith, K. G. 1958. Erosional processes and landforms in Badlands National Monument, South Dakota. *Geol. Soc. America Bull.* 69:975-1008.

15

Copyright © 1962 by the Geological Society of America

Reprinted from pp. 1042-1045 and 1028 of *Geol. Soc. America Bull.* 73:1025-1046 (Sept. 1962)

Quantitative Geomorphology of Some Watersheds in the Appalachian Plateau

MARIE E. MORISAWA *Route 202, Towaco, N. J.*

[*Editor's Note:* The bulk of this paper, a study of the quantitative morphology of fifteen drainage basins, has been deleted.]

LIST OF SYMBOLS AND DEFINITIONS

u — order of basin or stream segment denoting level of magnitude in drainage network hierarchy

s — highest order within a drainage network

N_u — number of streams of order u

R_b — bifurcation ratio, ratio of branching in a drainage network; ratio of number of streams of order u to number of streams of next higher order, $\dfrac{N_u}{N_{u+1}}$

$\bar{L}_u$ — mean length of stream channel segments of order u

ΣL_u — total length of all stream segments of order u

$\sum_{u=1}^{s} L_u$ — total cumulative length of all stream segments of all orders contained in basin of order u

R_L — stream-length ratio, ratio of mean length of streams of order u to mean length of streams of next higher order, $\dfrac{\bar{L}_u}{\bar{L}_{u+1}}$

$\bar{A}_u$ — mean area of basin of order u

R_A — basin-area ratio, ratio of mean area of basin of order u to mean area of basin of next higher order, $\dfrac{\bar{A}_u}{\bar{A}_{u+1}}$

D — drainage density, ratio of total length in miles of all streams in a basin to area of the basin in square miles

C — constant of channel maintenance, area in feet required to maintain 1 foot of drainage channel; reciprocal of D, in feet

F_u — frequency, or density, of stream of order u; number of streams of a given order per unit area

R_c — basin circularity, ratio of the area of a basin to the area of a circle with the perimeter of the given basin

S_u — channel gradient of order u, tangent of the vertical angle at point of measurement; or ratio of fall in feet to length of channel in feet

R_s — slope ratio, ratio of mean gradient of streams of order u to mean gradient of streams of next higher order, $\dfrac{\bar{S}_u}{\bar{S}_{u+1}}$

$\bar{H}_u$ — average total relief of basin of order u

R_R — total relief ratio, ratio of relief of a basin of order u to total relief of basin of next higher order, $\dfrac{\bar{H}_u}{\bar{H}_{u+1}}$

R_h — relief ratio, ratio of total relief of a basin to its longest dimension parallel to the principal drainage line

$\bar{Q}$ — mean discharge in cfs for a basin

Q_{max} — peak discharge in cfs for a basin

RELATIONS BETWEEN GEOMORPHOLOGY AND HYDROLOGY

Factors Controlling Runoff

Stream flow depends upon those factors which determine amount of rainfall excess and those which influence length of time for rainfall to travel through the basin. Rainfall excess is determined primarily by climate, vegetation, infiltration capacity, and surface storage. Geomorphic factors such as stream lengths, basin shape, and ground slope, as well as geologic characteristics such as rock type and structure, influence runoff intensity and discharge.

If other factors remain the same, discharge is proportional to the area of a watershed. Average discharge as well as peak discharge has a direct relation to size of watershed (Hack, 1957, p. 54; Eisenlohr, 1952, p. 143), average and peak discharge increasing with increasing area.

Sherman (1932, p. 339) concluded there was a definite relation between basin outline and the unit hydrograph. Jarvis and others (1936, p. 34) review maximum-discharge formulas which use basin shape as a factor. Morisawa (1958, p. 590) found regressions of circularity on runoff to be significant.

Stream density and bifurcation ratio affect discharge, as closely spaced, numerous tributaries would result in rapid runoff and probably a large volume of flow. Longer stream length would mean a longer lag from time of onset of rainfall excess to flood peak. However, as all these factors are hard to isolate and relate to discharge in a qualitative way, quantitative analyses must be carried out.

General Correlation of Geomorphometry and Hydrology

Many geomorphic characteristics are related to area, inasmuch as they change systematically with area. If discharge or runoff intensity are functions of area,

$$Q = f(A) ; \quad (11)$$

and if

$$A = f'(L, \Sigma L, N, S, H, R), \quad (12)$$

then Q will also be a function of the same morphologic elements in (12). Correlations of discharge and runoff intensity with those geomorphic properties that are clearly related to watershed area should be high.

Correlation coefficients for each morphometric property of all 15 total basins were determined for regressions on each hydrologic factor (Table 6). Regressions were tested for the probability that $\beta = 0$, and were found to be significant at the 0.05 level. Area and stream-length factors, as expected, are very closely related to stream flow. Basin shape is influential in determining discharge and runoff intensity. Close correlation with stream frequency reflects the fact that more streams per unit area enable runoff to proceed faster, because the stream network can carry a large amount of water and can discharge it quickly. Numerous

TABLE 6. CORRELATION COEFFICIENTS FOR GEOMORPHIC CHARACTERISTICS AND HYDROLOGY OF WATERSHEDS OF APPALACHIAN PLATEAU PROVINCE

	Peak flow (cfs.)	Mean runoff (in.)	Mean annual discharge (cfs.)
Area	.9480	.7828	.9865
Total length	.9574	.7484	.9838
Circularity	−.5378	−.6590	−.6997
Relief ratio	−.5724	−.2528	−.3475
Drainage density	−.2496	−.2992	−.2032
Constant of channel maintenance	.2498	.3022	.2131
Stream frequency	−.5872	−.4496	−.6132
First-order frequency	−.7214	−.6214	−.8386
Bifurcation ratio	.2630	.2064	.1000
Gradient	−.0327	−.5470	−.7912

first-order tributaries conduct stream flow out of a basin in a short time. A large number of channels also means more rainfall conducted out of the watershed by stream flow, rather than by infiltration through the soil.

Correlation coefficients for regressions of stream gradient on runoff are negative. This is explained by the fact that streams with steep gradients are shorter and carry less water than streams with low gradients which are long and deep, and thus have greater channel storage. Leopold and Maddock (1953, p. 15) showed that increase of depth in channel overcompensates for decreased gradient and tends to provide a net increase in stream velocity at mean annual discharge stations. Hence, if precipitation conditions are equal, a stream with steeper gradient will have a smaller mean annual runoff and lower peak flow than a stream with low gradient.

Relief ratio seems closely related to peak discharge and to runoff-rainfall ratio, but not to mean-discharge or runoff-intensity figures. Other parameters which might be expected to have a close relation to hydrology, such as drainage density and constant of channel maintenance, do not. This may be because greater channel storage downstream more than compensates for high discharge resulting from high-drainage density.

Thus, hydrology of a basin can be said to be directly related to total stream length, basin area, and first-order stream frequency, and is inversely related to stream gradient, circularity, and relief ratio. Stream flow, then, can be expressed as a general function of geomorphology of a watershed as follows:

$$Q = f(A, \Sigma L, \frac{1}{F}, \frac{1}{S}, \frac{1}{R_c}, \frac{1}{R_h}), \quad (13)$$

where Q can be average or peak discharge, or may be replaced by runoff-rainfall ratio. Since gradient and relief ratio are both measures of ratio of relief to length, or vertical to horizontal measurements, we can eliminate one of them from the equation. Stream flow, then, is geomorphologically related to drainage area, stream length, frequency of stream channels, basin shape, and basin relief.

Multiple Correlation with Peak Discharge

Runoff intensity and discharge result from the combined action of a number of factors: climate, vegetation, soil conditions, and topography. Hence a multiple regression, simultaneously taking a number of these factors into consideration, should be drawn up. Potter (1953, p. 69) computed a regression for peak discharge using factors of topography, area, and rainfall, determined empirically for 51 basins in the Appalachian Plateau. The vegetation factor was not introduced because it had been tested and found to be of no significance in these basins.

The rainfall-intensity factor, designated the P-ratio, is defined as ". . . the average ratio of rainfall intensity for 10-, 25- and 50-year recurrence intervals for any particular location to similar values derived from long-term rainfall records for Columbus, Ohio" (Potter, 1953, p. 68). Frequency of rainfall, designated the S-ratio, is defined as ". . . the average ratio of the product of annual rainfall and number of excessive storms for the 10-, 25-, 50-year recurrence records for any location to similar values derived from long-term rainfall records for Columbus, Ohio" (Potter, 1953, p. 69).

The topographic factor, or T-factor, chosen by Potter is ratio of longest length of principal stream to square root of average channel slope from head to mouth. Potter's equation for regression of these factors on 10-year peak intensity of runoff is

$$\log q_{max} = -1.421 + .170 \log A - .554 \log T \\ + .929 \log P + .449 \log S, \quad (14)$$

where

q_{max} = peak intensity of runoff, cubic feet per second, per acre, for a 10-year recurrence interval,
A = area in acres,
T = T-factor (topography),
P = P-ratio (rainfall intensity), and
S = S-ratio (frequency of rainfall).

Analysis of covariance showed all independent variables to be significant beyond the 0.001 level, and average standard error was 18.2 per cent.

The writer has substituted other geomorphic quantities for Potter's T-factor in an effort to explain even more of the variance. Because of their high correlation coefficients with peak-runoff intensity, and because by rational deductions they seem to have a relation to peak intensity of runoff, circularity, relief ratio, and frequency of first-order streams if each were substituted as the T-factor in a multiple regression on peak intensity of runoff. None of the length or slope measurements were used as Potter had considered these in his equation. As 10 of the basins chosen for this study were among those included in Potter's study, his values for peak-runoff intensity, area, P-ratio, and S-ratio were used in determining the multiple regression. The author's values for relief ratio, circularity, and first-order stream frequency were substituted in each case for the T-factor. New regression coefficients were determined by the Doolittle method of solution of normal equations for five variables (Ezekiel, 1941, p. 460).

The following equations resulted:

Relief ratio

$$\log q_{max} = -6.547 + 0.168 \log A + 3.352 \log P \\ + 0.135 \log S + 0.255 \log R_h \quad (15)$$

Circularity ratio

$$\log q_{max} = -4.062 + 0.077 \log A + 2.318 \log P \\ + 0.204 \log S - 0.002 \log R_c \quad (16)$$

Frequency of first-order streams

$$\log q_{max} = 17.044 - 2.177 \log A - 2.302 \log P \\ + 6.984 \log S - 6.682 \log F_1 \quad (17)$$

where q_{max}, A, P, and S are the same factors as in Potter's equation and

R_h = relief ratio,
R_c = circularity ratio,
F_1 = frequency of first-order streams.

Potter calculated the effect of addition of each independent variable as a per cent error. This is shown in Table 7, with per cent error for each substituted T-factor. Each variable substituted reduced the per cent error. Table 8 presents standard error of estimate and multiple and partial correlation coefficients and states whether regression coefficients are significant. Use of first-order stream frequency gives highest correlation coefficient, 0.9378, and lowest standard error of estimate at 0.099.

TABLE 7. EFFECT OF CHANGE OF T-FACTOR ON STANDARD ERROR

Variables	Average standard error per cent
Q_p vs A,P,S, and T	18.2
Q_p vs A,P,S, and R_h	11.6
Q_p vs A,P,S, and R_c	12.1
Q_p vs A,P,S, and F_1	9.1

Circularity gives a multiple correlation coefficient of 0.7120 and standard error of 0.201; relief ratio gives 0.7381 for correlation coefficient and 0.194 for standard error. An F test of multiple regression coefficients showed only the regression of first-order stream frequency to be significant. A test of partial correlation coefficients also showed only the correlation coefficient for first-order frequency to be significant.

Since relief ratio, circularity, and first-order stream frequency vary independently of each other and of area, they do not duplicate any other geomorphic factor. Together they show the shape, relief or steepness, and network composition of a drainage basin. Hence, the combination of all ought to be a determining factor of the hydrologic character of a watershed. Accordingly, a T-factor of the product of these three properties was substituted in a multiple regression with A, S, and P on peak intensity of runoff. Using the same figures for A, S, P, and q_{max}, and a new T factor,

$$T = F_1 \cdot R_c \cdot R_h,$$

a new regression equation was solved by the Doolittle method:

$$\log q_{max} = -8.958 + 0.542 \log A + 4.239 \log P - 0.290 \log S + 0.723 \log T. \quad (18)$$

The standard error of estimate for this regression was 0.064 and the multiple correlation coefficient equal to 0.9478, which an F test showed to be significant. The partial correlation coefficient was 0.9071 and was significant. The use, then, of a T-factor which is the expression of relief, shape, and network composition of a watershed considerably reduces standard error and results in a high correlation with peak intensity of runoff.

SUMMARY AND CONCLUSIONS

Analysis of quantitative geomorphic characteristics of watersheds in the Appalachian Plateau tends to confirm Horton's laws of drainage composition for stream numbers, lengths, and areas. Properties relating vertical to horizontal measurements, that is, height-length ratios such as relief ratio or stream gradient, for the most part do not show an exponential decrease with order. This indicates that horizontal or planimetric aspects conform to Horton's laws, but control by structure and lithology governs vertical or gradient aspects. Relief ratio and stream gradients do, however, decrease consistently with increasing order. Because each is an exponential function of order, the area, total stream length, longest length, and mean stream length are related to each other as power functions.

In a region of horizontally uniform lithology,

TABLE 8. STANDARD ERROR OF ESTIMATES, CORRELATION COEFFICIENTS, AND SIGNIFICANCE FOR MULTIPLE REGRESSIONS ON PEAK RUNOFF

T-factor	Standard error of estimates	Mult. correlation coefficients	Significant?	Partial correlation coefficient	Significant?
F_1	.099	−.9378	yes	−.7816	yes
R_c	.201	−.7120	yes	−.1072	no
R_h	.194	.7381	yes	.1758	no
$F_1 R_c R_h$	.064	.9740	yes	.9071	yes

larger basins tend to have longer stream segments with gentler gradients, are less nearly circular, and have a smaller relief ratio. Or, in a homogeneous region, a watershed with steep slopes will have a smaller drainage area, shorter streams, will be more nearly circular, and will have a greater relief ratio than a watershed with gentle stream gradients. Derived ratios such as stream-length ratio, bifurcation ratio, and area ratio are conservative and remain essentially constant for basins of the Plateau province. However, as geologic factors change, geologic controls exert a change in these ratios. A change in lithology will generally cause changes in the whole basin morphology. A more resistant bed results in basins with increased area, length, and stream gradient, and decreased relief ratio and drainage density. Total relief, while still increasing, does so at a lower rate. Hence, if laws of drainage composition do not hold within a basin, the basin is not homogeneous and we may look for the indicated geologic change.

Quantitative morphology permits objective comparison of similarity of form elements of watersheds to determine homogeneity of landforms as a basis for classification within a single physiographic province. An analysis of similarities and differences in geometry of watersheds in regions traditionally considered as distinct sections shows that the regions are different geometrically, but not conspicuously so. In particular, the Cumberland Mountain section and the unglaciated Allegheny Plateau section are similar except for scale ratio of stream lengths and drainage density. The Allegheny Mountain region is more distinctly different from these other two sections in all form elements.

A valuable practical application of quantitative geomorphology is demonstrated by correlation of topographic factors with hydrology of basins. Discharge and runoff intensity in these watersheds of the Appalachian Plateau are closely correlated with stream length, area, and frequency of first-order streams and inversely related to stream gradient, relief ratio, and circularity ratio. Analysis of simple correlations of hydrology and geomorphic features provided a basis for choice of properties for use in multiple regressions on peak intensity of runoff. Substitution of first-order stream frequency, relief ratio, and circularity ratio, each as topographic quantities in Potter's regression of area, rainfall and topography on peak flow, and calculation of new regression coefficients provided equations greatly reducing average standard error. An equation substituting a product of these three factors ($F_1 \cdot R_c \cdot R_h$) for T resulted in a regression with high correlation, significant at the 0.001 level. A multiple regression, then, of area, rainfall, and topography on peak intensity of runoff provides a means of predicting runoff intensity for watersheds on the Appalachian Plateau.

REFERENCES

Eisenlohr, W. S. Jr. 1952. Floods of July 18, 1942, in north-central Pennsylvania. *U.S. Geol. Survey Water-Supply Paper 1134-B.*

Ezekiel, M. 1941. *Methods of correlation analysis.* New York: John Wiley and Sons.

Fenneman, N. M. 1938. *Physiography of eastern United States.* New York: McGraw-Hill Book Co.

Hack, J. T. 1957. Studies of longitudinal stream profiles in Virginia and Maryland. *U.S. Geol. Survey Prof. Paper 294-B*:54–97.

Horton, R. E. 1945. Erosional development of streams and their drainage basins; Hydrophysical approach to quantitative morphology. *Geol. Soc. America Bull.* 56:275–370.

Jarvis, C. S. 1936. Floods in the United States, magnitude and frequency. *U.S. Geol. Survey Water-Supply Paper 771.*

Leopold, L. B., and Maddock, Jr. 1953. The hydraulic geometry of stream channels and some physiographic implications. *U.S. Geol. Survey Prof. Paper 252.*

Morisawa, M. E. 1958. Measurement of drainage basin outline form. *Jour. Geol.* 66:587–591.

Potter, W. D. 1953. Rainfall and topographic factors that affect runoff. *Am. Geophys. Union Trans.* 34:67–73.

Sherman, L. K. 1932. The relation of hydrographs of runoff to size and character of drainage basins. *Am. Geophys. Union Trans.* 13:332–339.

Hydrograph Synthesis for Small Arid-Land Watersheds

R. B. Hickok
Member ASAE

R. V. Keppel
Assoc. Member ASAE

and

B. R. Rafferty

A method of estimating hydrograph characteristics from physiographic features of a watershed under study

FOR many hydrologic design problems it is necessary not only to estimate the expected peak rate of runoff, but for flood-routing purposes to synthesize the entire inflow hydrograph. In areas where few runoff data are available, the designer must resort to some method of estimating hydrograph characteristics from physiographic features of the watershed. This is particularly true in the southwestern area of the United States where such data are extremely scarce. This paper presents a method of hydrograph synthesis developed especially for small arid land watersheds.

The method involves (a) estimation of a characteristic lag time from readily determined watershed parameters, (b) use of the watershed lag time to predict the hydrograph peak rate for an assumed total volume of runoff, (c) synthesizing the entire hydrograph using the lag time, the estimated peak rate, and a standard dimensionless hydrograph.

Development of the method is based on the analysis of rainfall and runoff records for 14 experimental watersheds in Arizona, New Mexico and Colorado (Fig. 1). The watersheds were established in 1938-39 by the Soil Conservation Service (USDA), and the studies have been continued since 1954 by the Agricultural Research Service in cooperation with the Soil Conservation Service and the agricultural experiment stations of Arizona, New Mexico, and Colorado.

The Experimental Watersheds

The 14 watersheds range in size from 11 to 790 acres. They are in four locational groups near Albuquerque and Sante Fe, N. M.; Safford, Ariz.; and Colorado Springs, Colo. All of the watersheds, except Colorado Springs W-I which has been cultivated for more than 50 years, are on arid or semiarid range lands at intermediate elevations (3500 to 7000 msl). Mean annual precipitation is about 8 in. for the Albuquerque and Safford watersheds, and 14 in. for Sante Fe and Colorado Springs watersheds. Over one-half of the precipitation and nearly all of the runoff results from intense convectional thunderstorms in the June to September period.

Runoff from each watershed is measured by a precalibrated triangular weir equipped with a water-level recorder giving continuous records of stage to 0.01 ft and time to 1 min. Each watershed has one or more 12-hr recording rain gages and standard gages, and each watershed or group of watersheds has a weekly recording gage. The periods of record are from 9 to 16 years. Detailed watershed characteristics are given in Table 1.

Determination of Watershed Lag Time

The term "lag time" has been used in the literature to denote various time relationships between rainfall excess and runoff characteristics. It was found in this study that the least variable and most readily determined time parameter was the time from the center of mass of a limited block of intense rainfall to the resulting peak of the hydrograph. For the convectional thunderstorms causing flood

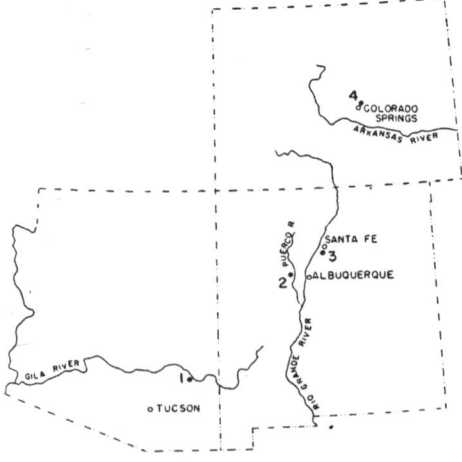

Fig. 1 Location map of experimental watersheds: (1) Safford, Ariz.; (2) Albuquerque, N. M.; (3) Santa Fe, N. M., and (4) Colorado Springs, Colo.

Paper presented at the annual meeting of the American Society of Agricultural Engineers at Santa Barbara, California, June, 1958, on a program arranged by the Soil and Water Division.

The authors — R. B. HICKOK, R. V. KEPPEL and B. R. RAFFERTY — are respectively, project engineer, Holmes and Narver, Inc., Los Angeles, (formerly project supervisor, SWCRD, ARS, USDA); agricultural engineer (SWCRD, ARS) U.S. Department of Agriculture; and teaching assistant, department of economics, University of Texas (formerly statistical clerk, SWCRD, ARS, USDA).

Acknowledgment: The authors are grateful to J. H. Dorroh, Jr. hydrologist, EWPU, SCS, Portland, Ore., for numerous helpful suggestions used in the preparation of this paper.

TABLE 1. WATERSHED CHARACTERISTICS

Watershed	A Size, acres	S Avg. slope, percent	S_{sa} Avg. slope of source area, percent	DD Drainage density, ft per acre	T_L Lag time, min	W_{sa} Width of source area, ft	L_{sa} Length to source area, ft	Cover	Soils
Safford, Arizona									
W-I	519	8.5	9.4	43	35	1690	9200	Shrubs-grass, 10-20%	Stony-sandy loam, moderate depth
W-II	682	12.4	19.7	70	19	1160	12300	Shrubs-grass, 10-35%	Stony-sandy loam, moderate depth
W-IV	764	2.6	2.6	106	46	1640	4800	Shrubs, 15-25%	Sandy loam, moderate depth
W-V	723	14.7	17.7	55	19	1500	12300	Grass, 10-35%	Stony-clay loam, moderate depth
Albuquerque, New Mexico									
W-I	97.2	16.5	17.2	100	9	1420	720	Brush-grass, 10-25%	Sandy-clay loam, shallow rock outcrops
W-II	40.5	14.3	16.4	86	12	615	1790	Grass, 5-30%	Fine-sandy loam, shallow crusted
W-III	183.0	6.9	9.7	75	19	1210	1350	Grass-shrubs, 5-35%	Fine-sandy loam, shallow crusted
Sante Fe, New Mexico									
W-I	50.0	9.9	10.4	89	16	680	2440	Grass-shrubs, 25-35%	Loam, shallow to moderate depth
W-II	790.0	4.3	5.4	42	41	1820	9180	Grass, 20-30%	Clay loam, moderate depth
W-III	51.6	18.9	19.4	143	7	880	620	Brush-grass, 10-35%	Sandy-clay loam, moderate depth
Colorado Springs, Colorado									
W-I	10.6	4.2	4.7	Cultivated	6	290	810	Cultivated	Clay loam, deep
W-II	39.7	6.0	6.8	57	20	540	1720	Grass	Clay loam, deep
W-III	35.4	5.6	5.9	100	20	520	1810	Grass	Loam, deep
W-IV	35.6	8.4	9.1	180	12	720	1720	Grass	Sandy loam to gravelly-clay loam, moderate depth

runoff from the watersheds, there was little difficulty in selecting the intense rainfall block responsible for the runoff peak. The minimum duration of the block causing the peak varied, of course, with watershed size. Fig. 2 shows a typical rainfall intensity-hydrograph plotting from which lag time was measured. About 130 such plottings (consisting of several of the largest peak flows from each watershed) were made for determination of characteristic watershed lag time.

Correlation of Lag Time with Watershed Characteristics

Snyder (1)*, using data from the Appalachian Mountain area, has related basin lag time (which he defined as the time from center of mass of rainfall excess to peak of the unit hydrograph) to watershed length parameters. Linsley (2) modified Snyder's equation and applied it to watersheds on the western slope of the Sierra Nevada. The U.S. Bureau of Reclamation (3) has used channel slope in addition to length parameters in estimating lag time. None of these combinations of watershed characteristics correlated very well with lag time for the 14 watersheds in this study. Therefore, multiple correlations of lag time with various combinations of watershed and channel slopes and lengths, drainage density, shape, and size were made. Of some 50

*Numbers in parentheses refer to the appended references.

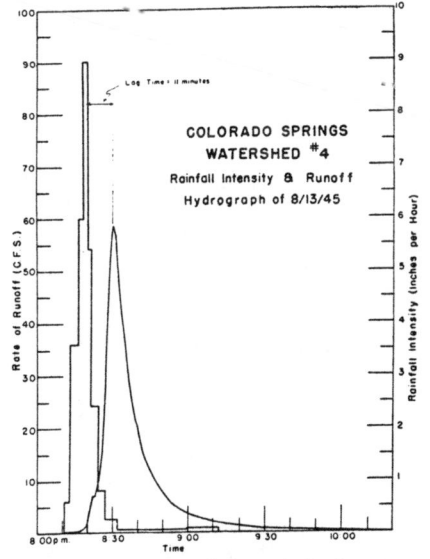

Fig. 2 Typical rainfall intensity (hydrograph plot shows lag time determinations)

such multiple correlations, two were decidedly superior. These relationships are described in equations [1] and [2].

$$T_L = K_1 \left[\frac{A^{0.3}}{S_a \sqrt{DD}} \right]^{0.61} \quad \quad [1]$$

Where:

T_L = lag time (time from limited block of intense rainfall to peak of hydrograph)
A = watershed area
S_a = average landslope of the watershed
DD = drainage density (total length of visible channels per unit area)

When length is in feet, area in acres, slope in percent, and lag time in minutes, $K_1 = 106$.

Equation [1] gave a standard error of estimate of 10.1 percent and a maximum deviation of 20.0 percent (Fig. 3).

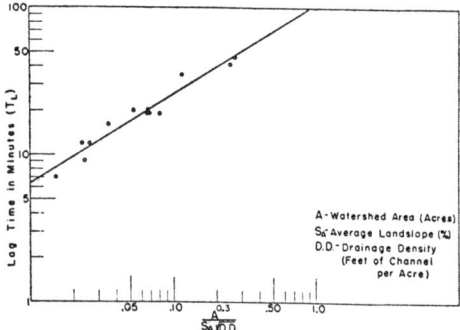

Fig. 3 Correlation of lag time with watershed parameters

In correlating lag time with watershed slope, it was noted that the slope of the half of the watershed having the largest average slope gave better correlation than did the slope of the entire watershed. This seemed to indicate that the steeper portion of the watershed may control the time of peak of the hydrograph even though rainfall excess occurs over the entire area. This gave rise to the concept of a controlling source-area for each watershed. The regression equation expressing lag time as a function of source-area parameters is:

$$T_L = K_2 \left[\frac{\sqrt{L_{sa} + W_{sa}}}{S_{sa} \sqrt{DD}} \right]^{0.65} \quad \quad [2]$$

Where:

L_{sa} = length from outlet of the watershed to center of gravity of source area
W_{sa} = average width of source area
S_{sa} = average land slope of the source area
DD = drainage density for entire watershed

When length is in feet, slope in percent, and lag time in minutes, $K_2 = 23$. Equation [2] was derived by considering the half of the watershed with the highest average landslope as the source area. Limited use of the equation in estimating lag times for larger heterogeneous watersheds indicates that the source area may be any important fractional part of the total watershed area which has distinctly greater slope or drainage density than the rest of the area.

Estimates based on equation [2] gave a standard error of estimate of 8.4 percent, and a maximum deviation of 12.6 percent (Fig. 4).

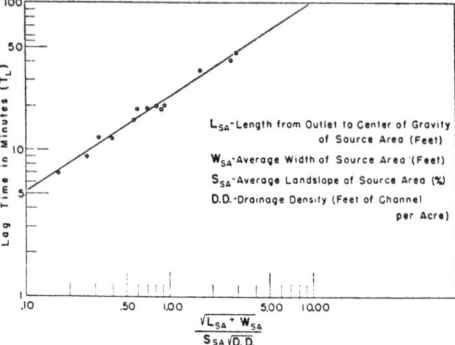

Fig. 4 Correlation of lag time with source area parameters

Equation [1] provides an estimate of lag time within practical confidence limits for reasonably homogeneous semi-arid rangeland watersheds up to about 1,000 acres in area. Very likely the size limitation is dependent on the character of rainfall and should be considered to be the upper limit of area from which the maximum runoff is expected to occur from rainfall excess over the entire area.

The source-area concept, equation [2], gives a more refined estimate of lag time for watersheds which differ widely in physiographic characteristics in some major portion of the area from the rest of the watershed, or which are large enough so that maximum runoff may occur as a result of rainfall excess occurring over only a portion of the watershed.

Correlation of Peak Rate — Volume Ratio with Lag Time

Lag time was found in this study to be a major determinant of hydrograph shape, and it is logical to express the ratio of the peak rate of runoff (q_p) to the total runoff volume (V) as a function of lag time. Fig. 5 shows this corre-

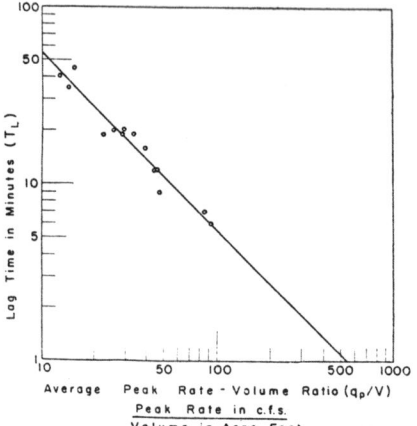

Fig. 5 Correlation of peak rate-volume ratio with lag time

lation of peak-volume ratio with lag time. The regression equation is:

$$q_p/V = K_3/T_L \qquad [3]$$

When q_p is in cfs, V in acre feet, and T_L in minutes, $K_3 = 545$.

Equation [3] gave a standard error of estimate of 14.6 percent with a maximum deviation of 29.0 percent.

Equation [3] is useful in estimating peak rate of runoff for any total volume of runoff. The relationship of peak-volume ratio to lag time also enables utilization of very limited hydrograph data for minor flows to predict peaks and estimate the hydrograph shape of much larger flows.

Development of Generalized Dimensionless Hydrograph and Mass Curve

For each of the 13 uncultivated watersheds included in the study, an average dimensionless distribution graph was prepared, using all suitable runoff hydrographs. The time base was made dimensionless in terms of the watershed lag time, and time increments of 20 percent of lag time were considered. The dimensionless distribution graphs for all watersheds were averaged, and a generalized dimensionless hydrograph and mass curve were computed (Fig. 6). The

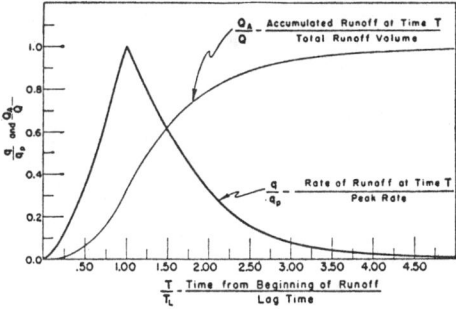

Fig. 6 Generalized dimensionless hydrograph and mass curve

synchronizing and averaging steps were more easily accomplished by working with distribution graphs rather than by working directly with hydrographs, since minor irregularities in the hydrographs were smoothed in preparing the distribution graphs. Coordinates defining the generalized dimensionless hydrograph and mass curve are given in Table 2.

TABLE 2. DATA FOR DIMENSIONLESS HYDROGRAPH AND MASS CURVE (Fig. 6)

T/T_L	q/q_p	Q_a/Q	T/T_L	q/q_p	Q_a/Q
0	0	0	1.6	0.545	0.671
0.1	0.025	0.002	1.7	0.482	0.707
0.2	0.087	0.007	1.8	0.424	0.742
0.3	0.160	0.020	1.9	0.372	0.773
0.4	0.243	0.036	2.0	0.323	0.799
0.5	0.346	0.063	2.2	0.241	0.841
0.6	0.451	0.096	2.4	0.179	0.875
0.7	0.576	0.136	2.6	0.136	0.900
0.8	0.738	0.180	2.8	0.102	0.917
0.9	0.887	0.253	3.0	0.078	0.932
1.0	1.000	0.325	3.4	0.049	0.953
1.1	0.924	0.400	3.8	0.030	0.965
1.2	0.839	0.464	4.2	0.020	0.973
1.3	0.756	0.523	4.6	0.012	0.979
1.4	0.678	0.578	5.0	0.008	0.983
1.5	0.604	0.627	7.0	0	0

Illustration of Method for Hydrograph Synthesis

The following example will illustrate the method of hydrograph synthesis. Walnut Gulch experimental watershed No. 4, located near Tombstone, Ariz., has been selected for the example because it is within the size range and climatic region represented by the watersheds from which the method was developed. The measured hydrograph having the largest peak flow for the four-year period of record was used for comparison with a computed hydrograph. Pertinent physical features for Walnut Gulch watershed No. 4 are:

Area (A) — 590 acres
Average land slope (S_a) — 9 percent
Drainage density $(D.D.)$ — 93 feet per acre

The average landslope was determined from a sample of slope measurements with an Abney hand level. Drainage density was determined by measuring the length of all drainage detail visible on an aerial photograph with a scale of 2 inches to the mile. Physiography of the watershed is relatively homogeneous; hence, equation [1] is used, which results in a value of 22.3 min for the lag time (T_L). This compares closely with an average measured lag time of 24 min for five of the major runoff events for this watershed.

The next step is to compute, from equation [3], the peak rate of runoff (q_p) for any assumed total volume of runoff (V). For a particular design application, V may be estimated from the assumed design storm rainfall volume and an estimated rainfall-runoff relationship, or a design storm runoff volume may be established directly. For this example, $V = 62.7$ acre-feet, the actual measured volume of runoff for the storm of July 19, 1955, was used for a realistic comparison of an actual and a computed hydrograph. Thus q_p is computed to be 1530 cfs. The measured peak was 1425 cfs. Fig. 6, is entered with values of T/T_L to obtain values of q/q_p, enabling the computed hydrograph to be determined in its entirety with any desired degree of detail. The computed and measured hydrographs are compared in Fig. 7.

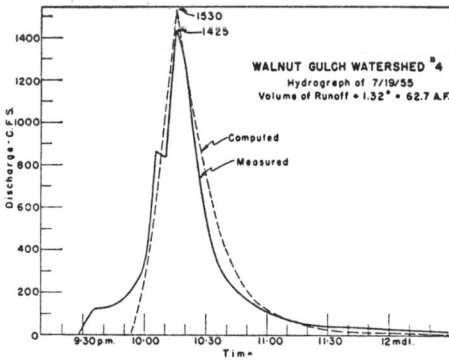

Fig. 7 Comparison of measured and computed hydrographs

Discussion and Conclusions

For the watersheds considered in this study, lag time as herein defined is the most significant time parameter in relating watershed influences to hydrograph shape. It was found that for a given watershed the rise time of the hydrograph (time from beginning of runoff to the peak) varied between much wider limits than did the lag time. Rise time is frequently affected by the duration of rainfall excess and minor variations in rainfall intensity—whereas lag time is relatively independent of the rainfall pattern. Rise time varied from 74 percent to 145 percent of lag time for the individual watersheds in this study. The average for all watersheds was 102 percent. Accordingly the generalized dimensionless hydrograph was constructed with lag time and rise time equal. However, it may not be concluded from this coincidence that the time of rise of the hydrograph affords a generally satisfactory index of watershed influence on the hydrograph shape. The measured hydrograph of Fig. 7 illustrates this point. The first portion of the rising limb of the hydrograph, from 9:38 p.m. to about 9:55 p.m., was caused by a minor burst of intense rainfall. That this portion is relatively insignificant in characterizing the overall hydrograph shape is indicated by the fact that only 8 percent of the total volume of runoff occurred during this 17-min period. The rise time of this hydrograph is 47 min compared with a measured lag time of 20 min. It is the latter time factor which is important in describing hydrograph shape and its relation to watershed physiography.

It should also be pointed out that lag time as defined in this paper does not correspond, except for very small watersheds with extremely simple drainage patterns, to the classical concept of "time of concentration." For natural watersheds of any size and complexity of drainage, runoff water originating from the most remote portion may and usually does arrive at the outlet too late to contribute to the flood peak. Accordingly, lag time will generally be less than the time of concentration for a given watershed.

For determination of lag time in the watersheds considered in this study, it is evident from equations [1] and [2] that the most important physiographic feature is land slope. All of the correlations involving channel slope were distinctly poorer than those involving land slope. Where flood peaks are almost exclusively the result of cloudburst-type convectional thunderstorms, it is highly probable that runoff water moves off the watershed and into the main channel in the form of abrupt translatory waves. Momentum effects might be expected to predominate over channel resistance effects in this type of flow. Consequently, from theoretical considerations, one would expect land slope to be of more importance than channel parameters in determining lag time, as was found in this case.

In equation [1] watershed area (A) provides an index of distance travelled, while in equation [2] the width of source area (W_{sa}) and the length of source area (L_{sa}) constitute a direct measure of travel distance. In both equations, the drainage density term (DD) provides a measure of what might be termed the hydraulic efficiency of the watershed i.e., the relative proportion of channel versus overland flow.

Within the range of conditions encountered in the present study, the shapes of the dimensionless hydrograph and mass curve (Fig. 6), were found to be independent of rainfall pattern and of soil and cover condition.

Experience has shown that estimates of runoff volume-frequency relations made from short-period records may be transferred to somewhat dissimilar watersheds with much greater confidence than is possible with similar estimates of peak runoff rates. Conversion of such a volume estimate to a corresponding estimate of peak rate on the basis of a lag time and correlated estimate of peak rate-volume ratio for the specific watershed being considered results in a substantially better estimate of the runoff peak.

The method of hydrograph synthesis presented is directly applicable to uncultivated arid-land watersheds of such size that major floods result from single thunderstorms producing runoff from the entire watershed. The authors feel that in order to satisfy this restriction the limit on size should be about 1000 acres for most areas of the Southwest. Work is currently under way to test the validity of this approach for watersheds of much greater area.

References

1 Snyder, F. F. Synthetic unit hydrographs, Trans. Am. Geophys. Union, vol. 19, pp. 447-454, 1938.

2 Linsley, R. K. Application of synthetic unit-graphs in the western mountain states, Trans. Am. Geophys. Union, vol. 24., part II, pp. 580-587, 1943.

3 U.S. Bureau of Reclamation—Unit graph procedures.

17

Reprinted from pp. 299-310 of *Fluvial Geomorphology*, ed. M. E. Morisawa, Publication in Geomorphology, State University of New York, Binghamton, 1973, 314 pp.

GEOMORPHIC THRESHOLDS AND COMPLEX RESPONSE OF DRAINAGE SYSTEMS

S. A. Schumm
Department of Earth Resources
Colorado State University
Fort Collins, Colorado

ABSTRACT

The alluvial and morphologic details of drainage systems are much too complex to be explained by progressive erosion alone. Within the constraints of the erosion cycle these complexities (terraces, alluvial deposits) must be explained by external variables such as climatic, tectonic, isostatic or land-use changes. However, field and experimental research into the details of fluvial landform development indicate that some abrupt modifications of such a system can be inherent in its erosional development and that two additional concepts are required for comprehension of drainage system evolution. These are 1) geomorphic thresholds and 2) complex response of drainage systems.

INTRODUCTION

Due to the complexity of Quaternary climatic and tectonic histories, topographic and stratigraphic discontinuities can be conveniently explained as a result of climatic and tectonic events. In this way the compulsion to fit geomorphic and stratigraphic details into a Quaternary chronology is satisfied, as is the basic scientific need to identify cause and effect. As the details of Holocene stratigraphic and terrace chronologies are studied, a bewildering array of changes are required to explain the behavior of a drainage system. In fact, it is now accepted that some major erosional adjustments can be induced by rather insignificant changes in the magnitude and frequency of storm events (Leopold, 1951).

The numerous deviations from an orderly progression of the erosion cycle has led many to discount the cycle concept completely. Current practice is to view the evolutionary development of the landscape within the conceptual framework of the erosion cycle, but to consider much of the modern landscape to be in dynamic equilibrium. There are obvious shortcomings in both concepts. For example, although the cycle involves continuous slow change, evidence shows that periods of relatively rapid system adjustment results from external causes. This, of course, is equally true of geomorphic systems in dynamic equilibrium. That is, for a change to occur in either the cycle or a system in dynamic equilibrium there must be an application of an external stimulus. Hence landscape changes and changes in rates of depositional or erosional processes are explained by the influence of man, by climatic change or fluctuations, by tectonics, or by isostatic adjustments.

One cannot doubt that major landscape changes and shifting patterns of erosion and deposition have been due to climatic change and tectonic influences and that man's influence is substantial. Nevertheless, it is the details of the landscape, the last inset fill, the low Holocene terrace, modern periods of arroyo cutting and gulleying, alluvial-fan-head trenching, channel aggradation and slope failure that for both scientific and practical reasons of land management require explanation and prediction. These geomorphic details are of real significance, but often they cannot be explained by traditional approaches.

Another aspect of the problem is that within a given region all landforms did not respond to the last external influence in the same way and, indeed, some have not responded at all. This is a major geomorphic puzzle that is commonly ignored. If land systems are in dynamic equilibrium conponents of the system should respond in a similar way to an external influence. Hence, the effects of hydrologic events of large magnitude should not be as variable as they appear to be.

The cyclic and dynamic equilibrium concepts are not of value in the location of incipiently unstable landforms because within these conceptual frameworks system change is always due to external forces. There is now both experimental and field evidence to indicate that this need not be true. The answer lies in the recognition of two additional concepts that are of importance for an understanding of landscape development. These may be termed 1) geomorphic thresholds and 2) complex response of geomorphic systems. Basically, they suggest that some geomorphic anomalies are, in fact, an inherent part of the erosional development of landforms and that the components of a geomorphic system need not be in phase.

These concepts are certainly not new, and one need not search long to find a geomorphic paper or book mentioning thresholds or the complexities of geomorphic systems (for example, Chorley and Kennedy 1971, Tricart 1965, Pitty 1971). However, these concepts have not been directed to the solution of specific problems nor has their significance been fully appreciated. The assumption that all major landform changes or changes in the rates and mechanics of geomorphic processes can be explained by climatic or tectonic changes has prevented the geologist from considering that landform instability may be inherent.

THRESHOLDS

Thresholds have been recognized in many fields and their importance in geography has been discussed in detail by Brunet (1968). Perhaps the best known to geologists are threshold velocities required to set in motion sediment particles of a given size. With a continuous increase in velocity, threshold velocities are encountered at which something begins and with a progressive decrease in velocity, threshold velocities are encountered at which something ceases, in this case, sediment movement. These are Brunet's (1968, pp. 14 and 15) "thresholds of manifestation" and "thresholds of extinction", and they are the most common types of thresholds encountered. When a third variable is involved, Brunet (1968, p. 19) identified "thresholds of reversal." An example of this is Hjulstrom's (1935) curve showing the velocity required for movement of sediment of a given size. The curve shows that velocity decreases with particle size until cohesive forces become significant, and then the critical velocity increases with decreasing grain size. Another example of this type of relationship is the Langbein-Schumm (1958) curve which shows sediment yield as directly related to annual precipitation and runoff until vegetative cover increases sufficiently to retard erosion. At this point there is a decrease in sediment yield with increased runoff and precipitation. Perhaps thresholds is not a good word to describe the critical zones within which these changes occur, but it is a simple and easily understood term.

The best known thresholds in hydraulics are described by the Froude and the Reynolds numbers, which define the conditions at which flow becomes supercritical or turbulent. Particularly spectacular are the changes in bed form characteristics at threshold values of stream power.

In the examples cited, an external variable changes progressively thereby triggering abrupt changes or failure within the affected system. Responses of a system to an external influence occur at what will be referred to as <u>extrinsic thresholds</u>. That is, the threshold exists within the system, but it will not be crossed and change will not occur without the influence of an external variable.

Thresholds can also be exceeded when input is relatively constant, that is, the external variables remain relatively constant, yet a progressive change of the system itself renders it unstable, and failure occurs. For example, it has been proposed that the progressive erosion of a region will cause short but dramatic periods of isostatic adjustment. With an essentially constant rate of denudation a condition is reached when isostatic compensation is necessary, and this probably takes place during a short period of relatively rapid uplift (Schumm 1963). Somewhat analagous to this is the long period of rock weathering and soil development required before a catalytic storm event precipitates mass movement. Following failure a long period of preparation ensues before failure can occur again (Tricart, 1965, p. 99). These <u>intrinsic thresholds</u> are probably common in geologic systems, but it is only geomorphic examples that will be considered here.

A geomorphic threshold is one that is inherent in the manner of landform change; it is a threshold that is developed within the geomorphic system by changes in the system itself through time. It is the change in the geomorphic system itself that is most important, because until the system has evolved to a critical situation, adjustment or failure will not occur. It may not always be clear whether the system is responding to geomorphic thresholds or to an external influence, but when a change of slope is involved, the control is geomorphic, and the changes whereby the threshold is achieved is intrinsic to the system.

EVIDENCE FOR GEOMORPHIC THRESHOLDS

Recent field and experimental work support the concept of geomorphic thresholds, and it has been used to explain the distribution of discontinuous gullies in the oil-shale region of western Colorado and to explain channel pattern variation along the Mississippi River.

Discontinuous gullies

Field studies in valleys of Wyoming, Colorado, New Mexico, and Arizona revealed that discontinuous gullies, short but troublesome gullied reaches of valley floors, can be related to the slope of the valley-floor surface (Schumm and Hadley, 1958). For example, the beginning of gully erosion in these valleys tends to be localized on steeper reaches of the valley floor. Carrying this one step farther, with the concept of geomorphic thresholds in mind, it seems that for a given region of uniform geology, land use and climate, a critical threshold valley slope will exist above which the valley floor is unstable. In order to test this hypothesis, measurements of valley-floor gradient were made in the Piceance Creek Basin of western Colorado. The area is underlain by oil shale, and the potential environmental problems that will be associated with the development of this resource are considerable.

Within this area, valleys were selected in which discontinuous gullies were present. The drainage area above each gully was measured on maps, and valley slopes were surveyed in the field. No records of runoff or flood events exist so drainage basin area was selected as a variable, reflecting runoff and flood discharge. When valley slope is plotted against drainage area, the relationship is inverse (Fig. 1), with gentler valley slopes being characteristic of large drainage areas. As a basis for comparison, similar measurements were made for valleys in which there were no gullies, and these data are also plotted on Fig. 1. The lower range of slopes of the unstable valleys coincide with the higher range slopes of the stable valleys. In other words, for a given drainage area it is possible to define a valley slope above which the valley floor is unstable.

Note that the relationship does not pertain to drainage basins smaller than about four square miles. In these small basins variations in vegetative cover, which are perhaps related to the aspect of the drainage basin or to variations in the properties of the alluvium, prevent recognition of a critical threshold slope. Above four-square miles there are only two cases of stable valley floors that plot above the threshold line, and one may conclude that these valleys are incipiently unstable and that a major flood will eventually cause erosion and trenching in these valleys.

Using Fig. 1, one may define the threshold slope above which trenching or valley instability will take place in the Piceance Creek area. This has obvious implications for land management for, if the slope at which valleys are incipiently unstable can be determined, corrective measures can be taken to artificially stabilize such critical reaches, as they are identified.

It seems possible that future work will demonstrate that similar relationships can be established for other alluvial deposits. For example, trenching of alluvial fans is common, and the usual explanation for fan-head trenches is renewed uplift of the mountains or climatic fluctuations. However, the concept of geomorphic thresholds should also be applicable to this situation. That is, as the fan grows it may steepen until it exceeds a threshold slope, when trenching occurs. Preliminary results from experimental studies of alluvial-fan growth reveal that periods of trenching do alternate with deposition at the fan head.

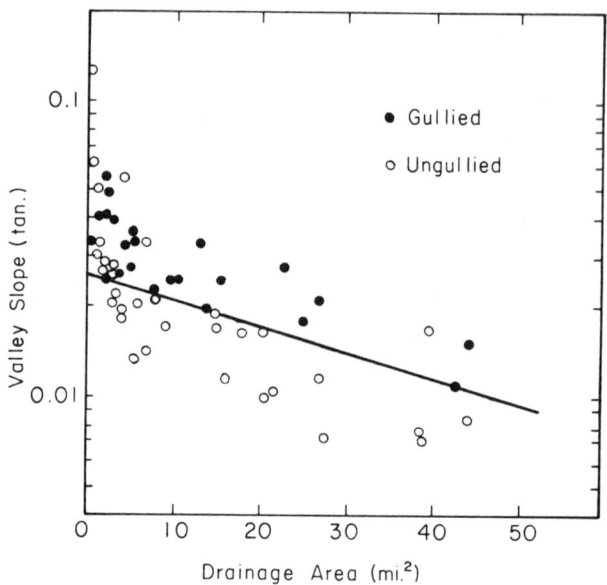

Figure 1. Relation between valley slope and drainage area, Piceance Creek basin, Colorado. Line defines threshold slope for this area (from P. C. Patton, 1973, unpublished M.S. thesis).

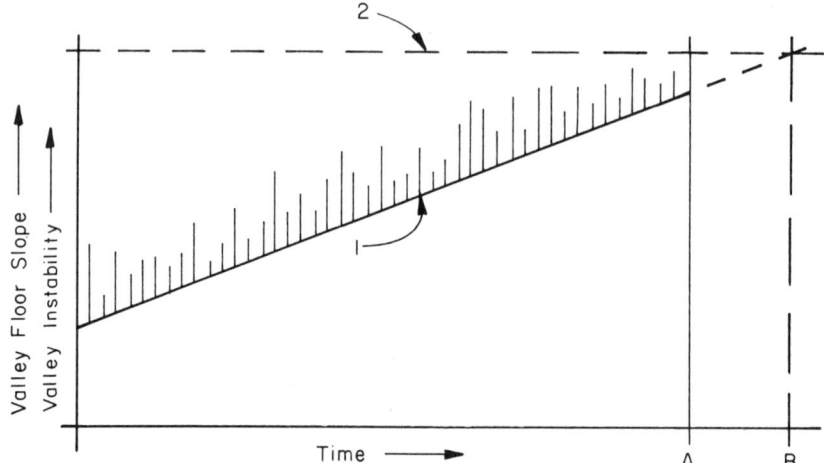

Figure 2. Hypothetical relation between valley-floor gradient and valley-floor instability with time. Superimposed on line 1, representing an increase of valley-floor slope, are vertical lines representing instability of the valley floor as related to flood events. When the ascending line of valley-floor slope intersects line 2 representing the maximum slope at which the valley is stable, failure or trenching of the valley alluvium will occur at time B. However, failure occurs at time A, as the apparent direct result of a major storm or flood event.

The concepts of thresholds as applied to alluvial deposits in the western U.S.A. is illustrated by Fig. 2, where the decreasing stability of an alluvial fill is represented by a line indicating increase of valley slope with time. Of course, a similar relation would pertain if, with constant slope, sediment loads decrease slowly with time. Superimposed on the ascending line of increasing slope are vertical lines showing the variations of valley floor stability caused by flood events of different magnitudes. The effect of even large events is minor until the stability of the deposit has been so reduced by steepening of the valley gradient that during one major storm, erosion begins at time A. It is important to note that the large event is only the most apparent cause of failure, as it would have occurred at time B in any case.

These studies of alluvial deposits in dry lands suggest that large infrequent storms can be significant but only when a geomorphic threshold has been exceeded. It is for this reason that high-magnitude, low-frequency events may have only minor and local effects on a landscape.

River patterns

Experimental studies of river meandering have been performed by both hydraulic engineers and geologists over many years. Such a study was designed to investigate the influence of slope and sediment loads on channel patterns (Schumm and Khan, 1972). It was found, during these experiments in which the water discharge was held constant, that if a straight channel was cut in alluvial material at a very low slope, the channel would remain straight. However, at steeper slopes the channel meandered. As the slope of the alluvial surface on which the model stream was flowing (valley slope) steepened, the velocity of flow increased, and shear forces acting on the bed and bank of the miniature channel increased. At some critical value of shear, bank erosion and shifting of sediments on the channel floor produced a sinuous course. The conversion from straight to sinuous channel at a given discharge occurred at a threshold slope (Fig. 3). As slope increased beyond this threshold, meandering increased until at another higher threshold the sinuous channel became a straight braided channel. The experiments revealed that there is not a continuous change in stream patterns with slope from straight through meandering to braided, but rather the changes occur at threshold slopes. Slope in this case is an index of sediment load and the hydraulic character of the flow in the channel; nevertheless, the relationship can be used to explain the variability of stream patterns.

If, in fact, the slope of the valley floor of a river varies due to the sediment contribution from tributaries or due to variations in deposition during the geologic past, then the river should reflect these changes of valley slope by changes in pattern. A comparison of the experimental results with Mississippi River patterns was made possible by data obtained from the Vicksburg District, Corps of Engineers. These data show that variations in channel pattern of the Mississippi River are related to changes of the slope of the valley floor (Fig. 4). Variations in valley slope reflect the geologic history of the river, and if today the valley slope exceeds a geomorphic threshold the river will show a dramatic change of pattern.

When the valley slope is near a threshold, major flood events will significantly alter the stream pattern. This conclusion has bearing on the work of Wolman and Miller (1960), concerning the geomorphic importance of events of high magnitude. They concluded that, although a major amount of work is done by events of moderate magnitude and relative frequent occurrence, nevertheless, the large storm or flood may have a major role in landscape modification. However, evidence on the influence of rare and large events on the landscape is equivocal. Major floods have destroyed the flood plain of Cimarron River (Schumm and Lichty, 1963), but equally large events have not significantly affected the Connecticut River (Wolman and Eiler, 1958).

These and other observations indicate that a major event may be of major or minor importance in landscape modification, and an explanation of the conflicting evidence requires further consideration of the threshold

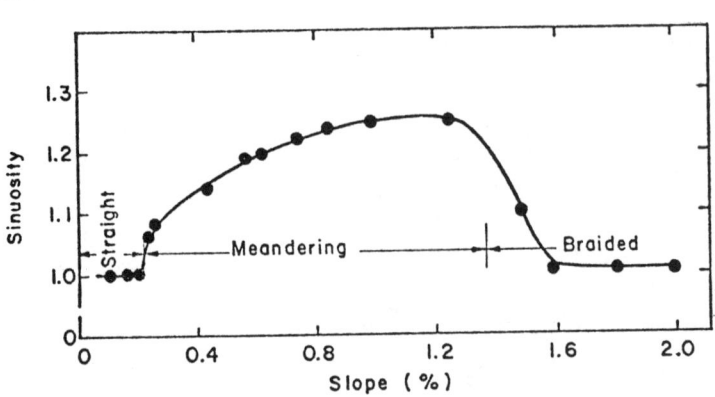

Figure 3. Relation between valley slope and sinuosity (ratio of channel length to length of flume or length of valley) during experiments. The change from a straight to a sinuous pattern, and from a sinuous to a braided pattern occurs at two threshold slopes. The absolute value of slope at which such changes occur will be influenced by discharge. Discharge was maintained at 0.15 cfs during the experiments. (After Schumm, S. A. and Khan, H. R., 1972.)

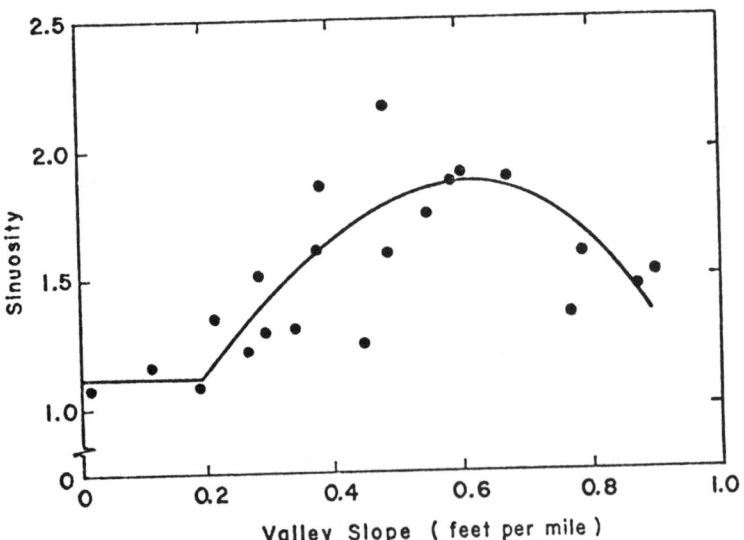

Figure 4. Relation between valley slope and sinuosity between Mississippi River between Cairo, Illinois and Head of Passes, Louisiana. Data from Potamology Section, U. S. Army Corps of Engineers, Vicksburg, Mississippi. Scatter about curve reflects natural variation of sinuosity. (After Schumm, et al., 1972)

concept. Some landscapes or components of a landscape have apparently evolved to a condition of geomorphic instability and these landforms fail; that is, depending on their development, they will be significantly modified by a large infrequent event whereas others will be unaffected. Therefore, there will be, even within the same region, different responses to the same conditions of stress.

When within a landscape some conponents fail by erosion whereas others do not, it is clear that erosional thresholds have been exceeded locally. One of the most significant problems of the geomorphologist or land manger is the location of incipiently unstable components of a landscape. The recognition of geomorphic thresholds within a given region will be a significant contribution to both an understanding of the details of regional morphology as well as providing criteria for identification of unstable land forms. At least one location, Piceance Creek basin, such a threshold has been identified. It remains to be seen how successful the concept can be applied elsewhere. It is very possible that major geomorphic thresholds associated with alluvial deposits will be identified most readily in subhumid, semiarid and arid regions, where the stabilizing influence of vegetation is least effective. However, slope stability and river pattern thresholds will exist in humid and tropical regions.

COMPLEX RESPONSE OF GEOMORPHIC SYSTEMS

Geomorphic histories tend to be complicated, and considering the climatic changes of the past few million years of earth history, one would expect them to be so. For example, throughout the world, geologists and archaeologists have studied the details of the most recent erosional and depositional history of valleys. This consists of identifying the sequence in which alluvial deposits were emplaced and then eroded. Because of worldwide climate changes during the Quaternary, it is reasonable to assume that alluvial chronologies applicable to large regions can be established. That is, a particular alluvial layer should be identifiable regionally and correlations of these deposits over large areas can then be made. There is no question that major climatic changes have affected erosional and depositional events, but when the alluvial chronologies of the last 10 to 15 thousand years are examined, it is not convincing that each event was in response to one simultaneous external change. In fact, investigations in southwestern United States reveal that during the last 10,000 years the number, magnitude and duration of erosional and depositional events in the valleys of this region not only varied from valley to valley but also varied within the same valley. For example, Kottlowski, Cooley, and Ruhe (1965) describe the situation as follows:

> "Late recent time is represented by epicycles of erosion and alluviation in the canyons and valleys of the southwest; however, the number, magnitude, and duration of the events differ from basin to basin and along reaches of the same stream."

This situation is so complex that correlations of terrace surfaces and alluvial fills over large areas seem impossible, but in the search for order, correlations are made. Haynes (1968) summarized the results of extensive radiocarbon dating of alluvial deposits in southwestern United States. His data demonstrate that, during the last 5000 years of record, there is significant temporal overlap among the three most recent alluvial deposits. This indicates that deposition was not in phase everywhere and that apparently deposition did not occur in response to a single event. Part of the complexity, at least as related to the most recent events, may be explained by the threshold concept with erosion occurring, as described for the Piceance Creek area, when a geomorphic threshold is exceeded. However, within one region not all valleys will achieve the threshold at the same time.

There is, in addition, another explanation. We know very little about the response of a drainage system to rejuvenation. Climatic change, uplift, or lowering of baselevel can cause incision of a channel. This incision will convert the flood plain to a terrace, the geomorphic evidence of an erosional episode. However, a drainage system is composed

Figure 5. Drainage evolution research facility at Colorado State University. Simulated storms can be generated by a sprinkler system over the 9 x 15 m container, and the erosional development of small drainage systems can be studied.

of channels, hillsides, divides, flood plains, and terraces; it is complex. The response of this complex system to change will also be complex. That this is true was demonstrated during an experimental study of drainage system evolution at Colorado State University (Schumm and Parker, 1973).

During experimentation, a small drainage system (Fig. 5) was rejuvenated by a slight (10 cm) change of base level. As anticipated, base-level lowering caused incision of the main channel and development of a terrace (Fig. 6a,b). Incision occurred first at the mouth of the system, and then progressively upstream, successively rejuvenating tributaries and scouring the alluvium previously deposited in the valley (Fig. 6b). As erosion progressed upstream, the main channel became a conveyor of upstream sediment in increasing quantities, and the inevitable result was that aggradation occurred in the newly cut channel (Fig. 6c). However, as the tributaries eventually became adjusted to the new baselevels, sediment loads decreased, and a new phase of channel erosion occurred (Fig. 6d). Thus, initial channel incision and terrace formation was followed by deposition of an alluvial fill, channel braiding, and lateral erosion, and then, as the drainage system achieved stability, renewed incision formed a low alluvial terrace. This low surface formed as a result of the decreased sediment loads when the braided channel was converted into a better-defined channel of lower width-depth ratio. The low surface was not a flood plain because it was not flooded at maximum discharge.

Somewhat similar results were obtained by Lewis (1949) in a pioneering experiment performed in a small wooden trough four m. long and 50 cm wide. Lewis cut a simple drainage pattern in sediment (four parts sand, one part mud). He then introduced water at the head of the flume into both the main channel and into two tributaries. The main channel debouched onto a "flood plain" before entering the "sea" (tail box of flume).

During the experiment, the break in slope or knickpoint at the upstream edge of the "flood plain" eroded back, rejuvenating the upstream drainage system. Initially, erosion in the head waters was rapid as the channels adjusted, and deposition occurred on the "flood plain," thereby increasing its slope. As the upstream gradients were decreased by erosion and the stream courses stabilized, sediment supply to the "flood plain" decreased. Because of the reduction of the sediment load the stream cut into the alluvial deposits in the upper part of the flood plain to form a terrace. Lewis concludes that "perhaps the most significant fact revealed by the...experiment is that terraces were built in the lower reaches without any corresponding change of sea levels, tilt or discharge." From our results and those of Lewis, it seems that an event causing an erosional response within a drainage basin (tilting, changes of base level, climate and/or land use) automatically creates a negative feedback (high sediment production) which results in deposition; this is eventually followed by incision of alluvial deposits as sediment loads decrease.

A similar sequence of events may be under way in Rio Puerco, a major arroyo in New Mexico, as well as in other Southwestern channels. For example, the dry channel of Rio Puerco, although previously trenched to depths of 13 m, is now less than four m deep near its mouth. This is due to deposition caused by very high sediment loads produced by the rejuvenated drainage system.

Within a complex natural system, one event can trigger a complex reaction (morphologic and/or stratigraphic) as the components of the system respond progressively to change. This principle provides an explanation of the complexities of the alluvial chronologies, and it suggests that an infrequent event, although performing little of the total work within a drainage system, may, in fact be the catalyst that causes the crossing of a geomorphic threshold and the triggering of a complex sequence of events that will produce significant landscape modification.

CONCLUSIONS

Although we continually speak and write about the complexity of geomorphic systems, nevertheless, we constantly simplify in order to understand these systems. The experimental studies discussed in this

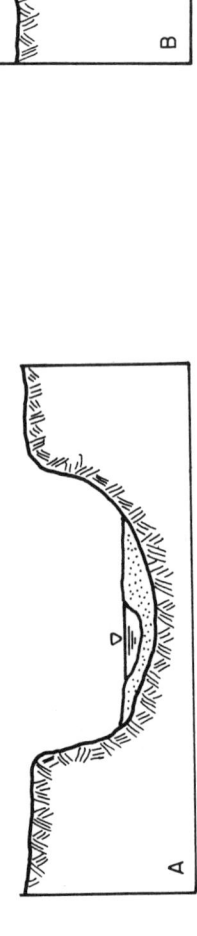
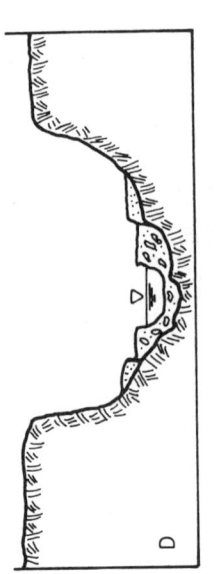

Braided Channel

Figure 6. Diagrammatic cross sections of experimental channel 1.5 m from outlet of drainage system (base level) showing response of channel to one lowering of base level.

A. Valley and alluvium, which was deposited during previous run, before base level lowering. The low width-depth channel flows on alluvium.

B. After base level lowering of 10 cm, channel incises into alluvium and bedrock floor of valley to form a terrace. Following incision, bank erosion widens channel and partially destroys terrace (Figure 2C).

C. An inset alluvial fill is deposited, as the sediment discharge from upstream increases. The high width-depth ratio channel is braided and unstable.

D. A second terrace is formed as the channel incises slightly and assumes a low width-depth ratio in response to reduced sediment load. With time, in nature, channel migration will destroy part of the lower terrace, and a flood plain will form at a lower level.

344

paper are good examples of this approach. Simplification and the search for order in simplicity caused instrinsic thresholds to be overlooked in preference to explanations based on external controls. For example, the explanation of the details of the Holocene record and the distribution of modern erosion and depositional features in the western United States may well depend on our understanding of thresholds and complex response. These concepts do not conflict with the cycle of erosion or with the concept of dynamic equilibrium; rather, they supplement them.

It need not be true that landscape discontinuities or what appear to be abrupt changes in the erosional evolution of drainage systems must always be related to external influences. The evolution of land forms, at least in semiarid and arid regions, need not be progressive in the sense of constant and orderly development; in fact, change may occur both progressively and by saltation; that is, by jumps from one dynamic equilibrium to a new one. Obviously, hydrologic and meteorological events are discontinuous (Tricart, 1962) but even if they occur at a constant rate, changes in the geomorphic system itself (changing landform morphology and sediment loads) will cause abrupt adjustments of the system due to the existence of intrinsic thresholds.

It is very possible that, without the influence of external variables and over long time spans progressive erosion reduction of a landscape will be interrupted by periods of rapid readjustment, as geomorphic thresholds are exceeded. Readjustment of the system will be complex as morphology and sediment yields change with time. The timing of these changes unquestionably will be related to major flood or storm events, but such events, as emphasized earlier, may be only the catalyst that induces the change at a particular time. That is, it is the existence of geomorphic thresholds, and the complex feedback response of geomorphic systems, that permit high magnitude events to play a major role in landscape evolution.

Newtonian physical principles are utilized by engineers to control the landscape and by geomorphologists to attempt an explanation for the inception, evolution, and character of geomorphic systems. These physical laws apply to natural geomorphic situations, but their predictive power is reduced by the complexity of the field situation. For example, an increase in gravitational forces would probably not everywhere cause an equal acceleration in erosional rates. That is, increasing stress may not produce commensurate strain, but local failures will occur. Thus, the application of stress over time will not everywhere have the same result especially as the system to which stress is applied is itself changing through time. The logical consequence of the above situation as outlined in this paper is that high magnitude events will not everywhere produce dramatic erosional events; rather the result depends on the character of the geomorphic system.

The importance of this approach to the investigation of landforms is in its potential for application to prediction of landform response to both natural and man-induced change. The fact that, at least locally, geomorphic thresholds of instability can be defined quantitatively indicates that they can be identified elsewhere and then used as a basis for recognition of potentially unstable landforms in the field. This approach provides a basis for preventive erosion control. Using geomorphic principles the land manager can spend his limited funds in order to prevent erosion rather than spending it in a piecemeal fashion to attempt to restore seriously eroding areas to their natural conditions.

ACKNOWLEDGMENTS

The research reviewed herein has been supported by the National Science Foundation, Army Research Office and the Colorado Agricultural Experiment Station. William Weaver reviewed an early draft of this paper.

REFERENCES

Brunet, Roger, 1968, Les phénomenes de discontinuité en geographie: Centre de Recherches et Documentation Cartographic et Geographique. v. 7, 117 p. Editions du Centre National de la Recherche Scientifique.

Chorley, R. J. and Kennedy, B. A., 1971, Physical geography, a systems approach: London, Prentice-Hall International, 370 p.

Hayes, C. V., Jr., 1968, Geochronology of late-Quaternary alluvium: in-Means of correlation of Quaternary succesions (editors: H. E. Wright, Jr., and R. B. Morrison), Salt Lake City, University of Utah Press, p. 591-631.

Hjulstrom, Filip, 1935, Studies of the morphological activity of rivers as illustrated by the River Fyris: University Upsala Geological Institute, v. 25, p. 221-527.

Kottlowski, F. E., Cooley, M. E. and Ruhe, R. V., 1965, Quaternary geology of the Southwest: in Quaternary of the United States (editor: H. E. Wright, Jr.), Princeton, Princeton University Press, p. 287-298.

Langbein, W. B. and Schumm, S. A., 1958, Yield of sediment in relation to mean annual precipitation: Amer. Geophys. Union, Trans., v. 39, p. 1076-1984.

Leopold, L. B., 1951, Rainfall frequency: An aspect of climatic variation: Amer. Geophys. Union, Trans., v. 32, p. 347-357.

Lewis, W. V., 1944, Stream trough experiments and terrace formation: Geol. Mag. v. 81, p. 241-253.

Pitty, H. F., 1971, Introduction to geomorphology: London, Methuen and Company, 526 p.

Schumm, S. A., 1963, The disparity between present rates of denudation and orogeny: U. S. Geol. Survey Prof. Paper 454-H, 13 p.

_____, and Hadley, R. F., 1957, Arroyos and the semarid cycle of erosion: Amer. Jour. Sci., v. 255, p. 161-174.

_____, and Lichty, R. W., 1963, Channel widening and floodplain construction along Cimarron River in Southwestern Kansas: U. S. Geol. Survey Prof. Paper 352-D, p. 71-88.

_____, and Khan, H. R., 1972, Experimental study of channel patterns: Geol. Soc. America Bull., v. 85, p. 1755-1770.

_____, et al., 1972, Variability of river patterns: Nature (Physical Science) v. 237, p. 75-76.

_____, and Parker, R. S., 1973, Implications of complex response of drainage systems for Quaternary alluvial stratigraphy: Nature (Physical Science) v. 243, p. 99-100.

Tricart, Jean, 1962, Les discontinuités dans les phénomenes d'erosion: International Assoc. Scientific Hydrology Pub. 59, Comm. d'erosion Continentale, p. 233-243.

Tricart, Jean, 1965, Principés et Méthods de Géomorphologie: Paris, Masson et Cie, 495 p.

Wolman, M. G., and Eiler, J. P., 1958, Reconaissance study of erosion and deposition produced by the flood of August 1953 in Connecticut: Amer. Geophys. Union, Trans., v. 39, p. 1-14.

_____, and Miller, J. P., 1960, Magnitude and frequency of forces in geomorphic processes: Jour. Geol. v. 68, p. 54-74.

AUTHOR CITATION INDEX

Abrahams, A. D., 232
Adams, G. F., 199
Atwood, W. W., 168
Avena, G. C., 71

Bailey, R. W., 167
Balchin, W. G. V., 199
Beckinsale, R. P., 7
Becraft, R. J., 167
Bennett, H. H., 167
Benson, M. A., 188
Beutner, E. L., 167
Bilham, E. G., 199
Brigham, A. P., 51
Brögger, W. C., 52
Brown, J. S., 47, 53, 62
Brunet, R., 346
Butts, C., 199
Butzer, K. W., 3

Cadell, H. M., 15
Cailleux, A., 180
Challinor, J., 304
Chamberlin, T. C., 253
Chapman, C. A., 176
Chorley, R. J., 3, 7, 180, 188, 199, 346
Christofoletti, A., 71
Coates, D. R., 176
Connaughton, C. A., 227
Cooke, W., 199
Cooley, M. E., 346
Cotton, C. A., 167

Dake, C. I., 47, 53, 62
Daly, R. A., 253
Daubrée, A., 63, 253
Davis, W. M., 7, 48, 51, 65, 167, 253, 304
Derbyshire, E., 180
Dewey, H., 199
Dobson, G. C., 167
Dodge, R. E., 253

Duley, F. L., 167
Dunn, A. J., 7
Dutton, C. E., 48

Ebright, J. R., 199
Eiler, J. P., 346
Eisenlohr, W. S., Jr., 329
Exekiel, M., 329

Feller, W., 312
Fenneman, N. M., 304, 329
Fletcher, J. E., 167
Fok, Y.-S., 180
Forsling, C. L., 167
Foster, C. L., 15

Gaebe, R. R., 167
Garner, H. F., 180
Geikie, J., 64
Ghose, B., 71
Gilbert, G. K., 56, 167, 253
Giuliana, G., 71
Glenn, L. C., 167
Glock, W. S., 199, 261, 304
Gravelius, H., 71, 167
Gray, D. M., 315
Green, A. H., 13
Gregory, K. J., 3, 315
Gumbel, E. J., 188

Hack, J. T., 227, 312, 329
Hadley, R. F., 304, 346
Haggett, P., 3
Hains, C. F., 324
Happ, S. C., 167
Harder, E. C., 63
Hayes, C. V., Jr., 346
Hayes, O. E., 167
Hirsch, F., 180
Hjulstrom, F., 346
Hobbs, W. H., 48, 62, 63, 253

Author Citation Index

Horton, R. E., 71, 167, 168, 176, 188, 199, 209, 304, 312, 324, 329
Howard, A. D., 8, 71, 233
Howe, E., 253
Hursh, C. R., 227, 304

Iddings, J. P., 253
Ingham, A. I., 199

Jacob, C. E., 188
Jagger, T. A., Jr., 54, 56, 59, 253
Jarvis, C. S., 329
Jeffreys, H., 168
Johnson, D., 63, 233, 304
Jukes, J. B., 13
Jukes-Brown, A. J., 15

Kahn, J. S., 228
Kelly, L. L., 167
Kemp, J. F., 51
Kennedy, B. A., 3, 180, 346
Khan, H. R., 346
King, F. H., 253
King, N. J., 324
Kirkby, M. J., 180
Kottlowski, F. E., 346
Krumbein, W. C., 304

Lamplugh, G. W., 52
Langbein, W. B., 72, 176, 188, 304, 312, 346
Lawson, A. C., 304
Leopold, L. B., 3, 180, 304, 312, 329, 346
Lev, J., 199
Lewis, W. V., 346
Lichty, R. W., 346
Linsley, R. K., 334
Lobeck, A. K., 8
Lueder, D. R., 8
Lupia-Palmieri, E., 71

Mace, M. M., 324
Maddock, T., Jr., 180, 304, 329
Malm, D., 199
Margerie, E. de, 233
Martin, L., 168
Maxwell, J. C., 176
Melton, M. A., 176, 228
Meunier, S., 253
Middleton, H. E., 168
Miller, C. F., 8
Miller, J. P., 3, 188, 304, 312, 346
Miller, R. L., 228
Miller, V. C., 8, 176, 199
Moore, R. C., 199
Morgan, M. A., 180, 188

Morgan, R. S., 168
Morisawa, M., 3, 233, 329

Nathorst, A. G., 253
Neal J. H., 168
Newman, M. H. A., 228
Noe, G. de la, 233

Olson, E. C., 228

Palmer, H. S., 59
Pandy, S., 71
Parker, R. S., 233, 346
Parvis, M., 8
Paulsen, C. G., 304
Penck, W., 304
Peterson, H. V., 324
Pettijohn, F. J., 304
Pitty, H. F., 346
Pogorzelski, H., 199
Potter, W. D., 329

Renner, F. G., 168
Rittenhouse, G., 167
Ruhe, R. V., 233, 346
Rzhanitsyn, N. A., 72, 180

Salisbury, N. E., 3
Salisbury, R. D., 168, 253
Scheidegger, A. E., 3, 72, 233
Schumm, S. A., 8, 176, 199, 233, 304, 315, 324, 346
Segerstrom, K., 304
Semmes, D. R., 199
Sheafer, A. W., 199
Sherman, L. K., 329
Sherwood, A., 199
Shreve, R. L., 72, 233
Siegel, S., 228
Slater, C. S., 168
Smart, J. S., 72, 233
Smith, K. G., 176, 199, 305, 324
Smith, W. S. Tangier, 253
Snyder, F. F., 334
Stall, J. B., 180
Stephenson, J. W., 199
Strahler, A. N., 72, 176, 188, 199, 228, 305

Tarr, R. S., 168, 254
Thornbury, W. D., 8
Thornthwaite, C. W., 199
Tipton, R. J., 168
Topley, W., 15
Tricart, J., 180, 346

Unwin, W. C., 168

Van Sickle, D. M., 324

Walker, H. M., 199
Walling, D. E., 3, 315
Ward, R. C., 315
Weaver, J. E., 199
Werritty, A., 72

Williamson, W. C., 254
Willis, B., 38, 48, 50
Wilson, L., 180
Woldenberg, M. J., 180
Wolman, M. G., 3, 188, 305, 346
Wooldridge, S. W., 168
Wright, S., 228

SUBJECT INDEX

Alluvial fan, 43
Alps Mountains, 41
Amazon River, 1, 92, 164
Angle, of tributary junction, 244, 290
Appalachian Mountains, 19, 22, 48, 58
Available relief, 296

Badlands, 235, 270
Base level, 23, 28, 37, 166
Belt of no erosion, 115, 118, 119, 153, 232
Bifurcation ratio, 88, 170, 276, 308, 309
Black Hills, 54, 56

Capture, 14
Channel maintenance, constant of, 173, 275, 280, 327
Cheyenne River, 314, 316
Circularity ratio, 214, 285
Climate-vegetation index, 197
Colorado River, 164
Complex response, 341

Delaware River, 100
Depression storage, 108
Discharge, 326
  base flow, 184, 186
  mean annual, 184
  mean annual flood, 186
  peak, 326
Discontinuous gullies, 318
Drainage density, 70, 81, 82, 92, 151, 165, 172, 179, 181, 187, 189, 191, 214, 223, 275, 317, 327, 332
Drainage patterns
  evolution
    abstraction, 13, 267
    extension, 261
    integration, 261, 263
    maximum extension, 265
  types
    anastomotic, 61
    angulate, 62

  annular, 54–56
  centripetal, 64
  colinear, 66, 68
  complex, 68
  dendritic, 46, 47
  deranged, 67
  irregular, 67
  parallel, 57, 58, 64
  pinnate, 59
  radial, 53, 54
  rectangular, 50, 51, 68
  subdendritic, 59, 60
  subparallel, 59, 60, 65, 68
  trellis, 48, 49, 55, 148

Elongation ratio, 285
Entrance angles, 147
Entropy, 306
Erosion cycle. See Geographical cycle
Exmoor, 189

Feedback, 212
Floodplain, 33, 34
Flow, type of
  laminar, 108, 113, 128
  mixed, 108
  overland, 82, 115, 122, 127, 188
  sheet, 104, 107, 129
  subdivided, 108
  surge, 111

Geographical cycle (of Davis), 6, 23, 25, 150
Geomorphic thresholds, 335
Grade, 28, 29, 36
Great Salt Lake, 41

Henry Mountains, 56
Hortonian Revolution, 2
Hydraulic geometry, 207
Hydrograph, 330
Hypsometry, 174, 191, 287, 288, 299

Subject Index

Infiltration capacity, 82, 104, 105, 106, 130, 139, 140, 152, 178, 179, 181, 185, 211, 215, 221, 326
Interbasin areas, 281

Jura Mountains, 22, 32

Lag time (hydrograph), 330
Laws of drainage composition (Horton)
  area, 280
  numbers, 89, 100, 137, 276
  slopes, 93, 278
  stream lengths, 89, 137, 276
Length of overland flow, 151, 184

Meanders, 33, 34
Micropiracy, 131–136, 306
Mississippi River, 1, 92, 164, 339
Monadnock, 53

Peneplain, 17, 37, 38, 40, 43, 165, 166, 251
Piracy, 164, 266
Playfair's Law, 1, 78, 91, 93, 206
Precipitation effectiveness (Thornthwaite), 196, 210, 215, 221

Rainfall intensity, 327
Rain-wave trains, 111
Random walk, 308, 310
Relative density (of streams), 214
Relief ratio, 174, 285, 317, 319, 326
Rills, 129–137, 232, 235, 306
Roughness number, 215
Ruggedness number, 214, 219

Seine River, 40
Sediment load, 127
Sediment yield, 174, 286, 314
Sheet erosion, 130

Sinuosity, 340
Slopes
  graded, 251
  valley-side, 42, 173, 214
Stage (geographical or geomorphic cycle)
  maturity, 15, 25, 27, 29, 37, 79, 165, 191
  old age, 25, 27, 29, 37, 79
  youth, 25, 27, 29, 79, 165
Streams
  morphology
    area, 201
    frequency, 83
    length, 89, 171, 202
    order, 70, 79, 170, 171, 179, 201, 275
    sinuosity, 340
  types
    adventitious, 139
    antecedent, 40
    beheaded, 11, 13, 245
    consequent, 10, 12–15, 27, 30–32, 46, 55, 58, 235
    insequent, 30, 31, 32, 46
    obsequent, 16, 235
    perched, 150
    rejuvenated, 150
    subsequent, 10, 12, 18, 25, 30, 32, 55
Surface detention, 104

Texture, 84, 210
Texture ratio, 172
Thames River, 43
Tiber River, 44
Transmissibility, 106, 181, 184

Var River, 43

Watchung Mountains, 52

Zambesi River, 52

About the Editor

STANLEY A. SCHUMM, a Fluvial Geomorphologist and a Professor of Geology at Colorado State University, received the B.A. from Upsala College in 1950 and the Ph.D. in Geomorphology from Columbia University in 1955. He spent twelve years as a geologist for the U.S. Geological Survey and has been a Visiting Lecturer at the University of California at Berkeley and at the University of Sydney, Australia. During 1972–1973, Dr. Schumm was Acting Associate Dean for Research at Colorado State.

Dr. Schumm received the Horton Award in 1957 from the American Geophysical Union, and in 1970, he received an honorable mention for his paper "Geomorphic approach to erosion control in semiarid regions," from the American Society of Agricultural Engineers. In 1974, he was designated "Outstanding Educator of America." Dr. Schumm has delivered lectures at universities in Australia, New Zealand, Europe, Canada, and Venezuela, and has been a Visiting Scientist at the Polish Academy of Science. He is a member of several professional societies and has served as an expert witness during lawsuits on the relation of property boundaries to rivers and as an environmental geologist for the Regional Transportation District (Denver), Cameron Engineering Company, and Atlantic Richfield Oil Company.

Dr. Schumm has had sixty papers and reports published and has edited three books on river morphology, slope morphology, and Quaternary research. Two of his papers appear in this volume.